T0139294

5G-Enabled Internet of Things

5G-Enabled Internet of Things

Edited by
Yulei Wu
Haojun Huang
Cheng-Xiang Wang
Yi Pan

CRC Press is an imprint of the
Taylor & Francis Group, an **informa** business

CRC Press
Taylor & Francis Group
6000 Broken Sound Parkway NW, Suite 300
Boca Raton, FL 33487-2742

Printed on acid-free paper

International Standard Book Number-13: 978-0-367-19010-1 (Hardback)

Visit the Taylor & Francis Web site at
http://www.taylorandfrancis.com

and the CRC Press Web site at
http://www.crcpress.com

Contents

SECTION III PRIVACY AND SECURITY ISSUES

SECTION IV EMERGING APPLICATIONS OF THE 5G-ENABLED INTERNET OF THINGS

Preface

The Internet of Things (IoT) is changing the way we live and work. Its success and real value come from the establishment of services on top of the connected IoT devices. According to the Ericsson Mobility Report, there will be more than 30 billion connected devices worldwide by 2023, of which around 20 billion will be IoT-related devices. Between 2017 and 2023, the number of IoT devices is expected to increase at a compound annual growth rate of 19%, driven by promising IoT use cases such as smart wearables, smart display, smart metering, smart power, robotic control/production automation, robotic surgery, autonomous driving cars, and drone surveillance. These applications are usually integrated with wireless mobile communications.

Currently, a number of smart IoT devices exploit cellular networks such as the third generation (3G) and 4G long-term evolution (LTE) to maintain their connectivity and their connection with the cloud data centers. With the exponential growth of data produced by increasingly large numbers of IoT devices, several burning issues remain to be solved in application environments. For example, the transmission latency and reliability of the current cellular networks cannot be guaranteed, which in turn limits the effectiveness and feasibility of many emerging IoT applications such as the tactile Internet, autonomous driving cars, and robotic surgery, all of which require ultra-low latency and ultra-high reliability.

The 5G mobile communication system has been introduced with the capabilities of high throughput, low latency, high reliability, and increased scalability. These capabilities can enable a massive number of devices, with best quality-of-service and quality-of-experience provision, of ubiquitous connectivity solutions to fulfil their diverse IoT application requirements. 5G has the potential to allow the deployment of more Internet-connected devices without concern that existing issues would be exacerbated by an overcrowded network. The high speed and reliable connectivity underpinned by 5G will create new possibilities for IoT services far beyond those available today. In addition, the enabling technologies of 5G—including network function virtualization (NFV) and softwarization, software-defined networking (SDN), massive multiple input–multiple output, mobile/edge computing, and ultra-dense networks—have great potential to usher in a new IoT era, aiming to smoothly and flexibly support heterogeneous IoT services with distinct business characteristics under a massive number of smart devices. Furthermore, the 5G IoT will bring a rich source of Big Data. The powerful role of Big Data analytics in 5G will undoubtedly benefit IoT advancement.

However, how the enabling technologies in 5G—whether integrated into the whole or as a part of a system—can seamlessly fuel the IoT revolution remain challenging. This raises new considerations of network design, resource deployment, management, quality of experience, standards, and policy and regulation of 5G-enabled IoT. It is therefore of critical importance to devise novel solutions by designing smart 5G-enabled IoT paradigms integrated with the enabling technologies of 5G. The IoT has been established as a new cross-discipline research topic, requiring the anticipation of the technical and practical challenges faced by mixed research studies that cross multiple disciplines.

We invited world experts to contribute chapters covering the following four sections of this book:

- *Section I: 5G-Enabled Internet of Things: Architecture and Related Technologies*: provides an understanding of the properties, characteristics, architecture, emerging challenges, and opportunities of the 5G-enabled IoT.
  - Chapter 1: *5G Cloud, Mobile, and Edge Computing for Internet of Things*: focuses on the cloud and edge context and presents the use cases driven by IoT application scenarios.
  - Chapter 2: *Emerging Challenges and Requirements for Internet of Things in 5G*: includes instruction on the evolution and history of IoT, the features in 5G that enhance IoT systems, system-related challenges, and other related emerging issues.
  - Chapter 3: *Network Functions Virtualization–Based Internet of Things in 5G Networks.*
  - Chapter 4: *Exploring the Next Generation of the Internet of Things in the 5G Era*: allows the reader a very clear understanding of how SDN and NFV can change the design and utilization of the IoT in 5G networks.
  - Chapter 5: *Achieving Scalability in the 5G-Enabled Internet of Things*: elaborates the scalability enablers in 5G and how they can be applied to provide the system scalability that is urgently required by the IoT platforms.
- *Section II: 5G Access Network for the Internet of Things*: presents new proposals and network architectures to ensure the performance and success of 5G access networks for IoT applications. This part covers
  - Chapter 6: *5G Small Cells: The Harbinger of the Internet of Things and Connected Living*
  - Chapter 7: *Mobile Edge Computing for the 5G Internet of Things*, which presents the architecture, challenges and key technologies of complementarily integrating mobile edge computing with 5G-enabled IoT services;
  - Chapter 8: *Millimeter-Wave 5G-Enabled Internet of Things*: discusses the foremost mmWave 5G IoT usage scenarios;

- Chapter 9: *Algorithms and Performance Analysis for Narrowband Internet of Things and Broadband Long-Term Evolution Coexisting System*: presents important design and performance analyses of this coexisting system;
- Chapter 10: *Internet of Things Wireless Spectrum Sharing for Radio Access*: shows the general engineering of 5G in light of range-sharing strategies that clarify distinctive methods for spectrum sharing;
- Chapter 11: *Random Access Modeling for the Cellular-Based Massive Internet of Things*: deals with the challenge of connection establishment between IoT devices and base stations, for the sporadic uplink transmissions of massive IoT devices.

■ *Section III: Privacy and Security Issues*: addresses how to perform privacy protection and ensure system security in the 5G-enabled IoT.
  - Chapter 12: *Privacy and Security Issues in the 5G-Enabled Internet of Things*: discusses security and privacy issues in the 5G-enabled IoT and suggests future research opportunities.
  - Chapter 13: *Privacy-Preserving Techniques for the 5G-Enabled Location-Based Services*: focuses on the privacy issues of location-based services (LBSs) in the IoT and provides a comprehensive overview on the state-of-the-art and the key fundamentals of LBS privacy-preserving techniques.
  - Chapter 14: *Blockchain Technology for the 5G-Enabled Internet of Things Systems: Principle, Applications and Challenges*: provides a comprehensive overview on how blockchain technology can tackle the critical challenges of 5G-enabled IoT systems such as IoT device management, Big Data analytics and storage, and privacy and security protection.

■ *Section IV: Emerging Applications of the 5G-Enabled Internet of Things*: This part covers some emerging applications on
  - Chapter 15: *Searching for Internet-of-Things Resources: Requirements and Outlook*: deals with the challenges faced by the data deluge in 5G-enabled IoT;
  - Chapter 16: *Applications of the Internet of Things and Fog Computing for Community Safety toward the 5G Era*: discusses the challenges of community safety by IoT and introduces a hierarchical fog computing architecture for community safety based on fog gateway and Big Data;
  - Chapter 17: *Tactile Internet over Fiber-Wireless–Enhanced LTE-A HetNets via Artificial Intelligence-Embedded Multi-Access Edge Computing*: discusses how to enable emerging delay-sensitive interactive tactile Internet applications;
  - Chapter 18: *Smart Power Management Internet of Things System with 5G and LoRa Hybrid Wireless Networks*: describes a smart power management IoT system with 5G and LoRa hybrid wireless networks that provides energy usage information and efficient controls for power system energy analysis and allocation planning.

This book presents state-of-the-art solutions to the theoretical and practical challenges stemming from the integration of 5G enabling technologies into IoTs in support of a smart 5G-enabled IoT paradigm in terms of network design, operation, management, optimization, privacy and security, and applications. In particular, the technical focus covers a comprehensive understanding of 5G-enabled IoT architectures, converged-access networks, privacy and security issues, and emerging applications of the 5G-enabled IoT.

This book targets readers in both academia and industry. Graduate students can select promising research topics from this book that are suitable for their thesis or dissertation research. Researchers will gain a deep understanding of the challenging issues and opportunities of 5G-enabled IoT and can thus easily find an unsolved research problem to pursue. Industry engineers from IT companies, service providers, content providers, network operators, and equipment manufacturers can get to know the engineering design issues and corresponding solutions after reading some of the practical schemes described in some chapters.

All chapter authors have provided as many technical details as possible. Each chapter also includes references for readers' further studies and investigations. If you have any comments or questions on certain chapters, please contact the chapter authors for more information.

Thank you for reading this book. We hope this book is helpful to you in your scientific research and in understanding the practical problems related to the 5G-enabled IoT.

Editors

Yulei Wu, PhD, is a senior lecturer in Computer Science at the University of Exeter. He earned his PhD degree in Computing and Mathematics and BSc (Hons) degree in Computer Science from the University of Bradford, UK, in 2010 and 2006, respectively. His main research focuses on network slicing and softwarization, future network/Internet architecture and technologies, intelligent networking technologies, green networking, wireless networks, network security and privacy, and analytical modeling and performance optimization.

His recent research has been supported by the Engineering and Physical Sciences Research Council of the United Kingdom, the National Natural Science Foundation of China, and University's Innovation Platform as well as by industry. He has published more than 70 research papers in prestigious international journals, including *IEEE Journal on Selected Areas in Communications*; *IEEE Transactions on Parallel and Distributed Systems*; *IEEE Transactions on Mobile Computing*; *IEEE Transactions on Communications*; *IEEE Transactions on Wireless Communications*; *IEEE Transactions on Vehicular Technology*; *IEEE Communications Magazine*; *ACM Transactions on Multimedia Computing, Communications, and Applications*; and *ACM Transactions on Embedded Computing Systems* as well as the proceedings of reputable international conferences. He was the recipient of Best Paper Awards from IEEE SmartCity 2015, IEEE CSE 2009, and ICAC 2008 international conferences.

Dr. Wu serves as an associate editor or a member of the editorial boards for several international journals including *IEEE Transactions on Network and Service Management* and Elsevier's *Computer Networks* as well as a guest editor for many international journals, such as the *IEEE Journal on Selected Areas in Communications, IEEE Transactions on Sustainable Computing, IEEE Transactions on Computational Social Systems, IEEE Transactions on Cognitive Communications and Networking, IEEE Transactions on Network Science and Engineering*, and Elsevier's *Computer Networks.* Dr. Wu has been the chair or vice-chair of more than 50 international conferences/workshops and has served as the program committee member of more than 90 professional conferences or workshops. He was awarded Outstanding Leadership Awards from IEEE HPCC 2018, IEEE CIT 2017, IEEE ISPA 2013, and IEEE TrustCom 2012 and Outstanding Service Awards from IEEE HPCC 2012 and IEEE CIT 2010. He has also been a reviewer for book proposals, grant proposals, and a number of international conferences and journals, including many IEEE/ACM transactions and magazines. He was awarded the Outstanding Contribution in Reviewing status by the Elsevier journal *Future Generation Computer Systems* in 2018. He was elected as a member of Youth Innovation Promotion Association of the Chinese Academy of Sciences in 2015 and Employee of the Year in 2013. He is a senior member of the Institute of Electrical and Electronics Engineers and a fellow of the Higher Education Academy.

Haojun Huang, PhD, is currently an associate professor within the School of Electronic Information and Communications, Huazhong University of Science and Technology, China. He earned his BSc degree from the School of Computer Science and Technology, Wuhan University of Technology, in 2005, and his PhD degree from the School of Communication and Information Engineering, University of Electronic Science and Technology of China, in 2012. He worked as a postdoctoral researcher in the Research Institute of Information Technology, Tsinghua University, from 2012 to 2015, an assistant professor in communication engineering within the College of Electronic Information Engineering at Wuhan University, from 2015 to 2017, and a professor in the Department of Network Engineering within the College of Computer science at the China University of Geosciences in 2018. His current research interests include wireless communication, ad hoc networks, Big Data, and software-defined networking.

Dr. Huang has published more than 50 journal papers and refereed conference papers with more than 23 papers published in various IEEE journals, and he has co-authored/co-edited 3 books. He serves as an executive associate editor for Springer *Peer-to-Peer Networking and Applications* from October 2013, guest editor for ACM/Springer *Mobile Networks and Applications, Special Issue on Deep Learning for Big Data Analytics*, and also served as a technical program committee (TPC) Vice Chair for IEEE IUCC 2017 and GreenCom 2017. Furthermore, he has served as a TPC member for several important conferences such as IEEE INFOCOM 2015/2017/2018, HPCC 2018, DSS 2018, ICC 2013/2015/2016, GLOBECOM 2016, CIT-2015/2016/2017, CPSCom 2017, SmartData 2017/2018, GreenCom 2017, IUCC 2017, and WNM-2015, and as a reviewer for more than 20 various prestigious journals, including *IEEE Journal on Selected Areas in Communications*, *IEEE/ACM Transactions on Networking*, *IEEE Transactions on Mobile Computing*, and *IEEE Transactions on Services Computing*.

Cheng-Xiang Wang, PhD, earned his BSc and MEng degrees in Communication and Information Systems from Shandong University, China, in 1997 and 2000, respectively, and his PhD degree in Wireless Communications from Aalborg University, Denmark, in 2004.

He was a research assistant with the Hamburg University of Technology, Hamburg, Germany, from 2000 to 2001, a research fellow with the University of

Agder, Grimstad, Norway, from 2001 to 2005, and a visiting researcher with Siemens AG-Mobile Phones, Munich, Germany, in 2004. He has been with Heriot-Watt University, Edinburgh, UK, since 2005, where he became a professor in Wireless Communications in 2011. In 2018, he joined Southeast University, Nanjing, China, as a professor. He has co-authored 3 books, 1 book chapter, more than 170 journal papers including 20 highly cited papers and more than 110 papers published in various IEEE journals, and approximately 180 conference papers including 3 highly cited papers. He has also delivered 17 invited keynote speeches and talks and 6 tutorials in international conferences. His work has been cited more than 9,700 times. His current research interests include wireless channel measurements/modeling and B5G wireless communication networks.

Dr. Wang is a fellow of the IEEE and IET, an IEEE Communications Society Distinguished Lecturer for 2019 and 2020, and a Highly Cited Researcher recognized by Clarivate Analytics in 2017 and 2018. He was a recipient of nine Best Paper Awards from the IEEE GLOBECOM 2010, IEEE ICCT 2011, ITST 2012, IEEE VTC 2013-Spring, IWCMC 2015, IWCMC 2016, IEEE/CIC ICCC 2016, and WPMC 2016. He has served as a TPC member, the TPC chair, and the general chair for more than 80 international conferences. He is currently an Executive Editorial Committee member for the IEEE Transactions on Wireless Communications. He has served as an editor for nine international journals including the *IEEE Transactions on Wireless Communications* from 2007 to 2009, the *IEEE Transactions on Vehicular Technology* from 2011 to 2017, and the *IEEE Transactions on Communications* from 2015 to 2017. He was a guest editor for the *IEEE Journal on Selected Areas in Communications, Special Issue on Vehicular Communications and Networks* (lead guest editor), *Special Issue on Spectrum and Energy Efficient Design of Wireless Communication Networks, and Special Issue on Airborne Communication Networks.* He was also a guest editor for the *IEEE Transactions on Big Data, Special Issue on Wireless Big Data.*

Yi Pan, PhD, is currently a Regents' professor and chair of Computer Science at Georgia State University, Atlanta, Georgia, USA. He has served as an associate dean and chair of the Biology Department during 2013–2017 and chair of Computer Science during 2006–2013. Dr. Pan earned his BEng and MEng degrees in computer engineering from Tsinghua University, China, in 1982 and 1984, respectively, and his PhD degree in computer science from the University of Pittsburgh, Pittsburgh, Pennsylvania, USA, in 1991. His profile has been featured as a distinguished alumnus in both Tsinghua Alumni Newsletter and University of Pittsburgh Computer Science Alumni Newsletter. Dr. Pan's research interests include parallel and cloud computing, wireless networks, and bioinformatics. Dr. Pan has published more than 200 journal papers with more than 80 papers published in various IEEE journals. In addition, he has published more than 150 papers in refereed conferences. He has also co-authored/co-edited 43 books. His work has been cited more than 10,000 times. Dr. Pan has served as an editor-in-chief or editorial board member for 15 journals including 7 *IEEE Transactions*. He is the recipient of many awards including *IEEE Transactions* Best Paper Award, several other conference and journal best paper awards, 4 IBM Faculty Awards, 2 Japan Society for the Promotion of Science Senior Invitation Fellowships, and the IEEE BIBE Outstanding Achievement Award, National Science Foundation Research Opportunity Award, and Air Force Office of Scientific Research Summer Faculty Research Fellowship. He has organized many international conferences and delivered keynote speeches at more than 60 international conferences around the world.

Contributors

Naveed A. Abbasi
Next-generation and Wireless Communications Laboratory (NWCL)
Department of Electrical and Electronics Engineering
Koç University
Istanbul, Turkey

Hamed Ahmadi
School of Computer Science and Electronic Engineering
University of Essex
Colchester, United Kingdom

Ozgur B. Akan
Next-generation and Wireless Communications Laboratory (NWCL)
Department of Electrical and Electronics Engineering
Koç University
Istanbul, Turkey

and

Internet of Everything Group
Electrical Engineering Division
Department of Engineering
University of Cambridge
Cambridge, United Kingdom

Ashwin Ashok
Department of Computer Science
Georgia State University
Atlanta, Georgia

David de la Bastida
EECS International Graduate Program
National Chiao Tung University
Hsinchu City, Taiwan

Joe Billingsley
College of Engineering, Mathematics and Physical Science
University of Exeter
Exeter, United Kingdom

Ravishankar Chamarajnagar
VMware Inc.
Palo Alto, California

and

Department of Computer Science
Georgia State University
Atlanta, Georgia

Yansha Deng
Department of Informatics
King's College London
London, United Kingdom

Amin Ebrahimzadeh
Optical Zeitgeist Laboratory
INRS
Montréal, Québec, Canada

Yasmin Fathy
Computer Science Department
University College London (UCL)
London, United Kingdom

Simon Fong
Department of Computer and Information Science
University of Macau
Macau, China

Xingli Gan
State Key Laboratory of Satellite Navigation System and Equipment Technology
The 54th Research Institute of China Electronics Technology Group Corporation
Shijiazhuang, China

Dongyu Guo
Tianjin Key Laboratory of Advanced Networking
School of Computer Science and Technology
College of Intelligence and Computing
Tianjin University
Tianjin, China

Akhil Gupta
School of Electronics and Electrical Engineering
Lovely Professional University
Phagwara, India

Jinkun Han
Department of Digital Media Technology
North China University of Technology
Beijing, China

Meng Han
Data-driven Intelligence Research (DIR) Laboratory
Kennesaw State University
Marietta, Georgia

Haojun Huang
School of Electronic Information and Communications
Huazhong University of Science and Technology
Wuhan, China

Muhammad Imran
School of Engineering
University of Glasgow
Glasgow, United Kingdom

Nan Jiang
School of Electronic Engineering and Computer Science
Queen Mary University of London
London, United Kingdom

Anca Delia Jurcut
School of Computer Science
University College Dublin
Dublin, Ireland

Raabia Kausar
School of Electronics and Electrical Engineering
Lovely Professional University
Phagwara, India

Ruibin Li
Tianjin Key Laboratory of Advanced Networking
School of Computer Science and Technology
College of Intelligence and Computing
Tianjin University
Tianjin, China

Tengyue Li
Department of Computer and
Information Science
University of Macau
Macau, China

Wen Li
Institute of Command and Control
Engineering
Army Engineering University of PLA
Nanjing, China

Bao-Shuh Paul Lin
Department of Computer Science
National Chiao Tung University
Hsinchu, Taiwan

Fuchun Joseph Lin
Department of Computer Science
National Chiao Tung University
Hsinchu, Taiwan

Yi-Bing Lin
Department of Computer Science
National Chiao Tung University
Hsinchu, Taiwan

Liyuan Liu
Graduate College
Kennesaw State University
Kennesaw, Georgia

Chunbo Luo
College of Engineering, Mathematics
and Physical Science
University of Exeter
Exeter, United Kingdom

Martin Maier
Optical Zeitgeist Laboratory
INRS
Montréal, Québec, Canada

Wang Miao
College of Engineering, Mathematics
and Physical Science
University of Exeter
Exeter, United Kingdom

Geyong Min
College of Engineering, Mathematics
and Physical Science
University of Exeter
Exeter, United Kingdom

Arumugam Nallanathan
School of Electronic Engineering and
Computer Science
Queen Mary University of London
London, United Kingdom

Yi Pan
Department of Computer Science
Georgia State University
Atlanta, Georgia

Kai Peng
School of Electronic Information and
Communications
Huazhong University of Science and
Technology
Wuhan, China

Deli Qiao
School of Information Science and
Technology
East China Normal University
Shanghai, China

Abhishek Roy
Mediatek USA Inc.
San Jose, California

Navrati Saxena
Department of Software
Sungkyunkwan University
Seoul, South Korea

Sukhdeep Singh
Samsung R&D
India-Bangalore (SRI-B)
Bangalore, India

Liangliang Song
ThingPark China
Beijing, China

Wei Song
Department of Digital Media Technology
North China University of Technology
Beijing, China

Yifei Tian
Department of Digital Media Technology
North China University of Technology
Beijing, China

and

Department of Computer and Information Science
University of Macau
Macau, China

Xu Tong
Tianjin Key Laboratory of Advanced Networking
School of Computer Science and Technology
College of Intelligence and Computing
Tianjin University
Tianjin, China

Li-Ping Tung
Microelectronics and Information Research Center
National Chiao Tung University
Hsinchu, Taiwan

Chen Wang
School of Electronic Information and Communications
Huazhong University of Science and Technology
Wuhan, China

Cheng-Xiang Wang
National Mobile Communications Research Laboratory
School of Information Science and Engineering
Southeast University
and
Purple Mountain Laboratories
Nanjing, China

Chenyang Wang
Tianjin Key Laboratory of Advanced Networking
School of Computer Science and Technology
College of Intelligence and Computing
Tianjin University
Tianjin, China

Xianbin Wang
Department of Electrical and Computer Engineering
Western University
London, Ontario, Canada

Xiaofei Wang
Tianjin Key Laboratory of Advanced Networking
School of Computer Science and Technology
College of Intelligence and Computing
Tianjin University
Tianjin, China

Yulei Wu
Department of Computer Science
College of Engineering, Mathematics and Physical Sciences
University of Exeter
Exeter, United Kingdom

Lina Xu
School of Computer Science
University College Dublin
Dublin, Ireland

Bowen Yang
School of Engineering
University of Glasgow
Glasgow, United Kingdom

Turker Yilmaz
Next-generation and Wireless Communications Laboratory (NWCL)
Department of Electrical and Electronics Engineering
Koç University
Istanbul, Turkey

Tianqi Yu
Department of Electrical and Computer Engineering
Western University
London, Ontario, Canada

Lei Zhang
School of Engineering
University of Glasgow
Glasgow, United Kingdom

Rui Zhang
School of Computer Science and Technology
Wuhan University of Technology
Wuhan, China

Ping Zhao
School of Information Science and Technology
Donghua University
Shanghai, China

Weixing Zhu
Institute of Command and Control Engineering
Army Engineering University of PLA
Nanjing, China

Yongxu Zhu
Wolfson School of Mechanical, Electrical and Manufacturing Engineering
Loughborough University
Loughborough, United Kingdom

I

5G-ENABLED INTERNET OF THINGS

Architecture and Related Technologies

Chapter 1

5G Cloud, Mobile, and Edge Computing for Internet of Things

Ravishankar Chamarajnagar and Ashwin Ashok

Contents

1.1 Introduction

Fifth generation (5G) wireless technology has had tremendous growth over the last few years. Given all the benefits it has to offer, connected, distributed and real-time Internet of Things (IoT) applications will start to materialize soon in various industries. Both the proliferation of applications in the cloud and the battle for the edge will have 5G offering the core network infrastructure in the background. While IoT is expected to dominate the landscape, a wide range of applications will benefit from 5G's ultra-fast networks and real-time responsiveness. 5G is expected to provide the following capabilities [1]:

- Faster network speeds: The data rates are way off the charts for 5G compared with all the previous technologies, benefiting all the multimedia applications and services that will be making their way into the market.
- Lowered latency: The latency in the network is extremely low, which will result in all the interactive applications gaining a huge increase in responsiveness and a smooth experience.
- Network support for massive increases in data traffic: Not only are we referring to an increase in data rates, but we can now support the increased data usage per user due to the increase in data generated and transmitted by interactive applications, online games and enterprise cloud-based services.

5G, with the above capabilities [2], benefits bandwidth-heavy applications across various industries. This includes distributed IoT applications that generate huge amounts of data. Consider these few examples highlighting the need for 5G:

- Consumer applications are becoming more multimedia based for a richer experience and, in turn, demanding higher network bandwidth.
- On industrial floors, machine-to-machine (M2M) connectivity could benefit from low latency and higher bandwidth to reduce any kind of data congestion and downtime.
- The automotive industry is transforming the driving experience for consumers. The streamlined applications that will realize this transformative experience are extremely bandwidth hungry because most actions are in response to

external and contextual stimuli. Multiple compute-paradigms will need to be supported; for instance, an automated braking system could benefit from edge computing, and a more streamlined driver profile analysis for personalized experience is more compute intensive and, hence, needs to be cloud based.

- Smart cities are driving huge infrastructures across the globe; and projects such as smart lighting, traffic control, traffic analysis, pollution monitoring and control are just some of the most common that have materialized over the last few years and will continue to consume tremendous network bandwidth and compute resources for propagation and analysis.
- While many of the above applications are bandwidth intensive, some industries have higher real-time requirements than others. In critical health-care operations, the use of automated robotics arms to assist doctors in their surgeries or, as another example, the use of smart hospital beds for patient care will require a reliable infrastructure at the edge and in the cloud.

Like any technology innovation, 5G has explored multiple architectures and deployment directions. Some of the standards developing organizations (SDOs) like the Third Generation Partnership Project (3GPP) and the Fifth Generation Public Private Partnership (5GPPP) have worked together to make huge strides in establishing a level of consensus on the overall architecture and deployment paradigms. We will review some of the evolving architectural paradigms in subsequent sections of this chapter.

1.2 5G Edge Computing

1.2.1 With 5G, Is Edge Computing Necessary?

To a large extent, the explosion of connected things, devices, sensors, activators and systems that are part of the IoT has driven the development of 5G in recent times, along with the rich multimedia applications. Apps and services that collect or generate huge amounts of data rely on the promise of 5G with speeds that are 10 times faster than those of 4G, making connectivity to the cloud a reality. However, even as there are groups of applications that rely on huge amounts of data collected from the connected things to be propagated to the cloud for processing, there are a whole slew of applications and use cases where the responses are expected to be in real time. Given that 5G adoption cannot be instantaneous due to all the additional infrastructure around antennas, switches, repeaters, and so on needed to be deployed, the lack of guarantees of connectivity irrespective of the technology, and the infrastructure and the Service Level Agreements (SLAs) around real-time processing, edge computing is real; it is inevitable.

Mobile edge computing (MEC) [6] is edge computing that supports mobile devices using data repositories positioned close to the wireless infrastructure to keep the latency low. As mobile bandwidth increases and demands for data increase, the existing edge infrastructure is expected to change dramatically to support newer devices and applications in industrial, healthcare, manufacturing and other domains. As things at the edge grow more intelligent, we will see changes not only to the wireless infrastructure but to every aspect of the edge network. The compute servers, storage servers and network equipment will all go through some form of virtualization to create a more flexible and streamlined software-defined data center (SDDC) and will use a software-defined wide-area network (SD-WAN) for communication or network function virtualization (NFV) for network functions. Furthermore, private or hybrid cloud services will present a seamless experience, app-enabled frameworks will integrate well with all the data from the edge, and security paradigms will evolve to better address security and privacy concerns in a distributed environment.

Typically, applications that have a need for low latency communication are interactive or access time-sensitive data and can benefit greatly from distributed architectures like Content Distribution Network (CDN). We are already starting to see many of the applications and services that were traditionally in the public cloud being made available on edge appliances called IoT gateways. It is key to note that the revolution that happened in the public cloud years ago is happening now at the edge. While backend information technology (IT) infrastructure in the cloud will continue to be pivotal for many reasons, the battle for supremacy in IoT will be won or lost at the edge of the network.

1.2.2 Evolution of the Edge/Core/Cloud Paradigm

The generally accepted distributed architecture for IoT is three-tiered, with things comprising the first tier, IoT gateways aggregating data from things comprising the second tier, and the cloud making up the third tier. However, the architecture [7] proposed recently by Dell describes the three tiers of distributed architecture a bit differently: edge, core, and cloud, as shown in Figure 1.1.

The edge is seen as a compute node like a gateway capable of data processing along with all the things connected to them. Their physical proximity to things and their compute capabilities make them a huge part of edge computing. For instance, vibration and temperature readings of robotic devices on a factory floor are collected on the gateway and alerts are raised locally if values go above a certain threshold or local actions are executed based on rules configured on the gateways. Edge computing involves primarily things and the local processing of data generated by them. Extremely time-sensitive and real-time operations are typical candidates for edge processing.

For any workload on-premises, but slightly removed from the physical world, or taking place in distributed micro data centers, for instance, real-time analytics based on aggregation of all devices and gateways from the factory floor to finding

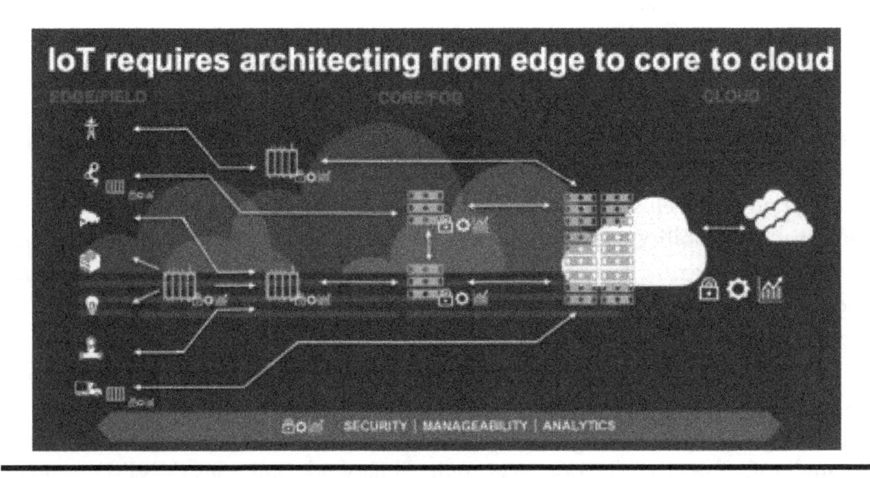

Figure 1.1 5G edge/core/cloud paradigm.

the optimal temperature for the floor or real-time promotion control of consumer products localized to a certain segment would not be edge computing; this would be core computing. The core has the benefits of edge computing as well as those of the cloud because it is both edge aware and cloud capable in terms of compute. If this ends up being the computing paradigm of the future, then the line between the core and the cloud will start to blur. "The edge" would no longer be the edge of the cloud, but with its own core, it would establish its own unique environment and possibly its own unique market.

Cloud computing was all about getting any workload to execute from any location. This would help keep the client extremely thin enabling sensors, actuators, and anything at the edge to run on minimal firmware and connect to the compute-intensive servers in the cloud through fat pipes and offloading all computation to the servers. However, although one can offload the computation, the response latency is a factor to consider for many applications. Latency or quality of service (QoS) is a key aspect that distinguishes the tiers. With demanding applications and services making their way into various verticals, IT would have to move quickly to support this architecture [8].

So, as the workloads start to be spread across the cloud, data centers or the edge network, their performance and QoS requirements will warrant where they are relegated: to the edge exclusively, or from the edge to the cloud, or to the core.

1.2.3 Cloud/Edge Computing Architecture

The 5G architecture caters to a wide range of services and applications with varying data rate and network bandwidth requirements. The services and applications belong to a wide range of industries—such as health care, aerospace, automotive, industrial, entertainment, home automation, smart cities—and the architecture would drive a flexible design to support all these domains with their myriad use cases. Binding them together loosely would be the key to a flexible design. Network slicing would guarantee a design that binds them loosely, helps isolate resources and at the same time helps reorganize components if requirements change. Achieving this level of flexibility and sophistication would require that all applications and services run on virtualized infrastructure. NFV is not new in the world of telecom and 5G. At the same time, SDDC, a newer paradigm, would be ideal to support the domain-based network slicing.

The NFV architecture framework provides a standardized model that moves away from proprietary, purpose-built hardware dedicated to a single service like load balancing or firewalling. Those services will be delivered as network functions through software virtualization as virtualized network functions (VNFs) on top of commercial off-the-shelf (COTS) hardware. The result is a network that is more agile and better able to respond to the on-demand, dynamic needs of telecommunications traffic and services, as shown in Figure 1.2. This is extremely relevant, as noted in [3], as we start to enable the next generation of services for the mobile edge computing.

An SDDC is an integrated abstraction layer that defines a complete data center by means of a layer of software that presents the resources of the data center as pools of virtual and physical resources and allows their composition into arbitrary user-defined services. SDDC is a data center where all the core elements

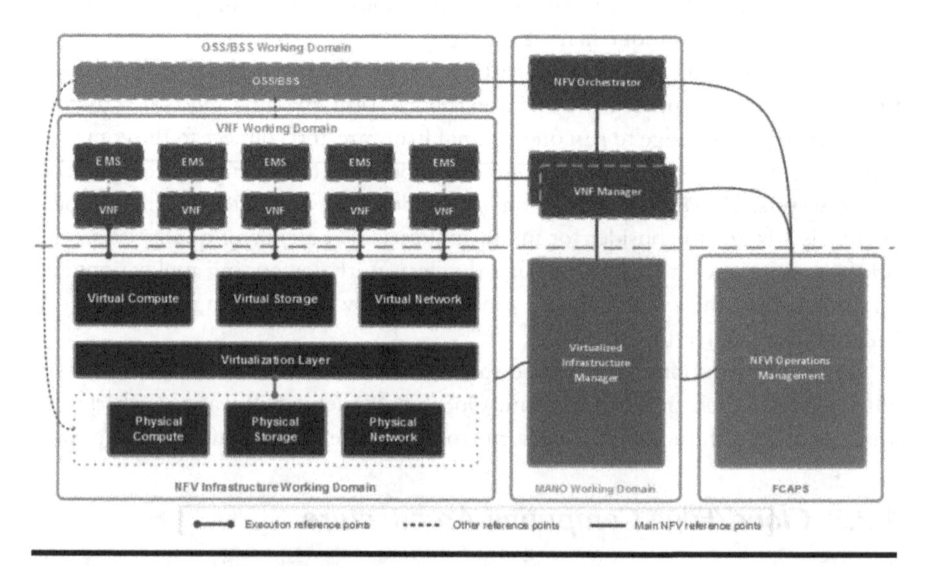

Figure 1.2 5G NFV architecture.

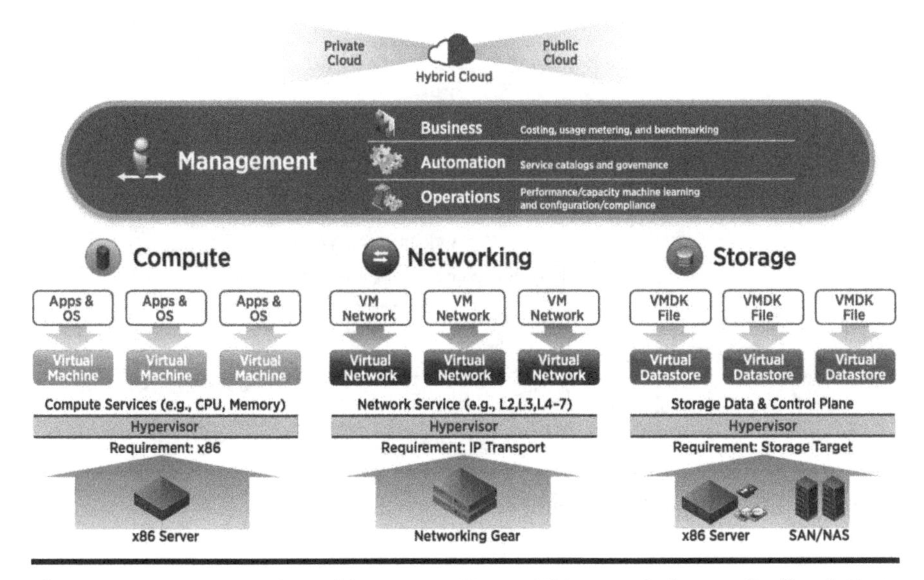

Figure 1.3 5G SDDC architecture. (From VMware Software-Defined Data Center, https://www.vmware.com/content/dam/digitalmarketing/vmwhitepaper-sddc-capabilities-itoutcomes-white-paper.pdf.)

(compute, networking storage, security, management) are virtualized and delivered as a flexible on-demand service, as shown in Figure 1.3. Refer to the SDDC whitepaper [4] from VMware for more details.

1. Faster deployment: People are delivering new software with new architecture and they need infrastructure, compute, storage, and networks that are more responsive and on-demand.
2. Increased security: Network virtualization can be a critical piece of a modern security policy.
3. Efficient and reduced cost: Scale-out architectures reduce capital expense to organizations and, through automation, can be more effective for them in how they use their employees.

Network slicing will help meet the demands of a flexible on-demand resourcing system. As we look at the needs of the various sections, their requirements and SLAs are very different and cannot be met by a monolithic resource pool, be it compute or storage or network. Slicing the stack to dedicate network resources along with the rest will help drive the customers independently, as depicted in Figure 1.4. This architectural paradigm has grown to be a huge part of all telecom implementations and has evolved into a prescribed standard, as outlined in [5].

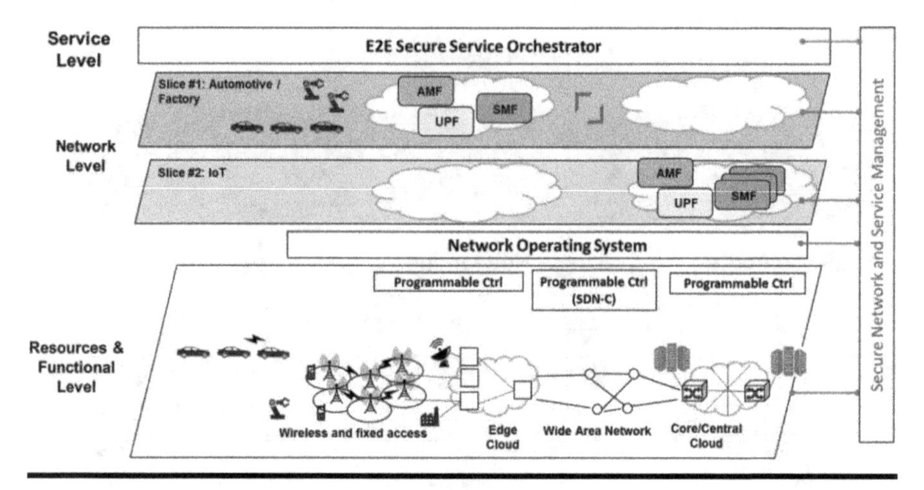

Figure 1.4 5G PPP standard architecture.

1.2.4 Platform: EdgeX Application Framework

As we start to focus on the edge workloads, either at the gateways or at the core, EdgeX Foundry [9] provides an open application platform for the edge. It is a vendor-neutral open-source project building a common open framework for IoT edge computing. At the heart of the project is an interoperability framework hosted within a full hardware- and Operating Systems (OS)-agnostic reference software platform to enable an ecosystem of plug-and-play components that unifies the marketplace and accelerates the deployment of IoT solutions. This is standards based and facilitates the execution of most kinds of applications in various tiers.

EdgeX Foundry proposes a loosely coupled tiered IoT architecture as shown in Figure 1.5 by allowing customers to deploy a mix of plug-and-play microservices on compute nodes at the edge, depending on the capability of host devices, and the targeted use cases. As an open-source, community-driven project, the current architecture scheme will evolve over time. It realizes the goal of assisting and unifying the developer community in driving a platform-agnostic approach to applications.

1.2.5 Applications at the Edge

Applications that have strong latency and QoS requirements are ideal candidates for edge computing, as outlined in the above section. As more compute capabilities are moving to the edge, applications are becoming more sophisticated as well.

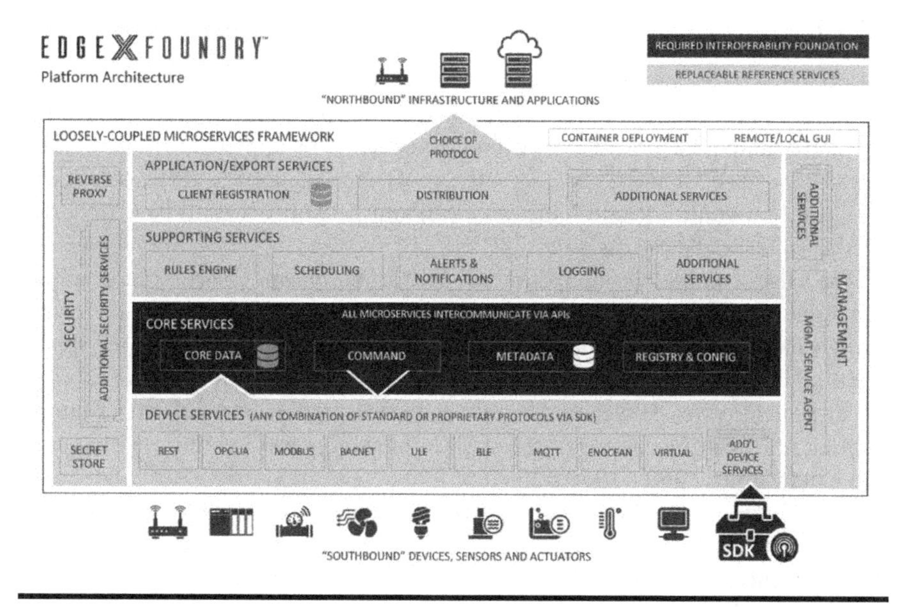

Figure 1.5 5G EdgeX architecture.

Outlined here are some examples in industrial, retail and automotive sectors, and we will dive deeper into the automotive sector:

- Industrial IoT: Quality assurance of integrated chips on an assembly line by analyzing chip images captured during the manufacturing process. With more powerful artificial intelligence (AI) capabilities on accelerator chips or Field Programmable Gate Arrays (FPGAs) available at the edge, it is now feasible to run most AI classification models in real time.
- Retail IoT: With local edge-based image analysis, extract demographics of the retail clientele and analyze their sentiment to drive sales pipelines for the various product lines:
  - Retail Realtime Crowd Insights to assess popularity of various product lines
  - Customer analysis at concession stands
  - Buying pattern analysis on cruise ships to drive localized promotions
- Automotive IoT: With the edge nodes in this case being the automobiles that are constantly in motion, the real-time and QoS requirements shoot up a notch higher. Here are a few examples using real-time image analysis at the edge to drive safety on roads:
  - Real-time driver monitoring through real-time image analysis to improve safety
  - Situational awareness with local maps and roadside hazards
  - Cooperative sharing of hazard data among nearby automobiles
  - Federation of user/cyclist discovery to ensure safety of pedestrians and cyclists

1.2.5.1 Automotive Edge

Connected vehicles and autonomous driving (AD) vehicles have emerged as key areas of the Automotive IoT, driving many innovations at the edge. The real-time service requirements in these areas cannot be achieved by services in the cloud but instead have to be done at the edge. The ecosystem supporting them, as shown in Figure 1.6, would include vehicles, roads, and networks that all have to come together to make the use cases possible.

The edge computing guidelines by ETSI (European Telecommunications Standards Institute) ISG (Industry Specification Group) MEC outlines the computing capabilities and the IT environment at the edge of the network for application and content providers. This environment is characterized by physical and logical proximity to clients, low latency between applications and data sources enabling near real-time access of the applications to context-rich information and high bandwidth for application traffic.

The 5G Automotive Association (5GAA) categorizes a comprehensive list of connected vehicle applications, categorized into four main groups of use cases: (1) safety, (2) convenience, (3) advanced driving assistance, and (4) vulnerable road user (VRU).

The following use cases provide a good view of the processing of all the exchanged information:

1. Realtime driver monitoring with edge-based image analysis of the driver to detect emotions, alertness, distractions that can help drive local actions and suggestions to improve safety on the road.
2. Situational awareness is essential for both driver-based and autonomous vehicles in cases where there may be unexpected turns or variable road conditions.

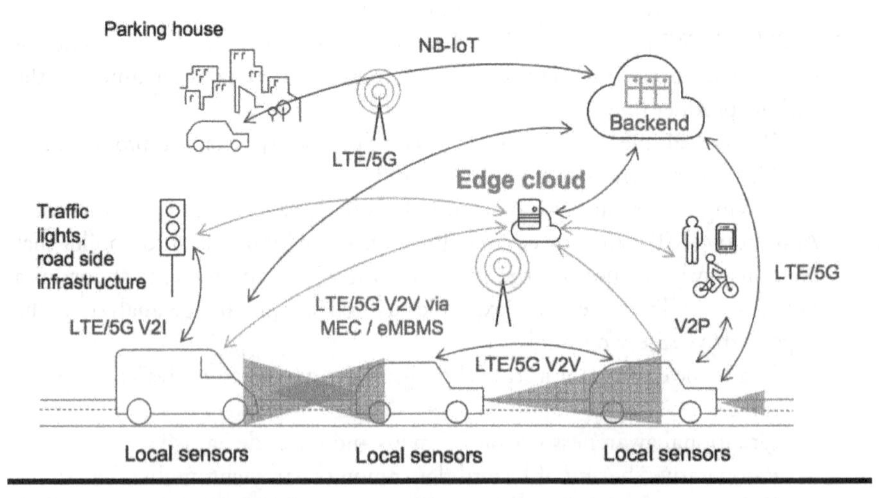

Figure 1.6 5G Auto vehicle-to-everything communication.

If these are captured and shared along with recent maps of the local area using the local edge server, they will improve safety in real time using the edge bandwidth effectively.
3. Cooperative sharing of hazard road conditions or other safety related data among nearby automobiles. This can assist drivers by warning them of either oncoming vehicles or upcoming hazard conditions on the road. This can effectively reduce accidents due to either road conditions or oncoming traffic, but this is only effective if it is real time and can be done completely at the edge.
4. Federation of pedestrian discovery to surrounding vehicles to make them aware of any pedestrians or cyclists on the roads or crossing a street at an intersection or jaywalking on a road. This can be extremely important for both driver-based as well as autonomous cars, and will rely heavily on the granular accuracy of the location.

1.3 Collaborative Edge Computing

The pervasiveness of the IoT is leading the way to a world of smart systems, applications and services. Mobile IoT powered by 5G is making this transformation possible with cloud and edge computing platforms. IoT devices that are mobile and portable within the bounds of wireless networks are all too common in enterprise deployments. It is easy to notice shared devices constantly in motion in retail locations, warehouse floors, or industrial assembly lines. Despite a strong 5G network backbone, one of the key challenges in realizing value with mobile IoT systems using current architecture is the heavy dependence on centralized cloud infrastructures. With data increasing exponentially, coupled with their access and storage, the network backbones will be inundated if they try to handle all application and service interactions using centralized data centers. Relying on data centers to download bits or to process data creates isolated points of failure, leading to unreliability on these systems. Given that 5G—with its flexible architectural paradigms like network slicing, virtual functions and software-defined networks and data centers—has made it possible to create and maintain an extremely reliable and scalable compute platform at the edge, IoT is taking advantage of compute at the edge.

Even as we move toward edge computing for IoT workloads to counter computational inefficiencies and network limits with cloud-based architectures, the fact that mobile IoT machines are becoming smarter and more powerful is encouraging. IoT devices are being packed with resources such as sensors, computing, and Input/Output (I/O). However, these resources are being completely underutilized given that most of them are used only for a certain duration for specific applications. For instance, a smartphone with quad-core processors has more computing power and storage than necessary for an average daily usage, which mostly involves access to text and emails; a smart printer in an office is largely in sleep mode and usually handles jobs in aperiodic chunks; and a smart TV is used more in the evenings at homes. The scenario applies to services available on devices as well. For instance, printer services and camera video capabilities are unused for a large percentage of the time. In essence, *there is excess resource and service capacity available in mobile IoT devices at the edge opening up a huge opportunity for sharing them across all available nodes, in a paradigm that we would term as the collaborative edge.* In this regard, we propose a system design for better utilization of resources and services in mobile IoT through opportunistic collaboration at the edge.

1.3.1 Toward Opportunistic Collaboration

As we look to address the above issues with network reliability and resource and service underutilization, we realize the need for opportunistic collaboration of IoT devices to share resources and services when they are available. Opportunistic collaboration becomes necessary because application offloading to the cloud may not always be possible due to bandwidth and latency concerns. In addition, the problem gets worse at scale. We seek to improve computational and network usage efficiency by mechanizing efficient ways to drive decisions and insights to execute applications and services closer to the devices, at the *edge*. Improving computation efficiency through collaboration coupled with the opportunistic usage of unused capacity in devices and at the edge, motivates our proposed *opportunistic collaborative mobile IoT* design. The proposed solution uses the well-known paradigm of blockchains to facilitate the collaboration, as illustrated in Figure 1.7.

1.3.2 Blockchain-Based Collaboration

We present a novel collaborative mobile IoT architecture that lets mobile IoT devices come together in an ad hoc manner and advertise their excess resource capacity and offered services using the blockchain framework. The collaborating devices are connected through a blockchain network that manages data dissemination in the network. The system uses *smart contracts* to advertise excess resources and network capacity, which are synchronized across a global network of blockchain nodes. These contracts are binding in nature and transparent to the network of nodes participating in the blockchain. Resources are made available to the seeking devices through containers (e.g., a docker) so that applications can be executed in a sandboxed fashion and do not need rooting the device. Services are made available through peer-to-peer (P2P) communication protocols. Here, services can be those initiated by apps or broad IT infrastructure-based services that cater to diverse applications and other dependent services.

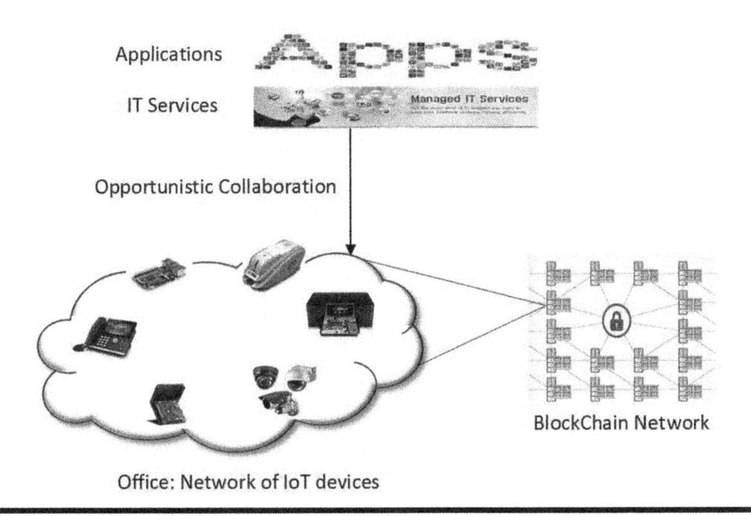

Figure 1.7 Opportunistic collaborative IoT using blockchains.

This approach lays the foundation for a distributed computing solution for mobile IoT that lets one opportunistically use excess capacity on resources and services. In essence, our proposal is a distributed middleware design that leverages blockchain technology for smart resource/service discovery, coordination and management.

In summary, the key contributions of this solution are as follows:

1. The design of a distributed mechanism to allocate resources and execute services through smart blockchain contracts and application execution on sandboxed containers
2. The implementation of a prototype collaborative mobile IoT system on Raspberry Pis and a Dell IoT edge gateway
3. The evaluation of network latency and benchmarking resource utilization under collaboration

1.3.3 Related Work

In this section, we discuss related works from the areas of resource abstraction platforms, resource lookup algorithms, distributed resource sharing platform and service invocation platforms.

Min Hong Yun et al. [14], proposed an architecture, *RIO*, for the abstraction of resources using a separation of the application layer from the operating systems services and kernel layer. Their approach addresses the specific area of resource abstraction with cross-memory mapping. Watanabe et al. [15], proposed a middleware, *ShAir*, for P2P resource sharing. It elaborates on the abstraction of resources using an event bus. Although the approach addresses resource abstraction, it does not specifically address the ad hoc networking and resource registry aspects. Karger et al. [16] discussed the approach of P2P lookup of Internet applications, which uses a distributed lookup protocol that helps map a given key to a node with the right content and efficiently handles the dynamic registration and exits of the nodes in the network. Salem et al. [17] proposed a mechanism for sharing resources at the edge that uses registries and a central mediation to choose the right resources based on demand. In addition, they introduced the concept of compensation for the resources used.

IBM and Samsung [18] have brought together in a proof of concept the blockchain as a repository of assets, their artifacts and related services, along with the P2P fabric using Telehash, which enables discovery and communication. This is testament to the fact that decentralized IoT is permeating the industry and there is a perceived need for it in the near term. One of the key aspects to fully realize is the flexibility to execute any apps or services on any device using any resource in the pool. SingularityNet [13] has brought together multiple AI services talking to each other to drive synergies from a decentralized AI. The underlying platform to support this level of communication is a decentralized system on blockchain that support services and apps. The group led by Prof. Bhaskar Krishnamachari in University of Southern California (USC) [19] explores blockchain technology for diverse areas. These are clear evidences that using the decentralized platform to advertise, discover and instantiate services and apps using blockchains is key to materializing highly sophisticated concepts in diverse areas.

Our proposed architecture helps provide a one-stop solution to the (1) sharing of system resources and services, (2) decentralized mediator and decision making, and (3) compensation for sharing. Although such features have been discussed in many technologies and research works before, combining all into a single entity framework has not been well explored.

1.3.4 Opportunistic Collaborative Platform Design

1.3.4.1 Blockchain Overview

Blockchain is a distributed database for an active list of records called *blocks*. Each block contains a timestamp and a link to a previous block, making blockchain a chronological sequence of blocks with all transactions recorded up to that point in time. Just as transactions are mined, validated and added to blocks to be synchronized with the rest of the blockchain network. A blockchain serves as an open, transparent distributed ledger that can record transactions between parties efficiently and in a verifiable and permanent way. The ledger hosts transactions and smart contracts that can be triggered and executed automatically in software. A smart contract is a computerized transaction protocol that executes the terms of a contract. All of its code and data are housed in the blockchain and synchronized across the network.

1.3.4.2 System Overview

The core of this design is a distributed middleware that enables collaboration of mobile IoT elements using the capabilities of the blockchain network. The blockchain framework helps to advertise, disseminate and make resources and services available to the network of IoT devices. Our design focuses on a localized blockchain network of nodes. We position that each network of IoT devices can form a small blockchain network that can expand organically as more devices are registered into this collaborative network. It also provides a virtual gateway to other blockchain networks, making information on resources and services available on a global scale.

As shown in Figure 1.7, there are two key elements that comprise this distributed architecture: (1) applications and services that require resources and (2) a distributed collection of IoT elements (devices and machines) with resources to share. Here, the IoT elements are termed the *collaborating nodes*. Discovery of services and resources (computing and storage), consent for collaboration, invocation of services and allocation of resources are done using *smart contracts* that are deployed by each of the collaborating nodes in the blockchain. These contracts contain information of the list of available services and resources on each node. Through the blockchain framework, these contracts are shared among all the nodes in the network.

A subset of the blockchain nodes, designated (through consensus) as *miners*, create and update smart contracts across the blockchain. For example, in an office room scenario with a collaboration network of a phone, printer, laptop, smart voice assistant and smart thermostat, the device with the highest computing power can be designated as the first miner. In essence, every device in the network can be designated as a miner. The downside is that the information dissemination time would increase because synchronization of each contract update will have to percolate through each miner node and arrive at a consensus. Depending on the number of miners, the convergence can take a few seconds to minutes. On the other hand, a reduced number of miners will result in heavy loads for a few miners. This lends itself to an interesting trade-off between miner count and resource availability.

1.3.4.3 System Workflow

The blockchain setup is a one-time process that involves all the mining and collaborating nodes. Here, we particularly focus on the collaborative execution process and not on the setup process. The setup process is analogous to setting up any IT infrastructure where registration of devices and information routing is checked through template test benches. In the office room example discussed above, the setup process is equivalent to registering the devices in the network of blockchain and designating the potential miners. In this section, we discuss the design details of the execution workflow of the process that happens once a blockchain is setup and a resource requirement is found in the network. How the network of nodes collaborate among themselves in a distributed manner to help each other achieve their tasks (applications and/or services) is the key notion of this workflow design. We discuss this workflow in more detail using the illustration in Figure 1.8. Note that this framework does not implement any new security measures, but, rather, it inherits the security guarantees provided by the blockchain and the underlying wired/wireless network.

(Step 1) Service and Resource Registration. Any node with resources or services to offer registers the resources and/or services with the smart contract deployed in its blockchain network. The contracts contain resources and services available for rent/lease along with their cost and node identity in an encrypted manner. When the resources are advertised, they are recorded along with the identity of the node offering the same. This minimizes the resource discovery time as it accounts for locality in addition to keeping fragmentation to a minimum. Once a resource or service becomes available, it goes through a dissemination cycle, beginning with the blockchain contract being invoked. As this request changes the state of the transaction, it is synchronized with the rest of the blockchain, which involves mining of the

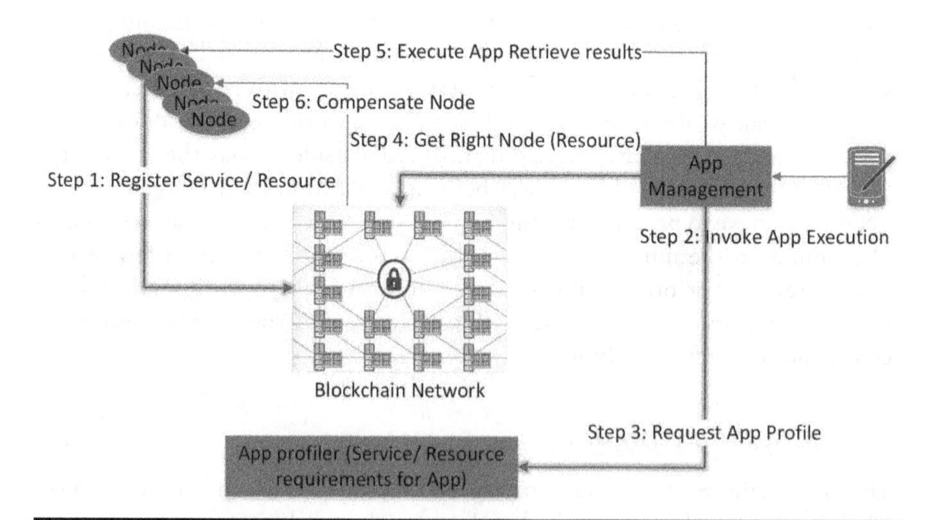

Figure 1.8 System workflow. Lines in bold are the contributions of this approach.

transaction, creation of a block and, subsequently, synchronization with the rest of the blockchain cluster. As the technologies move closer to autonomous operations, it is extremely important for resources and services to be discovered when needed. Having them on the blockchain makes them immediately available and easy to audit: two key aspects as we look into distributed autonomous systems.

(Steps 2 and 3) Application Profiling. Once an application is kicked off, a typical application management system profiles the apps for its required and excess resources. It also acquires the list of services that might be invoked as part of the execution cycle of the apps and some of these services might be spread across the network. The apps and services to profile are invoked through an automated triggering process during the contract preparation.

(Step 4) Resource Allocation. Once distributed or queued for execution, the edge computing units check for the availability of the resources and/or services necessary to execute the apps and services. These edge units can be a single edge computer or a collection of nodes in the blockchain at the edge of the network. If the edge units do not have sufficient computing resources as requested in the contracts, the request is made to the blockchain nodes to execute the apps and services. Selection of the blockchain nodes to execute the apps and services comprises the resource allocation phase. Any traditional resource allocation mechanism that optimizes for latency based on available computing resources works in this case. However, the ad hoc and distributed behavior of the blockchain approach simplify the resource allocation process to a simple *resource matching* process by comparing contracts. This is possible because the contracts are agreed upon based on consensus and all collaborating nodes are informed of every other node's resource requirement and availability. This execution is analogous to pairing and agreement among the collaborating IoT devices in the office scenario to *help* each other with resources to execute the apps pertinent in the network. In the event the nodes in the network do not have excess resources exposed for use, the nodes initiating the apps will fall back to the central edge or cloud computing framework.

(Steps 5 and 6) Execution and Compensation. Once the node with the available resources is acquired, the app is executed and the results gathered. App/service execution is done in sandboxed containers that are deployed in the blockchain nodes during the execution process. As we will describe later, our prototype system uses *docker* [10] containers; however, any type of sandboxing solution will work. In addition to satisfying the service and resource demands, it matches the rental price with the price the requesting entity is ready to pay for the resource or the service. Once either one is granted to the requesting entity, the node renting them is compensated for it with cryptocurrency (e.g., bitcoin [11], ethereum [12]).

Note the distinction between "initiation" and "invocation"; initiation of the app/service involves resource discovery, whereas invocation refers to execution, which is instantaneous. An example is deploying an HTTP service (initiation) versus accessing a webpage using HTTP urls (invocation). Service invocation takes on a lot more importance in solutions that tie them together by cascading them and getting results to be fed as inputs to the next set of services. SingularityNet [13] has set an example of cascading the AI services so that they no longer act as just speech or text AI services, but services that work in tandem to create a coherent solution. Such examples make a compelling case for the need of our decentralized collaborative platform.

1.3.5 Platform Efficiency

We conducted experiments using a prototype implementation of our proposed system to evaluate the feasibility and benefits of the collaborative approach. We use latency and resource utilization efficiency as the metrics for our evaluation, where

- *App execution latency* is defined as the time it takes for all steps in the workflow, including the latency for resource procurement, transfer and deployment, and initiation. This is equivalent to user response time or service delivery time.
- *Resource utilization efficiency* is defined as a combination of the dissemination efficiency to draw up a binding contract that makes the resource available on the blockchain network and the procurement efficiency to allocate resources to the requesting entity.

1.3.5.1 Experiment Setup and Methodology

We set up a blockchain network with four nodes forming a private network to mine transactions and blocks, each on a single thread. The setup comprises three Raspberry Pi3s and a Dell 5100 IoT gateway, all of which have resources to spare (x RAM, y CPU cycles, z storage) and services available that are advertised on the blockchain as part of a contract with appropriate usage costs. Any app that wants to use these resources and/or services will need to request and acquire a handle to them and compensate the offering node. The nodes each have (pseudo) accounts set up to receive compensation for mining, resources leased, and services invoked through pseudo cryptocurrency. Table 1.1 outlines the configuration of the Raspberry Pis and the Dell gateway. Four containerized apps on Docker [10] with different data and storage requirements and execution priorities are used as sandboxes for executing the apps and services. All experiments were conducted in an office room environment with the three Raspberry Pi nodes and the Dell edge gateway placed in the same local area network within a 5-m radius.

Table 1.1 HW Configuration of IoT Nodes

Nodes	*RAM (GB)*	*Storage (GB)*	*CPU*
Raspberry Pis	1-GB LPDDR2	32-GB microSD	4 × ARM 1.2 GHz
Dell5100 GW	2-GB DDR3L-1067 MHz	32-GB SSD	Intel E3825 1.33 GHz

We deploy four containerized apps using our architecture and they are set to be available as services in our prototype system. A Java program that simulates App Manager and Profiler invokes the blockchain contract to procure resources to invoke services. This setup helps evaluate the effects of sharing resources and services on the network. The containerized apps we use for our evaluation are

- Whoami: greets the invoking user; size: 2.1 MB
- Busybox httpd: a full-fledged web server; size: 3.1 MB
- Mysqld: a mysql database; size: 190 MB
- Dockerui: User Interface (UI) for docker management; size: 4.5 MB

1.3.5.2 App Execution Latency

We consider three aspects when evaluating app execution latency: (1) resource procurement, (2) app deployment, and (3) app invocation. We will compare these steps for our proposed *collaborative* approach, purely computing on *edge* and joint edge and collaborative when the system will *fallback to edge* when the resources are not available in the collaborating nodes. Here, edge implies a monolithic powerful edge with compute capabilities closer to the source of data, and our solution implies an "elastic edge" with multiple nodes collaborating. The edge is not opportunistic and all resources necessary for the execution of apps and services have to be provisioned at the edge when the app/service is initiated.

We observe from Figure 1.9 that our collaborative system's execution latency is comparable to that of a purely centralized edge computing system. This shows that

App Size (MB)	App Name	Collaborative (sec)	Edge (sec)	Fallback to Edge (sec)
2.1	whoami	18	19	21
3.1	busybox httd	19	21	24
190	mysqld	35	28	31
4.5	dockerui	21	18	20

Figure 1.9 App execution latency.

a collaborative distributive approach to mobile IoT is feasible. We also note that the collaboration may not necessarily provide all resources requested, in which case the system is able to fall back to the edge. We can observe that the fallback to edge approach has a minimal overhead given that resource discovery time is very small in the blockchain-based architecture due to the information availability on all nodes through smart contracts.

Note that in our setup the edge was placed in the same local network within a 5-m radius of the collaborating nodes. The latency for the app setup phase will scale with the distance between the requesting nodes and the edge unit. In a separate benchmark experiment, we observed a network latency (cellular) of the order of 50 ms for an edge computing device placed 10 miles away from the experiment office location. In the collaborative approach, the apps are present and available on the local network. It is important to note that once the resource is procured for a service app, it is available instantaneously for all other user apps that access the service.

The app execution latency, t_{AE}, is the cumulative sum of the resource procurement time, t_{RP}; app deployment time, t_{AD}; and app initiation time, t_{AI}. The app deployment time is a function of transmission rate and app size, and the app initiation time is transmission latency coupled with message processing time to procure a resource handle. The procurement time is the same as the time it takes for the replicas to reach consensus on when a resource has been released for use or reacquired into the pool. The slope is proportional to a few factors; proportional to the queuing delay at each replica in a bigger network (more than four replicas) and inversely proportional to the transmission delay (increased transmission delay offsets the queuing delay at each replica).

The three latency measures can be expressed as

$$t_{RP} = \left(slope * n_{replicas}\right) + b; slope = a * \frac{t_{queue}}{t_{trans}} \tag{1.1}$$

where a and b are empirically measured constants that account for network factors that impact performance and compute capacity of nodes in a heterogeneous network, respectively.

$$t_{AD} = \frac{size_{app}}{rate_{trans}}; t_{AI} = t_{trans} + t_{msgprocess} \tag{1.2}$$

1.3.5.3 Resource Utilization Efficiency

In contrast to our proposed collaborative approach, a purely edge computing environment does not take advantage of the device resource pool. Table 1.2 shows the baseline resource usage of the four nodes in our network, and Table 1.3 shows

Table 1.2 Resource Usage before Deploying Apps—Percentage of Total Capacity of the Node

Node	*RAM (GB [%])*		*Storage (GB [%])*		*CPU (%)*
RP1	0.381	(38)	5.922/29	(20)	2
RP2	0.329	(33)	7.271/29	(25)	2
RP3	0.421	(42)	8.372/29	(29)	2
Dell	0.597	(30)	11.383/25	(46)	3

Table 1.3 Resource Usage after Deploying Apps—Percentage of Total Capacity of the Node

Node (App)	*RAM (GB [%])*		*Storage (GB [%])*		*CPU (%)*
RP1 (whoami)	0.395	(40)	5.938/29	(20)	2
RP2 (httpd)	0.471	(47)	7.422/29	(26)	4
RP3 (dockerui)	0.489	(49)	8.391/29	(29)	3
Dell (mysqld)	0.723	(36)	12.123/25	(48)	9

the resource usage after deploying the apps using our collaborative model. We can observe from the usage statistics that the collaborative approach enables the allocation of resources through a proper matching of "available" with "requested." In our test case, we ensured that the resources required for the four apps are available in the collaborating nodes. In reality, the system will have to fall back to the edge or cloud if the resource is not available. However, the collaboration presents a first-hand opportunity to run the applications and services merely through mutual agreements and only approach the central edge/cloud units when no agreements can be reached. Due to the distributive nature of blockchains, such a case is rare because finding at least one node that has the available resource has a non-trivial probability.

To understand the resource utilization better, we discuss a use case to highlight the benefit of the collaborative approach: A gaming app needs resources from multiple nodes because a single node does not have the required resources. The request has to be supported by multiple nodes with partial resources.

Suppose the request is for 200 MB of memory, a camera and a microphone, it is possible to fulfill the request from multiple nodes by allocating 50 MB from four different nodes and camera and microphone from individual nodes. However, the caveat here is that the app must be able to accept resources from multiple sources. For example, an application that requires camera and microphone at the same location may not benefit from the collaborative approach; however, this can be resolved by finding a node that offers the RAM while using the camera and microphone from a different node. Our framework supports such a management of resources. Our distributed model allows for handling a heterogeneous set of resources from physically separated entities, which becomes complicated using traditional central resource allocation techniques. This is made possible through the use of blockchain.

1.3.6 Scalability

This model can be used further to explore larger-scale collaborative blockchain networks. Although a detailed empirical study of scale is out of the scope of this chapter, here is how a scaled collaborative edge-computing 5G mobile IoT platform is expected to evolve. The edge, in addition to collaboration with localized nodes as outlined here, will interact with fog or core nodes. These core nodes are analogous to small data centers and have the necessary compute, storage and network capabilities to run substantial workloads. These core nodes will also take advantage of architectural advances made with 5G, for example, NFV, network slicing and SDDC. To complete the end-to-end flow, these core nodes will connect back to the cloud for any central updates, giving this structure centralized and decentralized control. Blockchains, used for application and service coordination will evolve into a more sophisticated architecture with localized structural boundaries with more frequent synchronization that fits into a more globally distributed structure. On these blockchain nodes, although resource/service procurement times on a small network with little transmission delays are found to vary linearly with the number of nodes, queueing delays play a huge role in bigger networks, driving consensus times higher and making it important to profile and optimize the blockchain network for the workloads in which we are interested. Although empirical evidence from this chapter shows an ideal fit in localized networks like office buildings, store locations, or industrial floors, this scaled architecture over 5G with edge computing and blockchain-based collaboration would be ideal for any complex IoT workload (Figure 1.10).

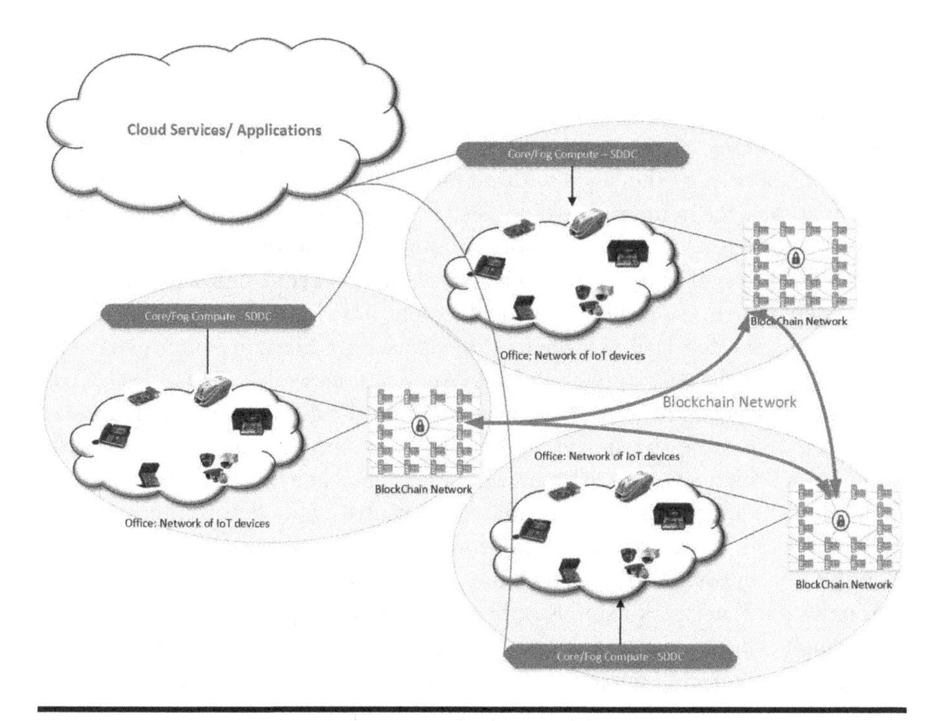

Figure 1.10 Collaborative edge scaling.

1.4 Conclusion

This chapter discussed IoT applications in the context of 5G, their characteristics and the reasons for their success with a 5G network backend. Even as all verticals and domains will benefit from 5G and IoT, the automotive sector is expected to be the biggest beneficiary in this technology prowess with connected car, smart vehicles and transportation infrastructure use cases.

As we start to see 5G deployments to support a slew of applications, we will see some architectural shifts in order to support a flexible and extensible infrastructure. With network slicing, we noticed various layers that were loosely bound so they could evolve separately, whereas with NFV, we had a way of virtualizing network functions moving away from dedicated hardware for individual functions and SDDC provided a programmable infrastructure so that all changes could be configured instead of hardware provisioned. We explored an application framework on top of this infrastructure, EdgeX, which made the deployment and management of this complex set of applications straightforward with a full application stack at the edge.

Although many of these applications across industries exist in the cloud, there is a huge shift to the edge, with many applications finding a home in the gateways at the edge in order to avoid huge amounts of data flowing to the cloud. It is predicted that while the cloud continues to be a prominent force, the battle for IoT will be fought at the edge. Even as the cloud and edge figure prominently in most deployments, another likely deployment paradigm that is taking shape is the core at the fog layer, closer to the edge with a compute-intensive stack that can support heavier applications. As applications evolve, and as we strive for a better utilization of resources and services on the devices, an opportunistic collaborative resource sharing for mobile IoT systems becomes extremely relevant. We designed a novel architecture that uses blockchains for collaboration. We developed a mechanism that enables opportunistically identifying available resources and advertised services, and invoking or utilizing them based on a collaborative consensus. We implemented a preliminary prototype of a four-node collaborative IoT network using Raspberry Pis and an edge computing gateway, and through experimental evaluation, showed the feasibility of the collaborative architecture with overall app execution latency and resource utilization as being comparable to the traditional centralized edge approach.

We expect to see an exciting next decade, with 5G and IoT driving innovations across industries and pushing the boundaries on technology and business.

References

1. Machina Prediction 5G IoT. https://machinaresearch.com/news/machina-research-predicts-10-million-5g-internet-of-things-connections-in-2024/. Accessed on March 15, 2018.
2. 5G Impact on IoT. https://internet-of-things-innovation.com/insights/the-blog/5g-means-internet-things/.w2padnhkjox. Accessed on March 15, 2018.
3. Enable Next-Gen Services with Multi-Access Edge Computing. https://www.vmware.com/content/dam/digitalmarketing/vmware/en/pdf/solutions/vmware-mec-solution-overview.pdf. Accessed on March 15, 2018.
4. VMware Software-Defined Data Center. https://www.vmware.com/content/dam/digitalmarketing/vm whitepaper-sddc-capabilities-itoutcomes-white-paper.pdf. Accessed on March 15, 2018.
5. 5G Telecom Public Private Partnership. https://5g-ppp.eu/wp-content/uploads/2017/07/5g-ppp-5g-architecture-white-paper-2-summer-2017-for-public-consultation.pdf. Accessed on March 15, 2018.
6. MEC: The Promise of Disruption. https://www.sdxcentral.com/mec/definitions/mec-5g/. Accessed on March 15, 2018.
7. Dell Edge Core Architecture. https://delltechnologiesworldonline.com/2017/connect/filedownload/ses. Accessed on March 15, 2018.

8. 5G IT Edge Cloud. https://www.pcmag.com/article/359036/it-needs-to-start-thinking-about-5g-and-edge-cloud-computing. Accessed on March 15, 2018.
9. Dell EdgeX Gateway Stack. https://www.edgexfoundry.org/. Accessed on March 15, 2018.
10. DockerOrg. https://www.docker.com/what-docker. Accessed on March 15, 2018.
11. BitcoinOrg. https://bitcoin.org/en/. Accessed on March 15, 2018.
12. Ethereum. https://ethereum.org. Accessed on March 15, 2018.
13. Cassio Pennachin Ben Goertzel, David Hanson. Singularitynet: A decentralized, open market and inter-network for AIs.
14. M. H. Yun, A. Amiri Sani, K. Boos and L. Zhong. Rio: A system solution for sharing I/O between mobile systems. *MobiSys'14 Proceedings of the 12th Annual International Conference on Mobile Systems, Applications, and Services*, 2014.
15. K. Watanabe, D. J. Dubois, Y. Bando and H. Holtzman. Shair: Extensible middleware for mobile peer-to-peer resource sharing. *ACM 978-1-4503-2237-9/13/08*, 2008.
16. D. Karger, M. F. Kaashoek, I. Stoica, R. Morris and H, Balakrishnany. Chord: A scalable peer-to-peer lookup service for internet applications. *SIGCOMM'01*, 2001.
17. A. Salem and T. Nadeem. Lamen: Leveraging resources on anonymous mobile edge nodes. *ACM 978-1-4503-4255-1/16/10*, 2016.
18. IBM.com. Adept: Empowering the edge; practical insights on a decentralized internet of things.
19. BlockChainUsc. http://blockchain.usc.edu. Accessed on March 15, 2018.

Chapter 2

Emerging Challenges and Requirements for Internet of Things in 5G

Lina Xu, Anca Delia Jurcut, and Hamed Ahmadi

Contents

2.1 A Brief History and the Classic Layer Structure

2.1.1 A Brief History of IoT

A WSN refers to a number of wired or wireless connected microdevices that are normally used for environmental monitoring and data collection. Those individual WSN devices can sense events, process the information, and store the data locally or transmit it throughout the network to a data center [1,2]. The concept of WSN was introduced nearly 30 years ago and it was originally proposed for military surveillance purposes [3].

Due to its first success, this technology was then adopted in many other fields, such as habitat [4], weather [5], agriculture [6], and wildlife monitoring [7]. However, still more effort was required before WSNs could truly make a dynamic improvement to our everyday life. According to the idea presented in [8], a similar and successful paradigm to compare against is the case of the Internet. The Internet was invented in the late 1960s. However, it did not become universally popular until 1995 when the Internet could be accessed more conveniently. The number of Internet users has dramatically increased over the last 10 years owing to the rich over-the-top (OTT) applications developed for it, such as the World Wide Web, electronic mail and social networking.[1]

The revolution that is necessary to ensure WSN technologies flourish in a similar manner as the Internet requires more effort in simplifying design, implementation, deployment and usability. The societal impact of WSNs is defined as the number of the users, which itself is determined by the quality and quantity of the available applications. Currently, it is extremely urgent to inspire developers to build more useful OTT applications to improve people's life quality or experience. Therefore, the concepts of the IoT [9] and the Internet of Everything (IoE) [10] were proposed, and now they are widely accepted and highly valued. Along with data mining and predictive analysis, a large body of applications includes smart homes, smart cities, smart transportation and smart health care [11].

WSNs were the former existence and one technology foundation for the IoT. However, there are several fundamental differences between WSNs and the IoT in terms of the dynamics and diversities. We will take the following aspects to discuss the differences in detail:

- **Objectives:** As mentioned, WSN systems are mainly used for monitoring the environment and collecting data. The user profile and system context are rarely under discussion. One of the main objectives for IoT systems is to improve people's lives and their personal experiences. User-oriented services and context-aware design are prioritized in IoT systems.

---

[1] http://www.internetworldstats.com/stats.htm.

- **Number of connected devices:** By the end of 2030, according to Cisco [12], there will be more than 500 billion connected devices. Other sources may have different predictions, though the predicted numbers are all extremely high. Intelligently managing systems with such a large number of devices along with all the generated data is a new thing belonging to IoT.
- **Applications:** Traditional WSN applications are normally developed for environment monitoring, only involving sensors deployed and fixed in the field. The volume of the data transmitted in the network is relatively small. In contrast, IoT applications can range from the smart kitchen to the smart city. The applications all have their own specific requirements on the lower layer network performance. For instance, smart road lighting in smart transportation systems requires short network delays. Smart monitoring in smart city systems demands high network bandwidth.
- **Communication ability:** Because the features of the connected devices in IoT systems vary greatly, their communication ability, including communication range, communication band, transmission rate, and power consumption, can all be different. This feature brings challenges to network management, and the physical architecture is critical in term of the quality of service (QoS).
- **Hardware capability:** WSN systems normally have homogeneous devices composed of the same sensors, the same micro-controller or processor and the same battery level, indicating identical sensing ability, computing capability and power supply. This feature can simplify the design of the relevant protocols, such as routing protocols. However, for IoT systems, this feature is no longer applied. For example, in a smart home system, the refrigerator, the vacuuming robot, and the light sensing devices all have their own hardware characteristics. Furthermore, the devices in today's IoT systems are no longer limited to those basic sensor motes.

2.1.2 The Classic Layer Structure

The traditional open systems interconnection (OSI) comprises seven layers. Because IoT systems normally have a large range of varieties, from the choice of the hardware to the types of the applications, the traditional seven network layers are simplified to three layers: the perception, networking, and application layers, as shown in Figure 2.1 [13]. The heterogeneous nature of IoT applications has prompted the requirement of a well-defined structure in order to support their various functionalities.

Perception Layer: Perception layer refers to the data-collecting end devices such as sensor tags, sensor nodes, actuators, and sensor gateways, which can support a range of communication technologies such as radio frequency identification (RFID), ZigBee, Bluetooth, or wireless fidelity

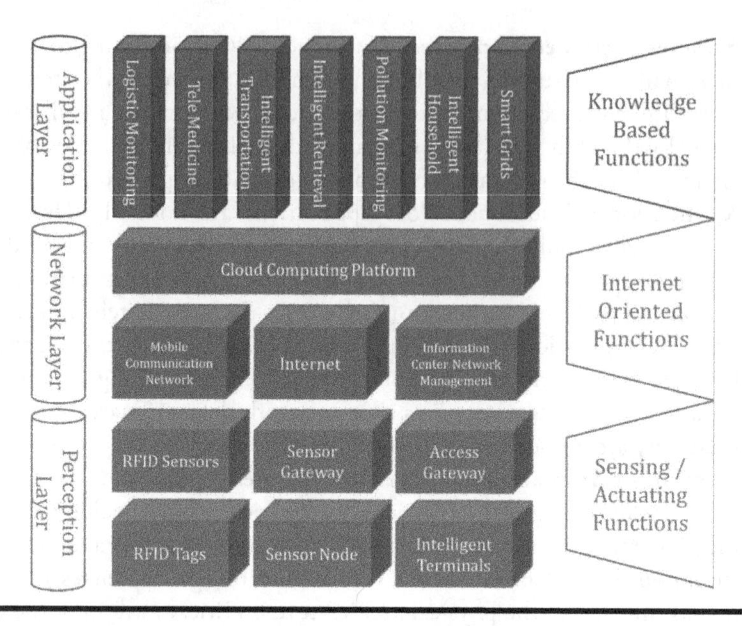

Figure 2.1 Classical three-layered IoT architecture.

(Wi-Fi). This layer represents the function of extracting the raw data from the environment and transmitting the data to higher layers or to other IoT devices for further processing or storage. According to the functionalities, the perception layer could be classified into two sublayers: the perception node layer and the perception network layer [14]. The node layer is responsible for data sensing and gathering. The network layer is in charge of interaction with the upper layer and provides necessary infrastructure for networking and data transmission. The communication ability of the networking layer will be greatly improved if the perception layer can support 5G technologies.

Network Layer: The network layer is the middle layer of the IoT three-layer architecture. It is designed for ensuring reliable transmissions of the data conveyed from the perception layer and facilitating the storage and processing functions while supporting access management. The entities in this layer vary from smart mobile devices to network devices to servers or the cloud computing platform. The infrastructure for the 5G mobile communication network can be utilized to enhance the connectivity among those IoT devices. Recently, many IoT devices have become quite powerful, with state-of-the-art battery and system-on-a-chip (SoC) technologies.

Application Layer: The application layer is a more rapidly updating and vastly diversified layer compared with the two lower layers. It enables a smart environment for users to interact with the entities in the other layers. In general, only the application layer is visible for the end users. This layer mostly involves data storage and analysis, data visualization, smart decision making, and control. For example, intelligent household, smart transportation, connected health, smart grids, and logistic monitoring are served based on the application layer. Because 5G can provide full connectivity and low-energy communication, a larger range of applications can be expected in the near future.

IoT systems are deployed in many domains, operating to fulfill a set of specific objectives and purposes. Based on the three-layer structure, there are two primary essential system-level functionalities required in IoT systems:

Data Collecting: The initial motivation to deploy IoT devices and systems into the field is that we need to know what is currently happening without physically being there. The systems can gather the data for background data mining and predictive analysis through machine learning techniques. The information and data first can help us understand our environment or business model better, and then actions based on those understandings can be more efficient and effective. The data collection process is referred to as data flow from the field to the server/cloud, replying on the perception and networking layer.

Interacting and Managing: IoT systems should be able to manage and control the designated equipment, space and areas intelligently. Collecting and analyzing the data are only part of the duties of IoT systems. We call an IoT system a real "smart system" only if the system can perform context-aware operations and proceed with live interactions [15]. Some of these systems can also make intelligent decisions because they can predict the future. Interacting with the sensing filed and managing the information collected are normally through instructions and commands injection. The implementation is replying on the application layer requirements.

The data and information exchanging between the field and the server/cloud are highly dependent on the connectivity of the system. If the field is extremely large, for example, smart cities, there will be high system requirements for scalability. The algorithms and protocols operating on different layers should consider scalability as one important design principle. When the systems are expected to operate for years, then energy-efficient protocols are essential in order to maintain system longevity. In many circumstances, data can only be meaningful if collected in a consistent manner, then the system availability and reliability will be mostly appreciated, such as fall detection systems in health care. Like any other cyber-physical system, privacy and security are also highly prioritized and many challenges need to be resolved when deploying such a system in the real world. There are many other requirements on IoT systems depending on the proposed architecture and their specific functionalities. For instance, a fatal signal detection system will have high requirements on the data accuracy. For smart transportation or auto driving, system delay is critical to making the right decision.

2.2 A New Era for IoT in 5G

For the first time in history long-term evolution (LTE) has brought the entire mobile industry to a single technology footprint, resulting in unprecedented economies of scale. The converged footprint of LTE has made it an attractive technology baseline for several segments that had traditionally operated outside the commercial cellular domain. There is a growing demand for a more versatile Machine-to-Machine (M2M) platform. The challenge for the industrial segment is the lack of convergence across the M2M architecture design that has not yet materialized. It is expected that LTE will remain as the baseline technology for wide-area broadband coverage also in the 5G area.

Figure 2.2 presents an overview of 5G network infrastructure. Mobile operators now aim to create a blend of preexisting technologies covering 2G, 3G, 4G, Wi-Fi, and others to allow higher coverage and availability and higher network density in terms of cells and devices with the key differentiator being greater connectivity as an enabler for M2M services [16]. New machine-type communication technologies have also been invented, such as long-term evolution–machine (LTE-M) and narrowband IoT (NB-IoT). An array of antennae supporting high-order MIMO is installed in a device and multiple radio connections are established between the device and the cellular base station, allowing simultaneous data transmission. Meanwhile, operators, vendors and academia are combining efforts to explore technical solutions for 5G that could use frequencies above 6 GHz and reportedly as high as 300 GHz. This platform will need to provide a network management and control layer to coordinate the activities from the application layer and the services from the underlying infrastructure.

This layer should be implemented between the network layer and the application layer. It provides functionalities/components as such network selection, traffic monitoring and user analysis. For example, smart network selection can be implemented

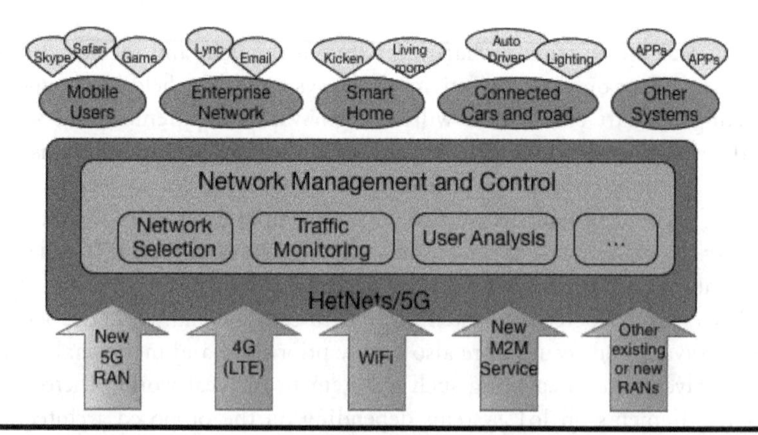

Figure 2.2 An overview of 5G network supporting IoT applications (RAN: Radio Access Network).

in the following way: The network selection component can match OTT usage to a suitable network interface based on the characteristics of the OTT application itself, the network conditions and the user profile. The traffic monitoring component can provide the network condition information. As we can see, those components will operate collaboratively to achieve intelligent 5G communication.

Many IoT systems can be supported and enhanced if implemented on top of the 5G network. For those systems, running multiple applications and usages at the same time is quite common. The network selection should know the applications and user scenarios in order to allocate suitable underlining network services for communication [17]. This information can be accomplished by the user analysis component. Besides, the network context itself, such as the congestion level, is also important when selecting the network interface [18]. Therefore, traffic monitoring is necessary. 5G aims to enhance the degree of automatic adaptation and configuration. User-oriented and quality of user experience (QoE)–aware design is one of the main tasks. The management layer should decouple software functions from the hardware resource layer. Meanwhile, it also needs to provide network performance analysis and optimization. The 5G networks aim to advance IoT technologies farther in the market [19].

Third Generation Partnership Project (3GPP) standard 5G-based backhaul has become popular as a great solution to connectivity problem for IoT systems. In recent years, many novel system architectures have been proposed and published [20], indicating that the next generation of mobile networks (i.e., the 5G network) will need not only to develop new radio interfaces or waveforms to cope with the expected traffic growth but also to integrate heterogeneous networks from end to end (E2E), with distributed cloud resources to deliver E2E IoT and mobile services. A novel backhaul architecture through the mobile network for smart building applications has been provided to improve services for users and will offer new opportunities for both service providers and network operators [21]. It has been revealed that the uplink for LTE network infrastructure remains idle most of the time and the resource can be used as the communication backhaul for IoT systems, which, on the other hand, rely on an uplink for data gathering [22]. An example of such a network can be a heterogeneous network in which a macro-cell tier is overlaid with a very dense tier of LTE-femtocells [23]. Optical wireless communication (OWC) can be considered as a mobile backhaul supporting Wi-Fi, LTE and 5G as a new access technology in IoT systems where it enables secure and reliable communications at low latency [24].

The 5G network is the ultimate means to connect everything together, building a great foundation for IoT systems [25,26]. Small cells take a huge role in such infrastructure [27,28]. In addition, software-defined networking (SDN) and network function virtualization (NFV) techniques are also widely used in such scenarios [29,30] to improve the scalability. Smart backhaul solutions have been proposed to improve users' utility and QoE [31,32]. Combining cloud technology with the cellular backhaul is also becoming dominant for IoT infrastructure

[33–35]. Network selection and resource allocation in 5G network is still a challenging research topic that urgently needs to be addressed [36].

In addition to the architecture design, resource management and load balancing in the backhaul is also critical to QoS, and several intelligent solutions have already been proposed for 5G [37–39]. Network reliability can also be improved in 5G through cognitive radio [40]. As we can see from the above evidence, 5G can provide a highly connected and reliable backhaul infrastructure for many IoT systems.

2.3 System-Related Challenges and Requirements

As discussed in Section 2.1.2, some general requirements such as energy efficiency, latency, availability and reliability, privacy and security exist in most IoT systems. In this section, the challenges and requirements in terms of system design and privacy and security are presented when deploying IoT systems in 5G scenarios.

2.3.1 System Architecture Design

Good system design and architecture can simplify the process when addressing the challenges in IoT and can also bring convenience when trying to meet the requirements.

2.3.1.1 Dust/Fog/Cloud Infrastructure

To deal with such a high complexity in IoT, a huge amount of industrial and academic effort has been put into developing gateway frameworks to facilitate device integration and application development. The dust/fog/cloud infrastructure [41] assisted by 5G backhaul has been commonly accepted as an effective and efficient way to manage the network and collect data. "Dust" refers to the end IoT devices. Fog is an architecture that aims to distribute computation, communication, control and storage resources closer to the end devices along the cloud-to-things continuum [42]. Sometimes the term fog is used interchangeably with the term edge. The data volume generated from the dust can be overwhelming and highly duplicated. The fog can significantly leverage the data processing and storage pressure from the cloud to the edge. Implementing fog/edge computing into IoT systems and performing intelligent resource allocation is still challenging. In addition, utilizing and integrating 5G backhaul to minimize the deployment cost of IoT infrastructure is another challenging research topic.

2.3.1.2 Cross-Layer Design

The perception layer can be seen as the combination of the traditional physical layer and the Media Access Control (MAC) layer. It can include two-dimensional barcode labels and readers, RFID tags and reader-writers, camera, GPS, sensors, terminals and the associated wireless infrastructure. The networking layer is responsible for the data transmission and communication inside the system and with the outside Internet. Reliability and energy efficiency are normally considered as the design principles [43]. The application layer is to analyze the collected data and provide services to the end users [44]. The perception layer and network layer together are considered as the foundation for the whole IoT system [45]. Together, these two layers provide the backbone and the fundamental infrastructure of an IoT system. However, the architecture design and detailed implementation normally can only be confirmed after knowing the application layer design. If the availability of any of the three layers (perception, network and application) fails, the availability of the whole system collapses. Where the system will be deployed, what size the field will be, and what kind data will be collected are all issues involved in the applications but highly affect the decision making on the perception layer and network layer. Cross-layer designs are widely applied for networking and communication protocol implementation [46,47] in order to improve reliability and energy efficiency. The concept of the "Web of Things" [48] has been introduced to enable interactions with lower-layer IoT devices with high visibility and performance for the system users. It aims to create an illusion of integrating the perception and networking layers into a whole [49].

It has been frequently argued that although layered architectures have been a great success for wired networks, they are not always the best choice for wireless networks. To address this problem, a concept of cross-layer design, based on an architecture where different layers can exchange information in order to improve the overall network performance, has been proposed and is becoming popular. There is a large body of recent state-of-the-art work for cross-layer protocols in the literature [50]. However, the 5G network is a 3GPP-defined cellular network with limited variations that can be made to support different protocol designs. Therefore, how to utilize existing 5G core network components to facilitate cross-layer implementation is the main challenge. Cross-layer technologies still require more multidisciplinary effort toward 5G application.

2.3.1.3 Middleware

Researchers from academia and industry are exploring solutions to enhance the development of IoT from three main perspectives: scientific theory, engineering design, and user experience [51]. Those activities can enrich the technologies for IoT, but they also increase the complexities when implementing such a system in the real world. For this reason, the concept of IoT middleware has been introduced and many systems are already available, such as Contiki [52], TinyOS [53], and OpenWSN [54]. However, the formal definition for IoT middleware is still missing. Researchers all have their own understandings. In some context, IoT middleware is equivalent to IoT Operating System (OS). In general, middleware can simplify and accelerate the development process by integrating heterogeneous computing and communications devices and supporting interoperability within the diverse applications and services [55]. Most existing implementations for middleware are designed for WSN, not for a service-oriented IoT system. Therefore, some IoT-specific middleware such as [56,57] have been proposed, focusing on supporting diverse IoT services.

In reality, middleware is often put in place to bridge the design gap between the application layer and the lower infrastructure layers. The requirements for middleware service for the IoT can be categorized into functional and nonfunctional groups. Functional requirements capture the services or functions such as abstractions and resource management [58]. Nonfunctional requirements capture QoS support or performance issues such as energy efficiency [59] and security [60]. When involving 5G as part of the network infrastructure, middleware has the potential to provide better functional and nonfunctional services by utilizing the great features of 5G. However, the line between middleware and lower network infrastructure has never been clarified. How to design a middleware architecture to enable smart interaction and management over 5G is where the challenge lies.

2.3.1.4 Adaptation to Multiple Input–Multiple Output

5G, which is an enabler for M2M-type communication, has features such as 1- to 10-Gbps speed, 1-ms latency, and 100% coverage and reliability [61]. It aims to provide QoE-aware services for a large range of applications and usages. MIMO technologies are introduced as one important approach to achieve those 5G goals [62,63]. This is where an array of antennae is installed in a device and multiple radio connections are established between a device and a cell [16]. IoT technology has been proposed with a higher degree of diversity in terms of communication abilities and user scenarios, supporting a large body of real-world applications. Cellular-involved backhaul is becoming a popular approach adopted by many IoT systems in order to provide full connectivity [64]. Many IoT devices now are designed with MIMO abilities [65].

Normally, there are two ways to increase the quality of network users' experience. Straight-forwardly, the first one is to increase the bandwidth. The second solution is to provide a wider variety of services than just a single class of best-effort service [66]. The relationship can be classified into three types of utility functions, which already have been well studied and include two types of inelastic traffic and one type of elastic traffic [67]. Based on the utility functions, it can be seen that different types of services should be implemented in the networks to cater to elastic and inelastic traffic, respectively. This is extremely useful in MIMO because the network itself is capable of providing various communication interfaces, and each interface has its specific features. The idea of MIMO provides a great platform for service classification. If the traffic from a user can be classified into a few categories and then mapped into suitable network access interfaces, the user's QoE will be significantly improved compared with that in a single-class service network. Service classification and network selection are necessary in order to improve user utility.

Because the 5G networks aim to cater to more diverse scenarios, it is highly possible to have several applications running on one single 5G-enabled device. When MIMO is available, utilizing all the available radio access networks interfaces and intelligently distributing the traffic flow to those interfaces can significantly improve users' utility. Smart network selection over MIMO is critical to improve QoE in the 5G environment. According to the regulations in 3GPP, Internet Protocol Flow Mobility (IFOM) are already supported by most mobile devices, many of which can be used as IoT devices. Based on the network selection rules defined in access network discovery and selection function (ANDSF) [68], the ForFlowBased rules in IFOM can map different network traffic/usages to preferred network interfaces. This component can even be extended to adapt to specific requirements from users [69]. However, the static policies in ANDSF fail to address unpredictable user behaviors. Some proposed traffic offloading protocols and approaches can be extended to support network selection in a MIMO scenario. Solutions for smart network selection following 3GPP 5G standards—which can reflect factors such as dynamic network changes, financial cost, user priority, and traffic characters—are still missing [70]. Without proper management, the profit coming from the extra bandwidth and cost may not meet our expectations.

2.3.1.5 Smart Applications

IoT technologies have been applied in many domains to advance the management and control of the traditional systems. The requirements and challenges in each system are different and application dependent. Herein we will demonstrate those challenges and requirements through several popular applications from the system design perspective.

IoT-based **intelligent transportation systems (ITS)** are introduced to improve transportation safety and degrade traffic congestions while minimizing the environmental pollution. In order to reduce average trip time and achieve energy/fuel efficiency, the management can be effective if the system has a global picture of a whole city including its transportation infrastructure and traffic flow for vehicles, cyclists and pedestrians. The entire system could be considered as a vehicular network, whereas the communications are established between vehicle-to-vehicle (V2V), vehicle-to-infrastructure (V2I), vehicle-to-pedestrian (V2P), and vehicle-to-grid (V2G). Therefore, supporting city-level full connections and data collections is the key challenge to providing intelligent management to enable the traffic to flow smoothly and efficiently. Establishing communication in an ITS requires a duplexer at each vehicular entity and roadside V2I hubs. However, deployment of such duplexing devices in practice would be quite challenging and the real-time data requirement that is generated from them could only be satisfied by the potentiality of 5G. Energy efficiency is also one of the critical requirements for smart transportation. In isolation, individual IoT devices are not generally big consumers of energy; however, given that IoT devices will number 50 billion by 2020, collectively they are a large consumer of electricity. The problem has been recognized by manufacturers of IoT hardware who have introduced protocols and technologies to support energy-efficient communication. Techniques to reduce power consumption of those devices—even just a 1% reduction—can make a huge impact when considering the whole picture. If we consider that the sources of the electricity can be gas or coal, electricity efficiency can contribute to the reduction of CO_2 emissions even more. This is the motivation for many related studies in this direction, such as energy-efficient routing and mobile edge computing. Similar challenges and requirements also exist in many other large-scaled IoT systems, such as **smart cities**, **smart agriculture**, and **smart energy/grid**.

IoT-based **telemedicine and-health care** systems are the most profitable and funded projects in the entire world. This is mainly due to the higher aggregate of aging people and that health is the most concerned aspect of human life. A sensory system embedded with actuators can be provided for individuals to use as a wearable device for tracking and recording vital measurements such as blood pressure, body temperature, heart rate, and blood sugar, along with exercises carried out by the individuals. This data would be conveyed and stored in a cloud as a personal health record (PHR) to be accessed by the user and the assigned physicians. Data accuracy and privacy are the most prioritized requirements in health care-related systems. Efforts should be made from all IoT layers to enable data accuracy and privacy. Failure to do so can cause physical damage, and sometimes loss of life. Privacy is also critical for other private smart systems, such as **smart homes**. Accuracy and availability of great concern in the Industrial IoT (IIoT).

2.3.2 Privacy and Security

Security is defined as a process to protect a resource against physical damage, unauthorized access, or theft by maintaining high confidentiality and integrity of the assets' information and by making information about that object available whenever needed. As for any other computer system, there are also high requirements for security for IoT systems. Most of the IoT systems collect various types of data, which sometimes can be highly sensitive and personal. Therefore, data confidentiality, privacy, and trust are three key concerns with IoT devices and services [71]. Privacy and security is considered one of the most critical QoS features in IoT systems [72].

Wireless broadcast communication suffers more security risks than others, and multihop wireless communication can be worse given that there is no centralized trusted authority to distribute a public key in a multihop network due to the nature of distribution. Even though 5G infrastructure can provide secured transmission, risks remain at the end devices and in the cloud server. Current proposed security approaches may be effective to a particular security issue in a specific layer. However, a strong need still exists for a comprehensive mechanism to prevent security problems in all layers [73]. Security issues, such as availability, need to be addressed and not only at each layer; a good cross-layer design and communication is encouraged. IoT systems are generally large and complex systems with many interconnections and dependency, such as in smart cities [74]. The heterogeneous nature of communication equipment deployed with IoT and the rapidly increasing population and industries would cause scalability issues for security. However, the security issues could be addressed at each layer. In general, the lower layer infrastructure (i.e., the perception and networking layers) must protect itself from bad behaviors from attackers and harmful control from unauthorized users. The application layer should be continuously available for authorized users without interruption from unauthorized users. A mutual authentication protocol with strong cryptographic primitives would ensure the trust, whereas practices such as homomorphic encryption would ensure the privacy and anonymity. Application layer security concerns vary from one application to another.

Privacy and security issues exist in almost every IoT application. The information circulated through the ambient assistant living (AML) application would pose a privacy concern for consumers for disseminating information regarding their habitual activities derived through power consumption patterns, but the impact could be severe for industries. Therefore, all data transmission should be secured through encryption. The smart health-care systems pose the most significant privacy threat for their users because extremely sensitive data can be collected and stored in the IoT cloud. Thus, the same privacy concerns also exist in cloud storage and computing. At the same time, wearable IoT devices face limited resource capability issues such as battery power, memory and processing ability. The main effort involves implementing a lightweight but highly secure system. In smart transportation systems, the privacy of drivers should be safeguarded from external observers where location details, driving patterns, and navigated routes would be exposed and allow them to take harmful initiatives. The security problem in ITS exist also due to the larger number of entry points that make it vulnerable to diverse attacks to be targeted from many sources [75]. The drivers, who are not engaging in any authentication activity when using the vehicles, might be subjected to privacy violation on their driving behavioral patterns. Privacy and security issues pose a significant intrinsic challenge toward 5G technologies and secured communication when various protocols such as secure socket layer (SSL)/transmission layer security (TLS) and Internet Protocol Security (IPSec) [76] are demanded.

2.4 Other Issues in IoT

2.4.1 General Data Protection Regulation

Because the new general data protection regulation (GDPR)[2] became enforceable on May 25, 2018, protecting user data and securing user privacy are urgent and predominant issues to be solved for any IoT applications. Users' data must not

---

[2] https://www.eugdpr.org.

be detected or captured without their knowledge. Privacy has the highest priority for all existing and future application development, including IoT systems. Users' identities must not be identifiable or traceable. Under the new legislation, data processing must be fair and transparent. The motivation and legitimate purpose for processing the information should be stated clearly in the first place. Any data collected during the entire operation must be adequate and relevant to the system's designed purposes. Organizations are required to capture the minimum amount of data needed to fulfill the specified tasks and objectives while guaranteeing the information to be accurate and valid. They should avoid replicated storages and redundant data collections. It is essential to protect the integrity and privacy of the data by making sure it is secure. All the data collected must be accountable.

Regulations and policies should be in place to protect user privacy and data. Increasing the system protection and security level is no doubt the most direct and effective way to follow those regulations. Other advanced technologies also should be implemented to facilitate lawful management and monitoring. Mobile edge computing or fog computing technologies supported in 5G can leverage part of the data processing into the edge of the network—the end IoT devices or the first gateway. Such an approach can greatly reduce the redundant data transmission and unrelated data storage in the data center. It also reduces the security risk over wireless transmission. Technologies that enable missing data prediction can reduce the amount of data collected while maintaining analysis accuracy. Intelligent data fusion and data analysis can present meaningful information and conclusions to the system manager without overexposing the users' private data.

2.4.2 Internet of "Too Many" Things

IoT technologies have been and will be deployed in many scenarios to provide better services and support advanced management, scaling from smart home to smart cities. Applied IoT technology can be seen in industrial predictive maintenance, connected health and translational medicine, smart transportation, asset tracking, smart cities and many other instances. For example, in the ITS domain, IoT contributes to smart parking, autonomous vehicles, smart traffic control, smart routing, traffic light sequencing, smart road lighting, bike sharing and public transportation. It is likely that the uses of IoT in this area will increase further as new application cases are required and full connectivity can be provided through 5G. The use of IoT in ITS demonstrates the quantity of IoT devices within a single domain.

Although the predictions for the number of IoT devices differ, all sources agree the number of devices will be large. For example, Gartner Research [11] expects to see 20 billion IoT devices by 2020. Statista[3] shows that by 2020, the installed base of the IoT devices is forecast to grow to almost 31 billion worldwide. There would be 75 billion

[3] https://www.statista.com/.

IoT devices by the end of 2025. Cisco [10] has also predicted that there would be 50 billion IoT devices by 2020. All the sources have strong confidence that the number of IoT devices will be extremely large and the total number is expected to keep increasing.

We propose the term "IoT flood" as shown in Figure 2.3 to describe the current situation of IoT. The side effects include problems from several aspects such as environmental concerns due to physical pollution and energy consumption, health concerns due to heavy radiation, and security risks and privacy issues. IoT devices and the infrastructures are like water, permeating our living environment even without our awareness. The increase in IoT devices has been supported by decreases in manufacturing costs, improvements to the reliability and accuracy of sensors, and great communication infrastructure such as 5G as well as cheaper data storage. This has led to IoT systems being deployed for many uses by industry, individuals and governments. Rivers, farmland, forests, oceans and urban environments are at risk of the IoT flood as new IoT systems are deployed to monitor and act in diverse situations.

As the proliferation of the IoT within the single domain of ITS demonstrates, it is now time for us to step back and carefully examine the IoT and question the problems that the IoT flood introduces while we consider ways to prevent the flood advancing. This is particularly relevant given that Gartner has predicted that the IoT is currently at the peak-inflated expectation in the hyper cycle for emerging technologies. The inflation will continue to increase and then dynamically drop due to the realization of the reality that the IoT is not the panacea for all the world's problems. Are so many systems

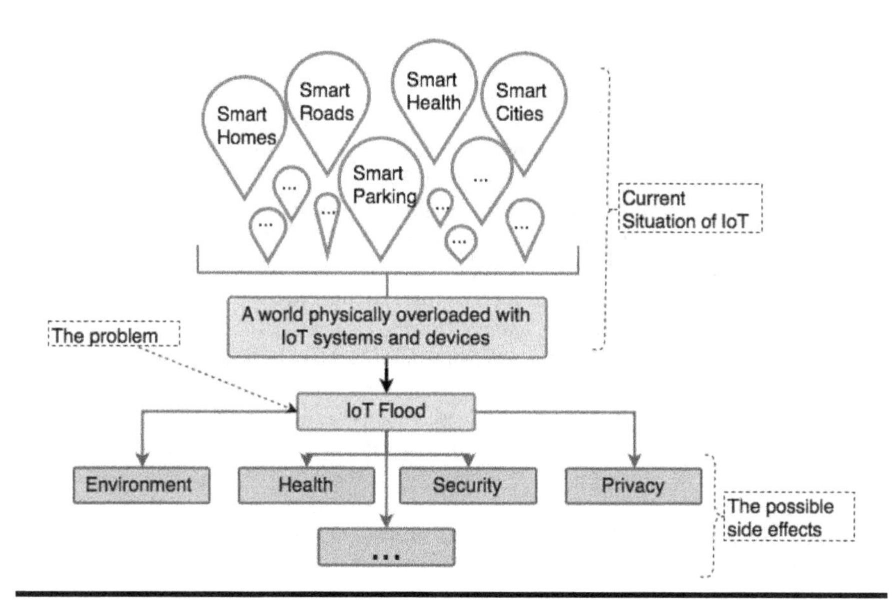

Figure 2.3 The causation of IoT flood and possible side effects. (From Xu, L. and Mcardle, G., *IEEE Access*, 6, 62840–62848, 2018.)

all necessary? What are they actually going to bring to us? The challenge emerging now is how to provide healthy and green IoT system design with manageable development and data collection. These are the keys to support real sustainable development.

References

1. C.-Y. Chong and S. Kumar, Sensor networks: Evolution, opportunities, and challenges, *Proceedings of the IEEE*, vol. 91, pp. 1247–1256, 2003.
2. I. Akyildiz, W. Su, Y. Sankarasubramaniam, and E. Cayirci, Wireless sensor networks: A survey, *Computer Networks*, vol. 38, no. 4, pp. 393–422, 2002.
3. K. Romer and F. Mattern, The design space of wireless sensor networks, *IEEE Wireless Communications*, vol. 11, pp. 54–61, 2004.
4. J. Polastre, R. Szewczyk, A. Mainwaring, D. Culler, and J. Anderson, Wireless sensor networks, In *Analysis of Wireless Sensor Networks for Habitat Monitoring*, pp. 399–423, Norwell, MA: Kluwer Academic Publishers, 2004.
5. G. Booch, R. Maksimchuk, M. Engle, B. Young, J. Conallen, and K. Houston, *Objectoriented Analysis and Design with Applications*, 3rd edition. Addison-Wesley Professional, 2007.
6. L. Ruiz-Garcia, L. Lunadei, P. Barreiro, and I. Robla, A review of wireless sensor technologies and applications in agriculture and food industry: State of the art and current trends, *Sensors*, vol. 9, no. 6, pp. 4728–4750, 2009.
7. K. H. Pollock, J. D. Nichols, T. R. Simons, G. L. Farnsworth, L. L. Bailey, and J. R. Sauer, Large scale wildlife monitoring studies: Statistical methods for design and analysis, *Environmetrics*, vol. 13, no. 2, pp. 105–119, 2002.
8. L. Xu, R. Collier, and G. M. P. O'Hare, A survey of clustering techniques in wsns and consideration of the challenges of applying such to 5G IOT scenarios, *IEEE Internet of Things Journal*, vol. 4, pp. 1229–1249, 2017.
9. A. Zanella, N. Bui, A. Castellani, L. Vangelista, and M. Zorzi, Internet of things for smart cities, *IEEE Internet of Things Journal*, vol. 1, pp. 22–32, 2014.
10. D. Evans, The internet of everything: How more relevant and valuable connections will change the world, *Cisco IBSG*, pp. 1–9, 2012.
11. M. Hung, Leading the IOT: Gartner insights on how to lead in a connected world, *Gartner Research*, pp. 1–29, 2017.
12. Internet of Things At a Glance, Cisco, https://www.cisco.com/.
13. P. Sethi and S. R. Sarangi, Internet of things: Architectures, protocols, and applications, *Journal of Electrical and Computer Engineering*, vol. 2017, 2017.
14. F. A. Alaba, M. Othman, I. A. T. Hashem, and F. Alotaibi, Internet of things security, *Journal of Network and Computer Applications*, vol. 88, pp. 10–28, 2017.
15. J. Gubbi, R. Buyya, S. Marusic, and M. Palaniswami, Internet of things (iot): A vision, architectural elements, and future directions, *Future Generation Computer Systems*, vol. 29, no. 7, pp. 1645–1660, 2013. Including Special sections: Cyber-enabled Distributed Computing for Ubiquitous Cloud and Network Services & Cloud Computing and Scientific Applications—Big Data, Scalable Analytics, and Beyond.
16. D. Warren and C. Dewar, Understanding 5G: Perspectives on future technological advancements in mobile, *GSMA Intelligence*, 2014.

17. L. Xu and N. Cao, Network selection in hetnets for advanced iot systems: A demonstration based on WiFi and LTE, *2017 IEEE International Conference on Computational Science and Engineering (CSE) and IEEE International Conference on Embedded and Ubiquitous Computing (EUC)*, vol. 2, pp. 301–304, 2017.
18. L. Xu, J. Xie, X. Xu, and S. Wang, Enterprise LTE and WiFi interworking system and a proposed network selection solution, *Proceedings of the 2016 Symposium on Architectures for Networking and Communications Systems*, ANCS'16, (New York, NY), pp. 137–138, ACM, 2016.
19. F. Boccardi, R. W. Heath, A. Lozano, T. L. Marzetta, and P. Popovski, Five disruptive technology directions for 5G, *IEEE Communications Magazine*, vol. 52, pp. 74–80, 2014.
20. R. Munoz, J. Mangues-Bafalluy, R. Vilalta, C. Verikoukis, J. Alonso-Zarate, N. Bart-zoudis, A. Georgiadis et al., The CTTC 5G end-to-end experimental platform: Integrating heterogeneous wireless/optical networks, distributed cloud, and iot devices, *IEEE Vehicular Technology Magazine*, vol. 11, pp. 50–63, 2016.
21. R. Fantacci, T. Pecorella, R. Viti, and C. Carlini, A network architecture solution for efficient IOT wsn backhauling: Challenges and opportunities, *IEEE Wireless Communications*, vol. 21, pp. 113–119, 2014.
22. E. Piri and J. Pinola, Performance of LTE uplink for IOT backhaul, *Consumer Communications & Networking Conference (CCNC), 2016 13th IEEE Annual*, pp. 6–11, IEEE, 2016.
23. M. Hindia, T. Rahman, H. Ojukwu, E. Hanafi, and A. Fattouh, Enabling remote health-caring utilizing iot concept over lte-femtocell networks, *PloS One*, vol. 11, no. 5, p. e0155077, 2016.
24. V. Jungnickel, D. Schulz, J. Hilt, C. Alexakis, M. Schlosser, L. Grobe, A. Paraskevopoulos, R. Freund, B. Siessegger, and G. Kleinpeter, Optical wireless communication for backhaul and access, *2015 European Conference on Optical Communication (ECOC)*, pp. 1–3, 2015.
25. M. Grob, The road to 5G: Providing the connectivity fabric for everything, *2015 IEEE Hot Chips 27 Symposium (HCS)*, pp. 1–26, 2015.
26. M. R. Palattella, M. Dohler, A. Grieco, G. Rizzo, J. Torsner, T. Engel, and L. Ladid, Internet of things in the 5G era: Enablers, architecture, and business models, *IEEE Journal on Selected Areas in Communications*, vol. 34, pp. 510–527, 2016.
27. A. Cimmino, T. Pecorella, R. Fantacci, F. Granelli, T. F. Rahman, C. Sacchi, C. Carlini, and P. Harsh, The role of small cell technology in future smart city applications, *Transactions on Emerging Telecommunications Technologies*, vol. 25, no. 1, pp. 11–20, 2014.
28. R. Fantacci, T. Pecorella, R. Viti, and C. Carlini, Short paper: Overcoming IOT fragmentation through standard gateway architecture, *2014 IEEE World Forum on Internet of Things (WF-IoT)*, pp. 181–182, 2014.
29. A. Hakiri, P. Berthou, A. Gokhale, and S. Abdellatif, Publish/subscribe-enabled soft-ware defined networking for efficient and scalable iot communications, *IEEE Communications Magazine*, vol. 53, pp. 48–54, 2015.
30. K. Wang, Y. Wang, D. Zeng, and S. Guo, An SDN-based architecture for next-generation wireless networks, *IEEE Wireless Communications*, vol. 24, pp. 25–31, 2017.
31. P. Huskov, T. Maksymyuk, I. Kahalo, and M. Klymash, Smart backhauling subsystem for 5g heterogeneous network, *The Experience of Designing and Application of CAD Systems in Microelectronics*, pp. 481–483, 2015.

32. A. Lea, K. Negus, H. Tapse, B. Varadarajan, and R. Vaughan, Spectrum options for wireless backhaul of small cells, in *Antennas and Propagation (EuCAP), 2014 8th European Conference on*, pp. 3310–3311, IEEE, 2014.
33. M. Peng, Y. Li, Z. Zhao, and C. Wang, System architecture and key technologies for 5G heterogeneous cloud radio access networks, *IEEE Network*, vol. 29, no. 2, pp. 6–14, 2015.
34. M. Peng, Y. Li, J. Jiang, J. Li, and C. Wang, Heterogeneous cloud radio access networks: A new perspective for enhancing spectral and energy efficiencies, *IEEE Wireless Communications*, vol. 21, pp. 126–135, 2014.
35. M. Hassanalieragh, A. Page, T. Soyata, G. Sharma, M. Aktas, G. Mateos, B. Kantarci, and S. Andreescu, Health monitoring and management using internet-of-things (IOT) sensing with cloud-based processing: Opportunities and challenges, *2015 IEEE International Conference on Services Computing*, pp. 285–292, 2015.
36. L. Xu, Context aware traffic identification kit (trick) for network selection in future hetnets/5G networks, *2017 International Symposium on Networks, Computers and Communications (ISNCC)*, pp. 1–5, 2017.
37. I. Loumiotis, E. Adamopoulou, K. Demestichas, and M. Theologou, *Optimal Backhaul Resource Management in Wireless-Optical Converged Networks*, pp. 254–261. Cham, Switzerland: Springer International Publishing, 2015.
38. C. Ran, S. Wang, and C. Wang, Balancing backhaul load in heterogeneous cloud radio access networks, *IEEE Wireless Communications*, vol. 22, pp. 42–48, 2015.
39. S. Andreev, M. Gerasimenko, O. Galinina, Y. Koucheryavy, N. Himayat, S. P. Yeh, and S. Talwar, Intelligent access network selection in converged multi-radio heterogeneous networks, *IEEE Wireless Communications*, vol. 21, pp. 86–96, 2014.
40. R. Bonnefoi, C. Moy, and J. Palicot, Advanced metering infrastructure backhaul reliability improvement with cognitive radio, *Smart Grid Communications (SmartGridComm), 2016 IEEE International Conference on*, pp. 230–236, IEEE, 2016.
41. G. Manogaran, R. Varatharajan, D. Lopez, P. M. Kumar, R. Sundarasekar, and C. Thota, A new architecture of internet of things and big data ecosystem for secured smart healthcare monitoring and alerting system, *Future Generation Computer Systems*, vol. 82, pp. 375–387, 2018.
42. M. Chiang and T. Zhang, Fog and IOT: An overview of research opportunities, *IEEE Internet of Things Journal*, vol. 3, pp. 854–864, 2016.
43. L. Xu, M. O'Grady, G. O'Hare, and R. Collier, Reliable multihop intra-cluster communication for wireless sensor networks, *Computing, Networking and Communications (ICNC), 2014 International Conference on*, pp. 858–863, 2014.
44. E. Borgia, The internet of things vision: Key features, applications and open issues, *Computer Communications*, vol. 54, pp. 1–31, 2014.
45. M. Wu, T.-J. Lu, F.-Y. Ling, J. Sun, and H.-Y. Du, Research on the architecture of internet of things, *2010 3rd International Conference on Advanced Computer Theory and Engineering (ICACTE)*, vol. 5, pp. V5–484–V5–487, 2010.
46. C. Han, J. M. Jornet, E. Fadel, and I. F. Akyildiz, A cross-layer communication module for the internet of things, *Computer Networks*, vol. 57, no. 3, pp. 622–633, 2013.
47. X. Chen, Y. Xu, and A. Liu, Cross layer design for optimizing transmission reliability, energy efficiency, and lifetime in body sensor networks, *Sensors*, vol. 17, no. 4, p. 900, 2017.

48. D. Zeng, S. Guo, and Z. Cheng, The web of things: A survey, *JCM*, vol. 6, no. 6, pp. 424–438, 2011.
49. D. Guinard, V. Trifa, T. Pham, and O. Liechti, Towards physical mashups in the web of things, *2009 Sixth International Conference on Networked Sensing Systems (INSS)*, pp. 1–4, 2009.
50. V. Srivastava and M. Motani, Cross-layer design: A survey and the road ahead, *IEEE Communications Magazine*, vol. 43, pp. 112–119, 2005.
51. M. A. Feki, F. Kawsar, M. Boussard, and L. Trappeniers, The internet of things: The next technological revolution, *Computer*, vol. 46, pp. 24–25, 2013.
52. Contiki, Accessed: May 17, 2018. http://www.contiki-os.org/.
53. Tinyos, Accessed: May 17, 2018. http://www.tinyos.net/.
54. Openwsn, Accessed: May 17, 2018. http://openwsn.atlassian.net.
55. M. A. Razzaque, M. Milojevic-Jevric, A. Palade, and S. Clarke, Middleware for internet of things: A survey, *IEEE Internet of Things Journal*, vol. 3, pp. 70–95, 2016.
56. C. Perera, P. P. Jayaraman, A. Zaslavsky, P. Christen, and D. Georgakopoulos, Mosden: An internet of things middleware for resource constrained mobile devices, *2014 47th Hawaii International Conference on System Sciences*, pp. 1053–1062, 2014.
57. H. Zhou, *The Internet of Things in the Cloud: A Middleware Perspective*, 1st ed. Boca Raton, FL: CRC Press, Inc., 2012.
58. L. Xu, D. Lillis, G. M. O'Hare, and R. W. Collier, A user configurable metric for clustering in wireless sensor networks., *SENSORNETS*, pp. 221–226, 2014.
59. L. Xu, G. O'Hare, and R. Collier, A balanced energy-efficient multihop clustering scheme for wireless sensor network, *7th IFIP Wireless and Mobile Networking Conference (WMNC)*, 2014.
60. A. H. Ngu, M. Gutierrez, V. Metsis, S. Nepal, and Q. Z. Sheng, IOT middleware: A survey on issues and enabling technologies, *IEEE Internet of Things Journal*, vol. 4, pp. 1–20, 2017.
61. Y. Mehmood, N. Haider, M. Imran, A. Timm-Giel, and M. Guizani, M2m communications in 5G: State-of-the-art architecture, recent advances, and research challenges, *IEEE Communications Magazine*, vol. 55, no. 9, pp. 194–201, 2017.
62. V. Jungnickel, K. Manolakis, W. Zirwas, B. Panzner, V. Braun, M. Lossow, M. Sternad, R. Apelfrojd, and T. Svensson, The role of small cells, coordinated multipoint, and massive mimo in 5G, *IEEE Communications Magazine*, vol. 52, pp. 44–51, 2014.
63. Y. Huang, Y. Li, H. Ren, J. Lu, and W. Zhang, Multi-panel mimo in 5G, *IEEE Communications Magazine*, vol. 56, pp. 56–61, 2018.
64. M. Chiang and T. Zhang, Fog and IOT: An overview of research opportunities, *IEEE Internet of Things Journal*, vol. 3, pp. 854–864, 2016.
65. S. Borkar and H. Pande, Application of 5G next generation network to internet of things, *2016 International Conference on Internet of Things and Applications (IOTA)*, pp. 443–447, 2016.
66. S. Shenker, Fundamental design issues for the future internet, *IEEE Journal on Selected Areas in Communications*, vol. 13, pp. 1176–1188, 2006.
67. F. Kelly, Charging and rate control for elastic traffic, *European Transactions on Telecommunications*, vol. 8, 1997.
68. G. A. N. Discovery and S. F. A. M. O. (MO), http://www.3gpp.org/dynareport/24312.htm. Accessed on March 20, 2019.

69. L. Xu, J. Xie, X. Xu, and S. Wang, Enterprise LTE and WiFi interworking system and a proposed network selection solution, *Proceedings of the 2016 Symposium on Architectures for Networking and Communications Systems*, ANCS'16, (New York, NY), pp. 137–138, ACM, 2016.
70. L. Xu and R. Duan, Towards smart networking through context aware traffic identification kit (trick) in 5G, *2018 International Symposium on Networks, Computers and Communications (ISNCC)*, pp. 1–5, 2018.
71. M. Abomhara and G. M. Koien, Cyber security and the internet of things: Vulnerabilities, threats, intruders and attacks, vol. 4, pp. 65–88, 2015.
72. R. Roman, J. Zhou, and J. Lopez, On the features and challenges of security and privacy in distributed internet of things, *Computer Networks*, vol. 57, no. 10, pp. 2266–2279, 2013.
73. Q. Zhang and Y. Q. Zhang, Cross-layer design for QOS support in multihop wireless networks, *Proceedings of the IEEE*, vol. 96, pp. 64–76, 2008.
74. A. Zanella, N. Bui, A. Castellani, L. Vangelista, and M. Zorzi, Internet of things for smart cities, *IEEE Internet of Things Journal*, vol. 1, pp. 22–32, 2014.
75. D. E. Kouicem, A. Bouabdallah, and H. Lakhlef, Internet of things security: A top-down survey, *Computer Networks*, vol. 141, pp. 199–221, 2018.
76. A. D. Jurcut, T. Coffey, and R. Dojen, Design requirements to counter parallel session attacks in security protocols, *2014 Twelfth Annual International Conference on Privacy, Security and Trust*, pp. 298–305, 2014.
77. L. Xu and G. Mcardle, Internet of too many things in smart transport: The problem, the side effects and the solution, *IEEE Access*, vol. 6, pp. 62840-62848, 2018. doi:10.1109/ACCESS.2018.2877175.

Chapter 3

Network Functions Virtualization–Based Internet of Things in 5G Networks

Haojun Huang, Kai Peng, Wang Miao, Geyong Min, Chunbo Luo, Wen Li, and Joe Billingsley

Contents

3.1 Introduction

With the global commercialization of the fourth generation (4G) long-term evolution (LTE) standard, the wireless community is now looking forward to the fifth generation (5G) mobile networks, which will be launched in 2020. Compared to 2G, 3G, and 4G, 5G is expected to have extended coverage, higher throughput, lower latency, and connection density of massive bandwidth, much higher energy efficiency, and spectrum efficiency [1], paving the way for the connection of billions of sensors over the networks. Global initiatives on 5G research investigations have been extensively carried out, starting with an investigation on user demands, application scenarios, technical trends, and potential solutions, for example, the work of Mobile and wireless communications Enablers for Twenty-twenty (2020) Information Society (METIS) in Europe, the 5G Forum in South Korea, and the IMT-2020 Promotion Group in China. Three types of 5G services have been defined by the International Telecommunication Union (ITU), including enhanced mobile broadband (eMBB), critical communications and network operations, and massive machine-type communications (mMTC), which are quite related to the Internet of Things (IoT) with different flavors compared with the current IoT services [2–4].

In order to fulfill the performance requirements of 5G, the telecommunication operators can install a large number of dedicated infrastructures to provide new network functions. This will increase CAPital EXpenditures (CAPEX) and OPerating EXpenditures (OPEX) for them, making low scalability and flexibility of network services. One efficient solution to these issues is the emerging network function virtualization (NFV) paradigm [5], which decouples network functions from customized hardware via full-blown virtualization technologies. It abstracts network services into software known as virtualized network functions (VNFs), and runs on commodity hardware.

IoT in 5G is a novel paradigm with access to wireless communication systems and artificial intelligence technologies that intend to revolutionize the whole world through connected "things" and enables machine-to-machine (M2M) communications. The development of 5G creates the possibility of deploying an enormous number of sensors in the network and to process large amounts of data, challenging the technologies of communications and data mining. Research investigations have indicated that roughly 7 billion smart devices will be connected through mobile networks such as 3G, 4G, and 5G, which, while currently being used for IoT, are not fully optimized for IoT applications [6]. The most obvious drawback of the current IoT is that it is domain specific and task oriented, tailored for particular applications with little or no possibility of reusing them for newer applications in 5G. This method is inefficient and leads to redundant deployments when new applications are contemplated. NFV will be a well-established concept that allows the abstraction of actual physical resources into logical units, enabling IoT efficient usage by multiple independent users [2,7–9]. It is an emerging and promising paradigm for the integration of 5G with IoT that can allow the efficient IoT utilization running on the same virtualized devices in 5G networks.

This chapter will present a basic overview of NFV-based IoT in 5G networks, with the focus on NFV-based IoT benefits, its key technologies, potential applications, and emerging challenges. This will allow the reader a very clear understanding of how NFV can change the design and utilization of IoT in 5G networks.

3.2 Network Functions Virtualization Benefits to 5G Internet of Things

Generally, the traditional IoT delivers network functions and improves network performances through installing hardware equipment for each new service, which increases CAPEX and additional OPEX. This will be resolved by the introduction of NFV, which decouples network functions from dedicated devices and deploys VNFs on commodity devices [10,11]. The benefits of NFV to 5G IoT mainly include, but are not limited to, the following:

1. Lower CAPEX and OPEX: IoT will generate massive data, with the features of larger size, higher velocity, more modes, higher data quality, and heterogeneity. Commonly, a large number of network functions are required at the edge and the core of IoT to process this data generated from the sensors or other devices [6]. These network functions are usually implemented at dedicated IoT devices, such as firewalls and load balancers, to provide network services. Once novel services are required, it is prerequisite to install additional devices and design dedicated software. By exploiting the technology of NFV, when service providers want to add new network functions or improve existing functions, they will only need to improve the software and instantiate them in virtual machine (VMs). In this way, both CAPEX and OPEX will be reduced because operators can use commodity hardware to replace the dedicated hardware.
2. More flexible and scalable: In the traditional IoT, different network functions tight coupling hardware are implemented on a variety of dedicated equipment in a specific order. However, this will not happen in the NFV-based IoT due to the decoupling of network functions from the underlying devices. The IoT services providers can instantiate different network functions in virtual machines

and build logical connections between them to implement network services for applications [12]. In addition, each VNF can upgrade independently without affecting the normal working of other VNFs. Furthermore, by exploiting NFV, when VM is under high traffic load, together with software-defined networking (SDN), VNFs can be easily migrated to other VMs to reduce traffic load. These features greatly enhance the flexibility and scalability of IoT.

3. Similar network performance: The goal of IoT NFV is to use full-blown virtualization technology to decouple the network functions from dedicated devices to commodity nodes to change the way the networks work [5], with the desired performance guaranteed in the whole network. The traditional dedicated IoT usually can fulfill the performance requirements in most applications, with additional CAPEX and OPEX. However, the NFV-based IoT will retain performance at least as good as that of purpose-built IoT device implementations without installing hardware equipment for each novel service. NFV industry specification group (ISG) had obtained encouraging results in the most stringent scenarios—such as broadband network gateway/broadband remote access server (BNG/BRAS), deep packet inspection (DPI), cloud radio access network (C-RAN), content delivery network (CDN) nodes—by a methodology based on tests with relevant IoT use cases and VNFs [10,11].

3.3 The Architecture of Network Functions Virtualization–Based Internet of Things in 5G

The NFV-based 5G IoT is mainly made up of various IoT and 5G networks that have decoupled network functions from underlying hardware via virtualization technologies. The NFV-based IoT will serve as the access and local communication networks and be responsible for data gathering and transmission in 5G networks. As opposed to traditional networks, the network functions of 5G IoT will have been virtualized into software and run on commodity IoT devices instead of dedicated nodes. A typical example of NFV-based 5G IoT applications includes the Internet of Vehicles (IoV), smart homes, smart cities, massive IoT systems, all integrating with 5G networks.

The architecture of NFV-based 5G IoT is illustrated in Figure 3.1. It presents the emerging technologies, referring to VNF, SDN, C-RAN, D-RAN, massive multiple input–multiple output (MIMO) networks, dense static small cell networks [6], and mobile small cell networks. Meanwhile, it also outlines the base stations, cloud, and mobile edge computing (MEC) in the 5G IoT architecture. Furthermore, it will incorporate the concepts of device-to-device (D2D) communications [13,14], small cell access points, and cross-technology communications [15].

The NFV-based 5G IoT consists of a large number of IoT devices that communicate with the virtual base stations through different protocols such as message queuing telemetry transport (MQTT) [15]. The virtual base station is composed of radio remote

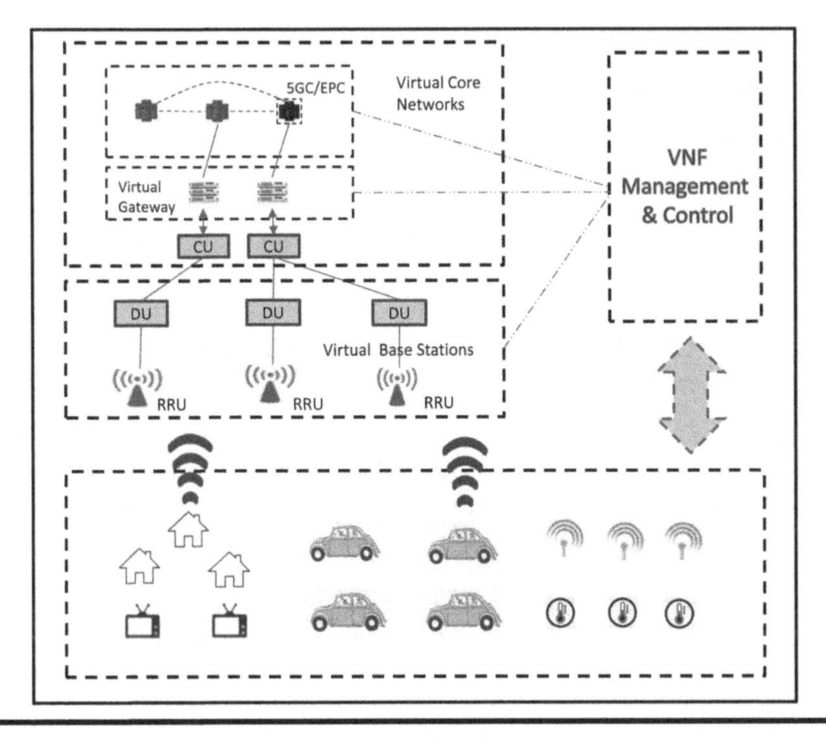

Figure 3.1 The architecture of NFV-based IoT in 5G networks.

unit (RRU), distributed unit (DU), and centralized unit (CU). The RRU is mainly used to receive and send signals, and perform a series of operations such as power amplifier and filtering. The DU mainly deals with services with high real-time requirements, whereas the services of CU have lower requirements for real-time performance. DU and CU are deployed on the virtual machines in the form of software.

In order to fulfill the ubiquitous communication requirements of 5G networks, the NFV-based IoT will actively construct different virtual networks, integrating with other infrastructure and mobile terminals. Furthermore, all IoT devices can communicate with each other with the same or different protocols—such as wireless fidelity (Wi-Fi), Bluetooth, ZigBee—through a gateway or through cross-layer communications [13,15,16]. This depends on the communication requirements and running state of networks. By exploiting NFV, the serving gateway (S-GW) and packet data network gateway (P-GW) will be virtualized to work with IoT. In fact, almost all 5G infrastructure will be virtualized to provide network services. All virtualized network functions in 5G IoT are managed and orchestrated by NFV (NFV management and orchestration [NFV MANO]), which will allocate the required network resources for diversified applications. In order to accelerate NFV performance, it will either dynamically migrate network functions and IoT or deploy new infrastructure in the network in a global view.

3.4 The Key Technologies in Network Functions Virtualization–Based 5G Internet of Things

In order to ensure that the NFV-based IoT can work well in 5G networks, a number of important technologies must be considered in its initial design and applications. These mainly include interconnection among heterogeneous IoT devices, network slicing, VNF migration, and NFV management and orchestration [6,13,17,18] and are presented as follows.

3.4.1 Interconnection Between Heterogeneous Internet of Things

There will be a variety of IoT devices that communicate with each other in a wireless way in 5G networks. In order to fulfill this requirement, a number of available protocols—for example Wi-Fi, ZigBee, or Bluetooth—have been proposed over the past years. However, none of the IoT devices installed these protocols can communicate directly with each other [6,11,19–21]. For example, a sensor installed with Wi-Fi is unable to communicate with another node working at ZigBee.

The heterogeneity of the networks—referring to devices and protocols—has become a hindrance to the development of IoT in 5G. Although the common belief that coexistence of wireless technologies results in harmful interference, it also offers new chances for them to collaborate [15]. A promising solution to this problem is to build a dedicated protocol like a multi-radio gateway that can process protocol conversion [8]. This gateway should have the following characteristics:

1. It can support the interconnection of different protocols and communication technologies.
2. It can translate data obtained from different sensors into a common format and convert commands sent by user terminal into a control message that controls a particular actuator.
3. It can exploit a lightweight and optimal protocol to transfer information on a device with limited resources.

Another promising solution is to exploit cross-technology communications—defined as direct communications among heterogeneous wireless devices [15]—to translate one protocol into another without additional traffic overhead while providing best performance guarantee. The research in [15] proposed a method called FreeBee for cross-technology communication without installing a gateway that can establish communication between devices that use different protocols. FreeBee modulates symbol messages by changing the timing of periodic beacon frames used in wireless communication to enable cross-technology communications. Such a generic cross-technology consumes zero additional bandwidth and incurs zero extra traffic.

The interconnection of different devices has become an important concept in 5G IoT. The investigation in [15] provided a promising solution to this issue by offering direct connectivity compared to a multi-radio gateway and ought to be taken into consideration in 5G IoT design and applications.

3.4.2 Network Slicing Built on Internet of Things

IoT systems can maintain the normal operation with the available compute, storage and network resources in 5G networks, but these resources are limited in quantity and the performances of these resources are limited by the underlying devices. Hardware resources are shared by all IoT applications under the same network.

The technology of network slicing is proposed to appropriate allocate network resources for network services with different network requirements, shown in Figure 3.2. Network Slicing is an emerging technology in 5G networks that will provide dedicated network services for users with different network requirements by dividing the physical network into multiple virtual networks and providing end-to-end services on these virtual networks. Different IoT applications have different service requirements for networks; using network slicing can provide customized network service to IoT applications [22,23]. The process of creating network slicing is actually the process of resource allocation. In the process of

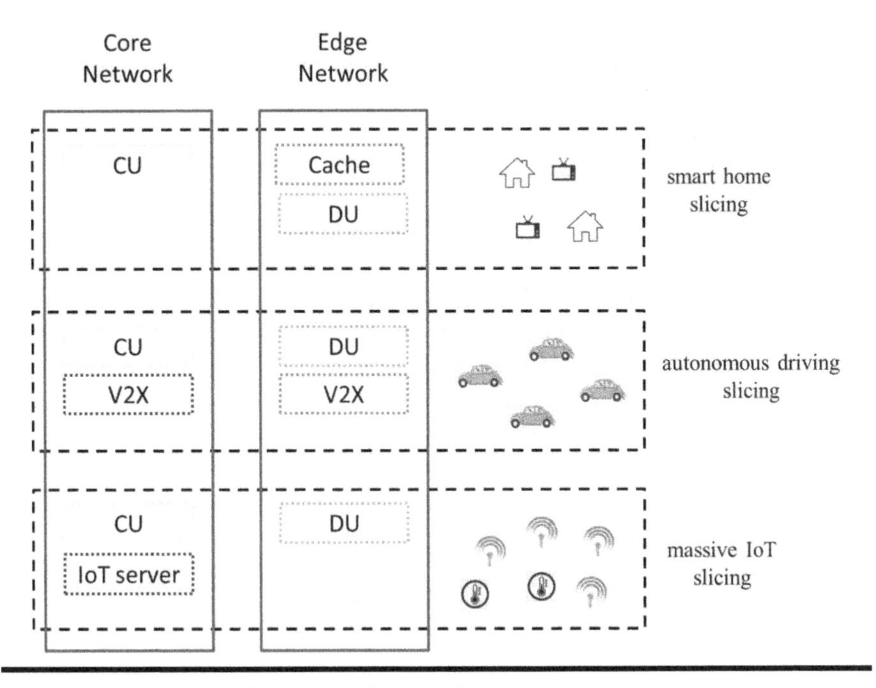

Figure 3.2 Network slicing paradigm in the IoT.

creating network slicing according to different network requirements and network conditions, a variety of factors such as data volume, data rate, transmission latency, mobility, and remaining resources on VMs need to be considered.

In order to provide high data rate services, the DU and cache function should be deployed at the edge network together. The IoV, as a part of the IoT, is a kind of mission-critical IoT task that needs high reliability, low latency and secure network connection, so deploying all proprietary functions in the edge network would be a nice solution. When a massive IoT application that doesn't require location update proposes a network request, we can just virtualize a simple 5G core without the mobility management function in the virtual core network and deploy DU at the edge network to handle the massive sensor data to balance the workload of data centers [23].

3.4.3 Virtualized Network Function Migration in the Internet of Things

It is necessary for 5G to dynamically adjust the network configuration to fulfill the requirements of network service and ensure the flexibility and scalability of the networks. This means that we ought to migrate VNFs from one generic device to another [24]. VNF migration solves the problem of unbalanced traffic distribution and load imbalance in the network.

The VNF migration for NFV-based 5G IoT is to remove the VNFs from the original virtual machine and redeploy it on a better node when the traffic distribution or workload is unbalanced. In order to maintain the continuity and consistency of the internal flow state of the VNF, the information of the VNF internal flow must be transmitted and redeployed together with the VNF, and finally the VNF migration is achieved.

VNF migration will bring the following benefits to the NFV-based 5G IoT:

1. Load balance of network traffic: VNF migration will alleviate overloaded network traffic or aggregate light-loaded network traffic.
2. Fast failover: When one VNF node breaks down or service is unreachable, a new VNF will quickly instantiate on another VM to take over it.

In light of the direction of migration, IoT VNF migration can be divided into merge migration and load balancing migration, as shown in Figures 3.3 and 3.4. Merge migration generally occurs when the sensor node is in a light load state. If the resource utilization of the node is too low, VNFs on this node will be merged with VNFs on other nodes. All VNFs on this node will be migrated to other nodes and the original node will enter standby mode to reduce energy consumption [25].

Load balancing migration generally occurs when a sensor node works at overload state, the resource utilization of this node is too high and results in a decreased

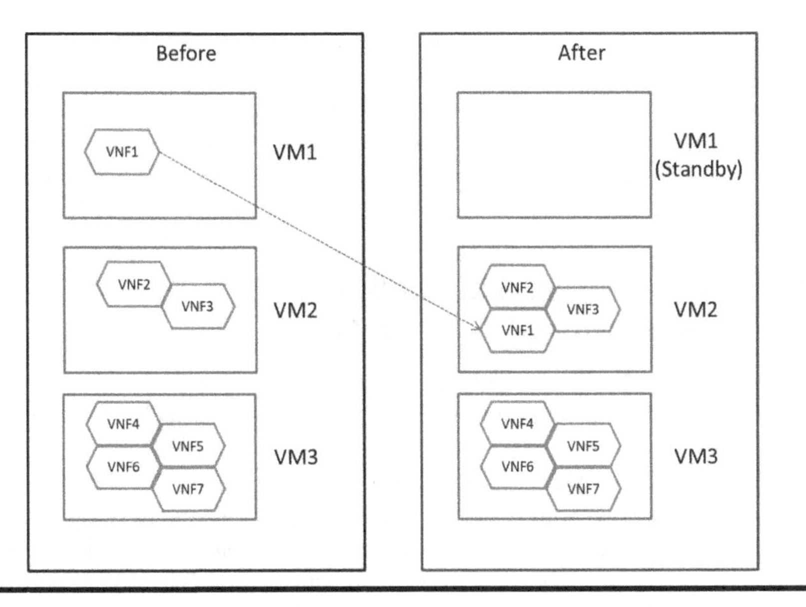

Figure 3.3 Merge migration in the IoT.

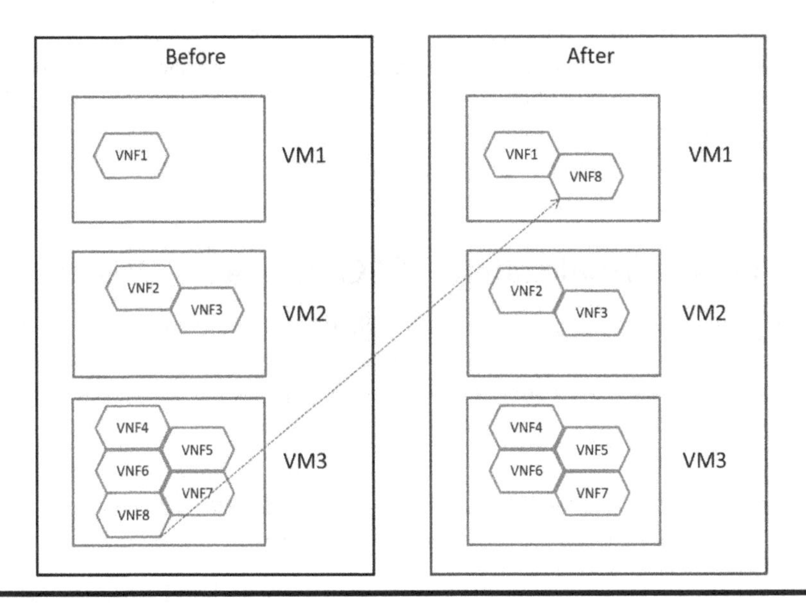

Figure 3.4 Load balancing migration the IoT.

efficiency [26,27]. Part of the VNFs on this node will be migrated to other nodes so that the resource utilization of this node is kept within a reasonable range.

Once it is required to migrate VNFs, we should consider the type of each VNF instance and the end-to-end performance requirements of the network slicing. The high access volume and data processing requirements of IoT can easily lead to an imbalance of workload, so it is necessary to introduce VNF migration into NFV-based 5G IoT.

3.4.4 Virtualized Network Function Management and Orchestration

In the near future, NFV will significantly change the way we deploy, operate and manage IoT and 5G networks and bring an unprecedented challenge to network and service solutions and flexible management. The NFV-based IoT will be partially or fully virtualized into the software to work with 5G networks. The ever-growing communication requirements of quality of experience (QoE) and the plug-and-play feature require that the virtualized IoT devices can be rapidly deployed to support network services. This can be achieved through efficient IoT NFV MANO [9,28].

Specifically, IoT NFV MANO focuses on flexible on-boarding and sidesteps the chaos that can be associated with rapid spin-up of network components, such as computing, networking, storage, and VM resources. Thus, a variety of applications can benefit from it. For example, the deployment of new services in traditional networks is equivalent to deploying new dedicated hardware, which may take several days and even tens of days to be finished, whereas the technology of NFV can reduce deployment time to hours. In the early stages when the technology of NFV is not widely used in IoT, the orchestration algorithm should manage the dependency management and information flow between virtualized and non-virtualized functions simultaneously rather than just managing and orchestrating resources that already have been virtualized [9].

3.5 Potential Applications of 5G Internet of Things

The integration of NFV-based 5G with IoT [29,30] will support the development of smart city, home and smart traffic on the basis of information sharing. In this section, we will illustrate two representative potential applications of 5G IoT, referring to NFV-based smart home and NFV-based IoV [10,11].

3.5.1 Network Functions Virtualization–Based Smart Home Built on 5G Internet of Things

Smart home is a typical example of the IoT integrating with 5G networks, which have been partially or fully virtualized through full-blown virtualization technologies. As shown in Figure 3.5, the common NFV-based smart home architecture includes a variety of components: home networks, with distributed heterogeneous

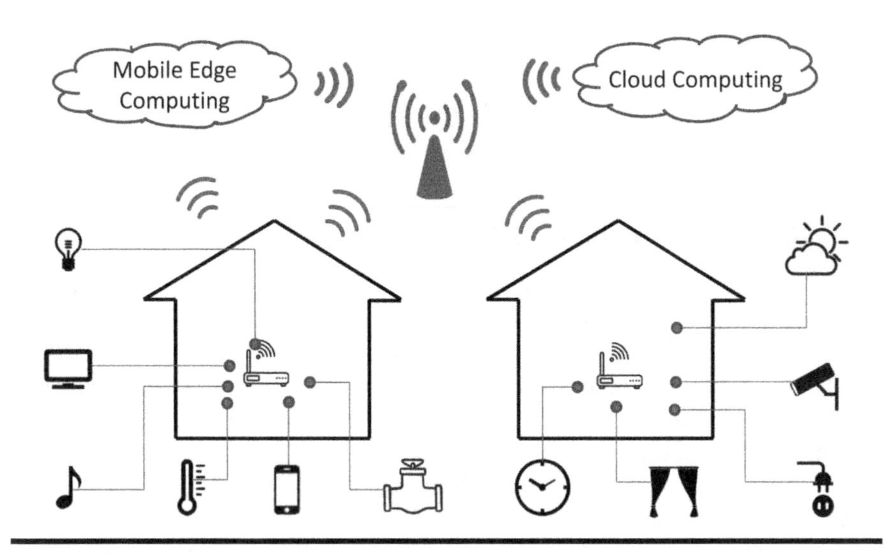

Figure 3.5 NFV-based 5G IoT in the smart home.

devices, such as mobile phones, PCs, TVs, refrigerators, washing machines, and various sensors and actuators; a gateway that collects the information from the sensor nodes installed in all home devices and that can be connected together and/or to the 5G; a cloud-hosted platform that receives the information from the gateway to store and analyze; and the mobile devices that can connect to the gateway or the cloud server to receive information and notification about the home while away.

This architecture allows an incremental deployment of new technologies and protocols for the network embedded within the home given that it is separated by the gateway from the wider, global Internet. The available protocols for ubiquitous communications among them include RS485, Ethernet, X-10, PLC-BUS, Wi-Fi, ZigBee, and Bluetooth [31,32]. The gateway acts as a translator between different protocols and the existing infrastructure.

In order to support ubiquitous communications among different devices, the converged gateways and cross-technology communications have been taken into consideration in smart home design and deployment. Smart home should be a coherent and dynamic system that can help individuals make decisions and identify its behavior through the data that is generated by IoT devices. For example, when the occupants leave the home, the sensor or gateway can identify this action (the human body signal is not detected or there is no mobile phone access to the gateway) and then intelligently control the operations of home appliances if needed.

Commonly, more smart home systems will coexist in one community, and they may access the same NFV-based base stations in the 5G edge networks. In order to provide high-definition video services, IoT NFV MANO should create one VNF in the edge networks to cache resources for use by multiple smart home systems.

In addition, NFV MANO should create their own VNFs and configure them in different network slicing. Once the edge network fails or the VMs overload, a smart home can allow the migration of IoT VNF.

3.5.2 Network Functions Virtualization–Based Internet of Vehicles Built on 5G

The IoV, as one of the revolutions mobilized by the IoT, is a promising paradigm to the future of automobiles and will undoubtedly boost the automobile market and accelerate innovation in 4G/5G applications. It is evolving from vehicular ad hoc networks (VANETs) to achieve the vision of "from smartphone to smart car" [3]. It will provide support for real-time communications among vehicles, vehicles and traffic lights, and improve traffic safety and efficiency with or without the help of roadside units (RSUs) [33]. For instance, sensors embedded in cars can detect an ongoing or upcoming accident, automatically call the emergency services, such as hospital or police, and deliver the onboard cameras and the medical records of the vehicle passengers to the emergency response department and then combine this with other data sets.

The network architecture of IoV includes five types of vehicular communications: vehicle-to-vehicle (V2V), vehicle-to-roadside unit (V2R), vehicle-to-infrastructure (V2I) of mobile networks, vehicle-to-personal devices (V2P), and vehicle-to-sensors (V2S). The vehicles equipped with NFV-based devices can easily exchange information with other vehicles and infrastructure through 5G networks in mobile environments and can make the best decision with the knowledge of road conditions to improve traffic safety and efficiency.

As shown in Figure 3.6, the NFV-based IoV consists of a lot of vehicles, things, and networks that have been equipped with a large number of NFV-based devices

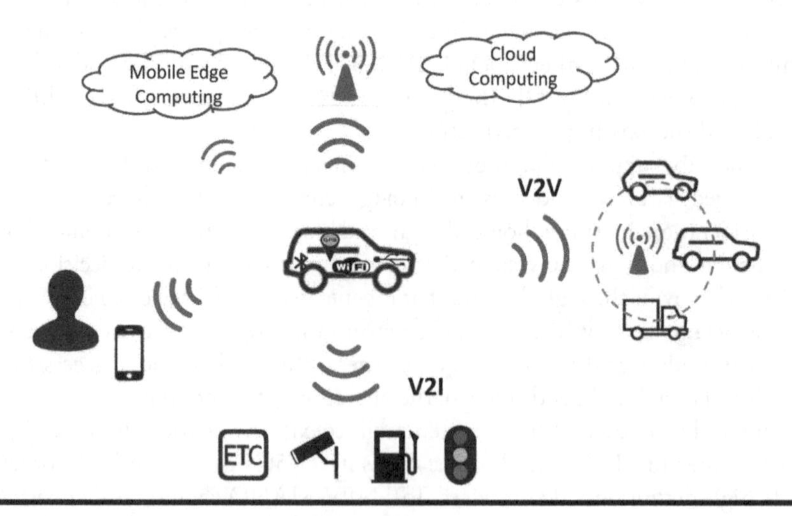

Figure 3.6 NFV-based 5G IoT in the IoV.

such as sensor nodes and mobile terminals. Built on 5G, this IoV is expected to have low latency, high speed, and intelligent scheduling in various applications. In order to achieve these goals, the IoT devices will be deployed on all roads and connect to intelligent traffic information systems through 5G networks. All devices embedded in cars and deployed in roads will be virtualized to dynamically provide enough resources in data storage, computing, and communications among vehicles, vehicles, and traffic lights [34].

3.6 Emerging Challenges in the Network Functions Virtualization–Based 5G Internet of Things

The NFV-based IoT brings the potential benefits of decreasing expenditures, more flexibility and scalability, and encouraged innovation in 5G network architecture and operations, but some emerging challenges remain to be addressed before this can be achieved. In this section, we highlight the existing challenges for further research and the potential directions on the NFV-based 5G IoT.

3.6.1 *Suboptimal NFV Performance*

The initial goal of NFV in 5G IoT is to decouple network functions from dedicated hardware and deploy VNFs on commodity devices [10,11]. This will greatly decrease the CAPEX and OPEX of IoT, while it cannot achieve better or similar NFV Performance. For example, data traffic in the NFV-based IoT may go through indirect paths rather than the direct path depending on the network topology in the traditional network, leading to a potential delay of packets [6,33]. Furthermore, it is challenging to isolate the fault in VNFs and recover from them because the IoT and its VMs will become more fluid. This is because the VNF orchestration cannot be executed in a global view and the additional operations on NFV will increase unwanted performance expenditures. Although the performance of some VNFs has been proven to have a high speed close to dedicated hardware, how to continuously improve the performance of virtualized network functions in the case of massive data growth in the NFV-based 5G IoT is still a challenge.

3.6.2 *Energy Shortage of the Internet of Things*

There will be dozens or hundreds of devices in the 5G IoT. Similar to the traditional IoT, the NFV-based 5G IoT still suffers from the lack of energy resources [6]. Most IoT devices can work well only if enough energy is provided. NFV has greatly reduced the CAPEX and OPEX of the IoT, but it cannot overcome an energy shortage though device reuse. In order to resolve this problem, the energy efficiency of IoT devices in communications ought to be improved on the one hand and to replenish energy for IoT devices on the other hand [35]. For example, we can design energy-aware Media Access Control (MAC) and routing protocols for data delivery [6]. The work in [6] has proposed an efficient data storage approach for energy conservation in IoT.

Recently, there have been a number of research efforts on wireless charging and simultaneous wireless information and power transfer (SWIPT) [36,37]. These can be regarded as a promising to resolve this issue for IoT. A sensor node, which intends to send information, can send a message containing the remaining energy to the destination. Built on this receiving message, the node can determine whether the required network function can be created or not in it. However, due to the attenuation of the SWIPT signal, the effective distance of SWIPT is only several tens of meters [36]. This challenge of energy supplementation in the NFV-based 5G IoT ought to be resolved in the future.

3.6.3 Security and Privacy of Internet of Things

IoT data is becoming related to almost all aspects of human network activities from just recording events to research, design, production, and digital services over networks. 5G mobile crowdsensing applications will collect detailed information from sensors and their owners during task management procedures. Most of the time, this information is sensitive and endangered if intercepted by a third-party malicious program. It should be kept secret from the networks for ethical, security, or legal reasons. However, in reality, IoT data is still more vulnerable to security and privacy attacks [10], such as the vulnerability to architecture and infrastructure defects, dishonest users, and unfaithful service providers [38].

The mass data has bought great challenges to both 5G and IoT [30]. Such distributed data are associated with the individual information of users but can hide their network privacy. Once data have has been gathered, the users will face potential security risks due to individual information leakage through data analysis. Almost all IoT services and 5G infrastructures will suffer from serious threats to privacy, integrity, reliability and availability [39,40]. Currently, most users fight against data collection in the network. If all or most of such data is hidden or neglected, its value could not be extracted for 5G networks and various applications. However, the current solutions only provide incredible security with users, and furthermore, mainly focus on static data rather than the time-variant network data. These make data privacy and security in the acquisition, storage, processing, and use still a formidable challenge to be addressed in 5G IoT [6].

3.7 Conclusion

In this chapter, we introduced the NFV concept, NFV-based IoT benefits, its network architecture, key technologies, potential applications, and emerging challenges in 5G networks. Being an emerging network paradigm, the NFV-based IoT is a quite important element to achieve the desired performance requirements of 5G networks. For example, it is efficient to improve the flexibility of network functions and reduce the cost of deploying new network functions, and it effectively isolates different network services but with the guarantee of the necessary network resource. However, the NFV-based IoT still faces many challenges, including efficient VNF management and orchestration, network security, and privacy protection. In the future, it will be necessary to introduce emerging network technologies like SDN, cloud computing, MEC and blockchain to further improve the NFV-based IoT for 5G networks. These will greatly improve the quality of the service and experience for all 5G scenarios.

References

1. S. Sun, M. Kadoch, L. Gong, and B. Rong, What Will 5G Be? *IEEE Journal on Selected Areas in Communications*, vol. 32, no. 6, pp. 1065–1082, 2014.
2. G. A. Akpakwu, B. J. Silva, G. P. Hancke, and A. M. Abu-Mahfouz, A Survey on 5G Networks for the Internet of Things: Communication Technologies and Challenges, *IEEE Access*, vol. 6, no. 99, pp. 3619–3647, 2018.
3. A. Aldaej and U. Tariq, IoT in 5G Aeon: An Inevitable Fortuity of Next Generation Healthcare, *2018 1st International Conference on Computer Applications & Information Security* (*ICCAIS*), pp. 1–4, 2018.
4. A. Baz, A. A. Al-Naja, and M. Baz, Statistical Model for IoT/5G Networks, *2015 Seventh International Conference on Ubiquitous and Future Networks*, pp. 109–111, 2015.
5. ETSI, Network Functions Virtualisation—White Paper #3, https://portal.etsi.org/NFV/NFV White Paper3.pdf, 2014.
6. A. Gupta and R. K. Jha, A Survey of 5G Network: Architecture and Emerging Technologies, *IEEE Access*, vol. 3, pp. 1206–1232, 2015.
7. S. Sun, M. Kadoch, L. Gong, and B. Rong, Integrating Network Function Virtualization with SDRr and SDN for 4G/5G Networks, *IEEE Network*, vol. 29, no. 3, pp. 54–59, 2015.
8. P. Wan, F. Ye, and X. Chen, A Smart Home Gateway Platform for Data Collection and Awareness, *IEEE Communications Magazine*, vol. 56, no. 9, pp. 87–93, 2018.
9. S. Abdelwahab, B. Hamdaoui, M. Guizani, and T. Znati, Network Function Virtualization in 5G, *IEEE Communications Magazine*, vol. 54, no. 4, pp. 84–91, 2016.
10. E. Dutkiewicz, B. A. Jayawickrama, and Y. He, Radio Spectrum Maps for Emerging IoT and 5G Networks: Applications to Smart Buildings, *2017 International Conference on Electrical Engineering and Computer Science* (*ICECOS*), pp. 7–9, 2017.
11. G. P. Fettweis, 5G and the Future of IoT, *ESSCIRC Conference 2016: 42nd European Solid-State Circuits Conference*, pp. 21–24, 2016.
12. J. Hwang, K. K. Ramakrishnan, and T. Wood, NetVM: High Performance and Flexible Networking Using Virtualization on Commodity Platforms, *IEEE Transactions on Network and Service Management*, vol. 12, pp. 34–47, 2015.
13. Y. Lu, P. Richter, and E. S. Lohan, Opportunities and Challenges in the Industrial Internet of Things Based on 5G Positioning, *2018 8th International Conference on Localization and GNSS* (*ICL-GNSS*), pp. 1–6, 2018.

14. J. Navarro-Ortiz, S. Sendra, P. Ameigeiras, and J. M. Lopez-Soler, Integration of LoRaWAN and 4G/5G for the Industrial Internet of Things, *IEEE Communications Magazine*, vol. 56, no. 2, pp. 60–67, 2018.
15. S. M. Kim and T. He, FreeBee: Cross-technology Communication via Free Side-channel, *Proceedings of the 21st Annual International Conference on Mobile Computing and Networking*, pp. 317–330, 2015.
16. C. X. Mavromoustakis, G. Mastorakis, and J. M. Batalla, Internet of Things (IoT) in 5G Mobile Technologies, *Modeling & Optimization in Science & Technologies*, vol. 8, pp. 371–397, 2016.
17. A. Shehab, M. Elhoseny, K. Muhammad, A. K. Sangaiah, P. Yang, H. Huang, and G. Hou, Secure and Robust Fragile Watermarking Scheme for Medical Images, *IEEE Access*, vol. 6, pp. 10269–10278, 2018.
18. K. E. Skouby and P. Lynggaard, Smart Home and Smart City Solutions Enabled by 5G, IoT, AAI and CoT Services, *2014 International Conference on Contemporary Computing and Informatics (IC3I)*, pp. 874–878, 2014.
19. M. Tavares, D. Samardzija, H. Viswanathan, H. Huang, and C. Kahn, A 5G Lightweight Connectionless Protocol for Massive Cellular Internet of Things, *2017 IEEE Wireless Communications and Networking Conference Workshops (WCNCW)*, pp. 1–6, 2017.
20. D. Wang, D. Chen, B. Song, N. Guizani, X. Yu, and X. Du, From IoT to 5G I-IoT: The Next Generation IoT-Based Intelligent Algorithms and 5G Technologies, *IEEE Communications Magazine*, vol. 56, no. 10, pp. 114–120, 2018.
21. J. Wang, C. Zhang, R. Li, G. Wang, and J. Wang, Narrow-Band SCMA: A New Solution for 5G IoT Uplink Communications, *2016 IEEE 84th Vehicular Technology Conference (VTC-Fall)*, pp. 1–5, 2016.
22. W. Guan, X. Wen, L. Wang, Z. Lu, and Y. Shen, A Service-oriented Deployment Policy of End-to-End Network Slicing Based on Complex Network Theory, *IEEE Access*, vol. 6, pp. 19691–19701, 2018.
23. H. Zhang, N. Liu, X. Chu, K. Long, A. Aghvami, and V. C. M. Leung, Network Slicing Based 5G and Future Mobile Networks: Mobility, Resource Management, and Challenges, *IEEE Communications Magazine*, vol. 55, no.8, pp. 138-145, 2017.
24. J. Zhang, D. Zeng, L. Gu, H. Yao, and M. Xiong, Joint Optimization of Virtual Function Migration and Rule Update in Software Defined NFV Networks, *IEEE Global Communications Conference (GLOBECOM 2017)*, 2017.
25. D. Cho, J. Taheri, A. Y. Zomaya, and P. Bouvry, Real-Time Virtual Network Function (VNF) Migration Toward Low Network Latency in Cloud Environments, *2017 IEEE 10th International Conference on Cloud Computing (CLOUD)*, 2017.
26. J. Xia, Z. Cai, and M. Xu, Optimized Virtual Network Functions Migration for NFV, *2016 IEEE 22nd International Conference on Parallel and Distributed Systems (ICPADS)*, 2016.
27. J. Xia, D. Pang, Z. Cai, M. Xu, and G. Hu, Reasonably Migrating Virtual Machine in NFV-featured Networks, *2016 IEEE International Conference on Computer and Information Technology (CIT)*, 2016.
28. G. Bernini, E. Kraja, G. Carrozzo, G. Landi, and N. Ciulli, SELEFNET Virtual Network Functions Management: A Common Approach For Lifecycle Management of NFV Applications, *2016 5th IEEE International Conference on Cloud Networking (Cloud-net)*, 2016.

29. S. Borkar and H. Pande, Application of 5G Next Generation Network to Internet of Things, *2016 International Conference on Internet of Things and Applications (IOTA)*, pp. 443–447, 2016.
30. A. Costanzo and D. Masotti, Energizing 5G: Near-and Far-Field Wireless Energy and Data Trantransfer as an Enabling Technology for the 5G IoT, *IEEE Microwave Magazine*, vol. 18, no. 3, pp. 125–136, 2017.
31. M. R. Palattella, M. Dohler, A. Grieco, G. Rizzo, J. Torsner, T. Engel, and L. Ladid, Internet of Things in the 5G Era: Enablers, Architecture, and Business Models, *IEEE Journal on Selected Areas in Communications*, vol. 34, no. 3, pp. 510–527, 2016.
32. L. J. Poncha, S. Abdelhamid, S. Alturjman, E. Ever, and F. Al-Turjman, 5G in a Convergent Internet of Things Era: An Overview, *2018 IEEE International Conference on Communications Workshops (ICC Workshops)*, pp. 1–6, 2018.
33. F. Yang, S. Wang, J. Li, Z. Liu, and Q. Sun, An Overview of Internet of Vehicles, *China Communications*, vol. 11, no. 10, pp. 1–15, 2014.
34. J. Cheng, J. Cheng, M. Zhou, F. Liu, S. Gao, and C. Liu, Routing in Internet of Vehicles: A Review, *IEEE Transactions on Intelligent Transportation Systems*, vol. 16, no. 5, pp. 2339–2352, 2015.
35. C. Kappor, H. Singh, and V. Laxmi, A Survey on Energy Efficient Routing for Delay Minimization in IoT Networks, *2018 International Conference on Intelligent Circuits and Systems (ICICS)*, pp. 320–323, 2018.
36. H. Lee and J. Lee, Resource and Task Scheduling for SWIPT IoT Systems with Renewable Energy Sources, *IEEE Internet of Things Journal*, 2018. doi:10.1109/JIOT.2018.2873658.
37. A. S. Toor, and A. K. Jain, A New Energy Aware Cluster Based Multi-hop Energy Efficient Routing Protocol for Wireless Sensor Networks, *2018 IEEE International Conference on Smart Energy Grid Engineering (SEGE)*, pp. 133–137, 2018.
38. J. Ni, X. Lin, and X. S. Shen, Efficient and Secure Service-Oriented Authentication Supporting Network Slicing for 5G-Enabled IoT, *IEEE Journal on Selected Areas in Communications*, vol. 36, no. 3, pp. 644–657, 2018.
39. M. A. Ferrag, L. Maglaras, A. Argyriou, D. Kosmanos, and H. Janicke, Security for 4G and 5G Cellular Networks: A Survey of Existing Authentication and Privacy-Preserving Schemes, https://arxiv.org/abs/1708.04027.
40. M. Pattaranantakul, R. He, Q. Song, Z. Zhang, and A. Meddahi, NFV Security Survey: From Use Case Driven Threat Analysis to State-of-the-Art Countermeasures, *IEEE Communications Surveys & Tutorials*, 2018. doi:10.1109/COMST.2018.2859449.

Chapter 4

Exploring the Next Generation of the Internet of Things in the 5G Era

Bao-Shuh Paul Lin, Yi-Bing Lin, Li-Ping Tung, and Fuchun Joseph Lin

Contents

4.1 Introduction

Until now, most of the systems of Internet of Things (IoT) applications and services have relied on fourth generation/long-term evolution (4G/LTE) or even 3G as the gateway to transport data to the destination for computing and applications creation and execution. Figure 4.1 illustrates wireless sensor network (WSN) which collects data and/or signals from devices and use 4G/LTE and future 5G as the gateway to network and services platform for further applications creation.

The development trends of networking and wireless communications technologies have evolved toward open networking (ON), software-defined networking (SDN), and networking function virtualization (NFV). By integrating SDN,

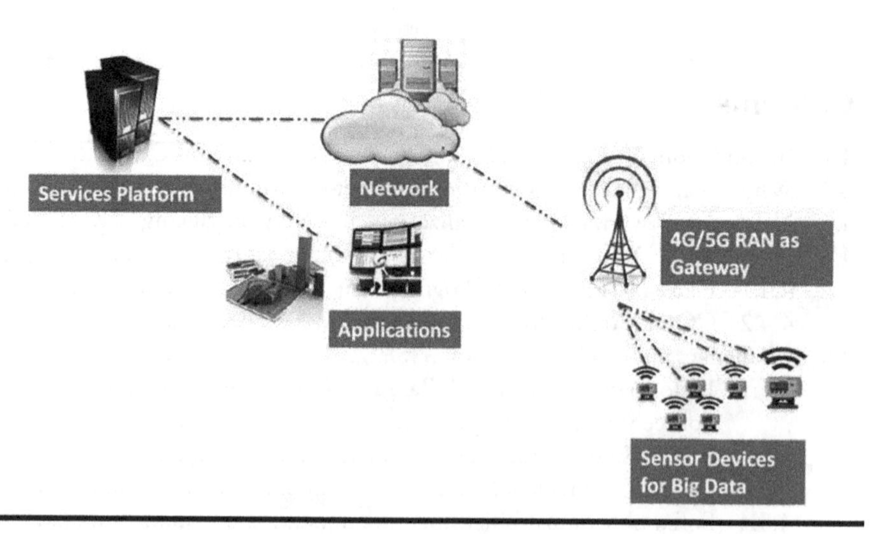

Figure 4.1 IoT conceptual and reference architecture.

NFV, and cloud technologies and networking industry paradigm shift to ON, are playing key roles for the softwarization and virtualization of next generation networks such as the fifth generation (5G) mobile network and future IoT applications/-services in 5G era.

It has been a paradigm shift in the networking industry, from a proprietary networking with vendor-dependent software/hardware to vendor-independent open source and white box switches as illustrated in Figure 4.2. The white box is the commodity hardware that is available from network equipment original design manufacturer (ODM) vendors. On the other hand, the open source software executed on the commodity hardware in the control plane (with controller) of SDN can be very complex. Currently, there are three well-developed open-source (OS) SDN controller operating systems available: open network operating system (ONOS), open daylight (ODL), and RYU [17,19]. Those OSs were developed by various open-source foundations.

At National Chiao Tung University (NCTU), with great research and development (R&D) effort by applying SDN/NFV, and open-source software to 5G and IoT, we have encountered and identified six challenging issues: system performance, end-to-end interoperability, software compatibility, global/local deployment, integration of SDN/NFV open source networking modules or software, and a shortage of talent. In this chapter, we also address how to tackle those challenges (Figures 4.3 and 4.4).

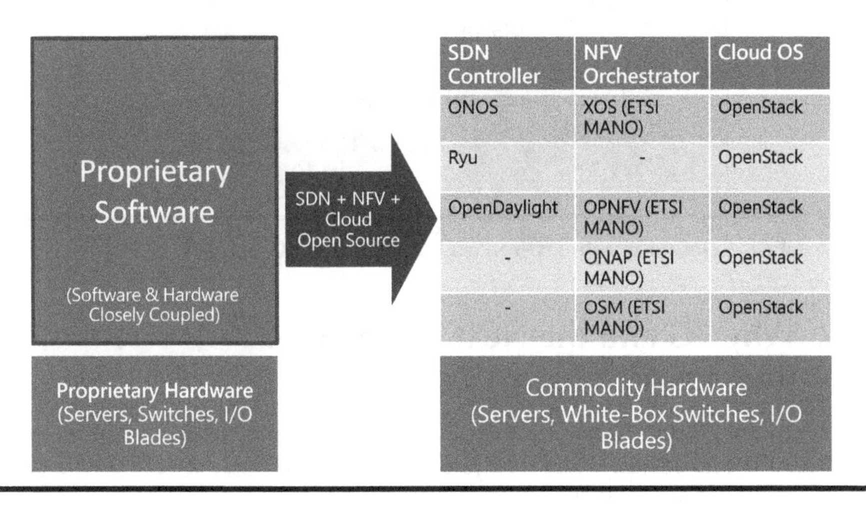

Figure 4.2 Open networking (ON) trends.

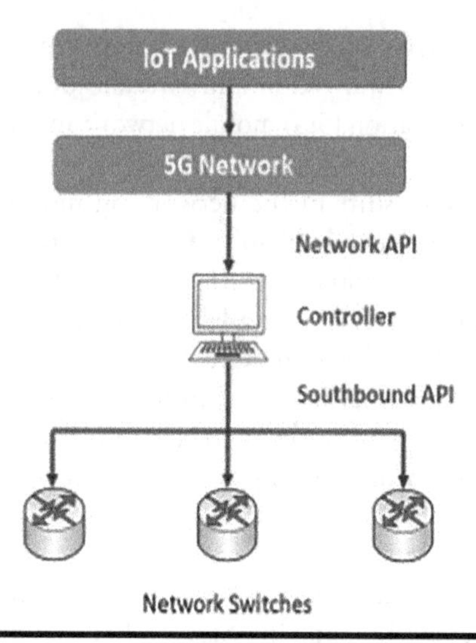

Figure 4.3 SDN architecture-based IoT applications.

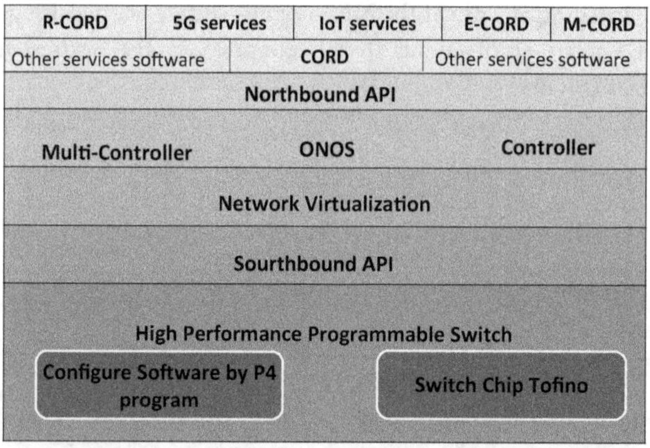

R-CORD	5G services	IoT services	E-CORD	M-CORD
Other services software	CORD	Other services software		
Northbound API				
Multi-Controller	ONOS	Controller		
Network Virtualization				
Sourthbound API				
High Performance Programmable Switch				
Configure Software by P4 program	Switch Chip Tofino			

Figure 4.4 SDN/NFV with layered architecture for 5G and IoT.

4.2 Why 5G and the Next Generation Internet-of-Things?

4G made possible fast Internet access, video reception and creation, apps that rely on location and identity, and always-on behavior; however, it still suffers with numerous limitations [29,30,31]:

1. For some applications: too-low data rate
2. For some applications: too-high delay rate

3. Too few simultaneous connections (insufficient density)
4. Weak (if any) quality of service (QoS) guarantees
5. Too-high price per bit
6. Too-high power consumption (and thus too-low battery)
7. Poor support for new applications/market (e.g., IoT, augmented reality (AR)/virtual reality (VR), connected cars)

5G is being developed to address 4G limitations. Its key advantages and impacts include that [30]

1. 5G impacts all network segments
2. 5G exploits SDN-based network slicing to handle different needs
3. 5G's Radio Access Network (5G RAN) defines multiple functional splits
4. 5G is the foundation for realizing the full potential of IoT

Figure 4.5 shows the key features of the current 4G-based IoT versus future 5G-enabled IoT. For example, the major differences between next generation IoT and the current IoT include data rate (throughput), latency, capability of intelligence (machine learning and artificial intelligence [AI]), protocol (narrowband IoT [NB-IoT] vs. machine-to-machine [M2M]), power consumption, application domains and complexities, and so on. So 5G is more than just fast downloads; its unique combination of very high connectivity, very low latency, and ubiquitous coverage will support smart vehicles and transport infrastructure such as connected cars, buses, and trucks.

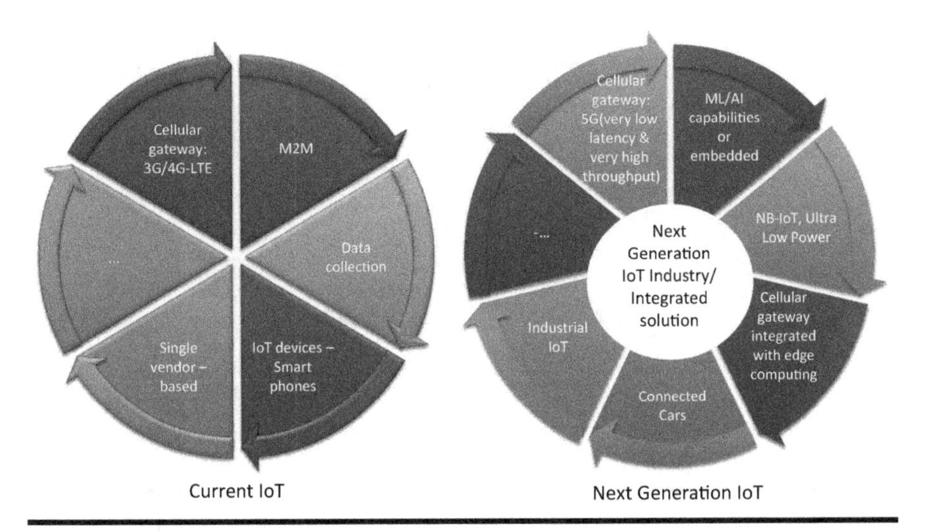

Figure 4.5 Key features or applications of current IoT versus future IoT.

4.3 Network Softwarization, Virtualization, and Network Slicing

Several studies [1,2,7,8] have focused on the realization of SDN technique for IoT and the virtualization of network functions of network elements of IoT by applying the NFV technique. Other studies [6,9–11] have investigated how to do softwarization and virtualization of 5G or mobile broadband networks by using SDN/NFV techniques. Recently, more studies evaluated the relationships between 5G and the IoT [27,28,35]. In this chapter, we address both IoT and 5G from the softwarization and virtualization points of view. We also discuss how the relationship between 5G and IoT from the 5G mobile broadband deployment impacts the future of IoT in terms of technology, application, and service.

To convert 5G and IoT to provide programmability, flexibility, and scalability, the steps to realize softwarization and virtualization are needed. The best techniques available are SDN and NFV. Furthermore, to perform network slicing of massive IoT applications and services, programmability, flexibility, and modularity are required to create multiple virtual networks on top of a common network, including 5G RAN and core network evolved packet core (EPC).

The international standard organization, Next-Generation Management Network (NGMN), has proposed three real, existing categories of network slicing for 5G, including smart phones, autonomous driving, and massive IoT, as shown in Figure 4.6. To zoom into the massive IoT category, it can be further divided into as many network slicing as shown in Figures 4.7 and 4.8. Figure 4.9 indicates the realization of 5G network slicing [22] by SDN/NFV technology through the virtualization of RAN and EPC.

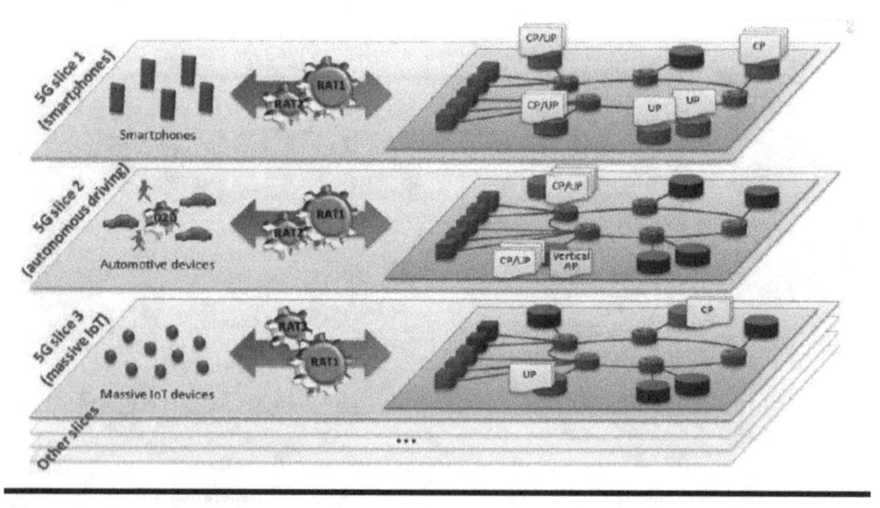

Figure 4.6 Three categories of network slicing proposed by NGMN. (Source: NGMN)

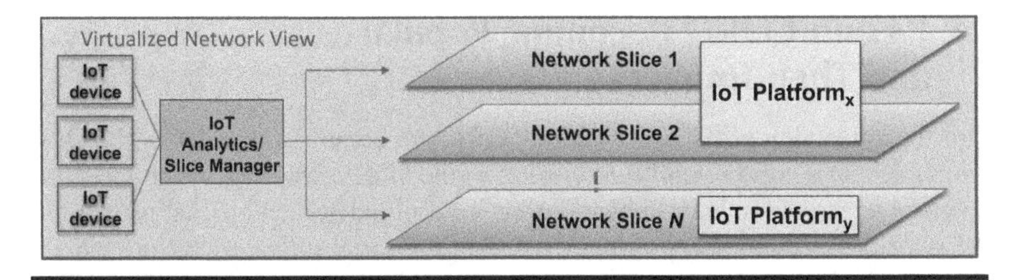

Figure 4.7 Application of SDN/NFV to IoT services to realize dynamic deployment of IoT services.

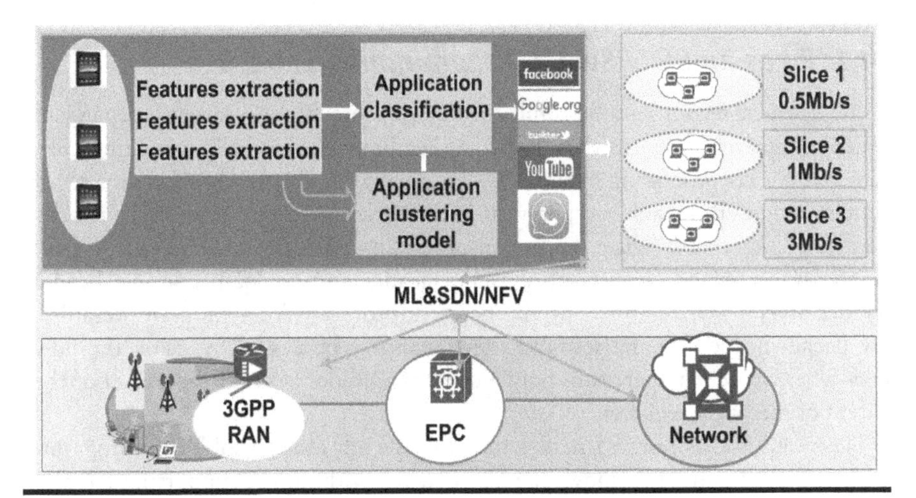

Figure 4.8 Network slicing for mobile broadband traffic.

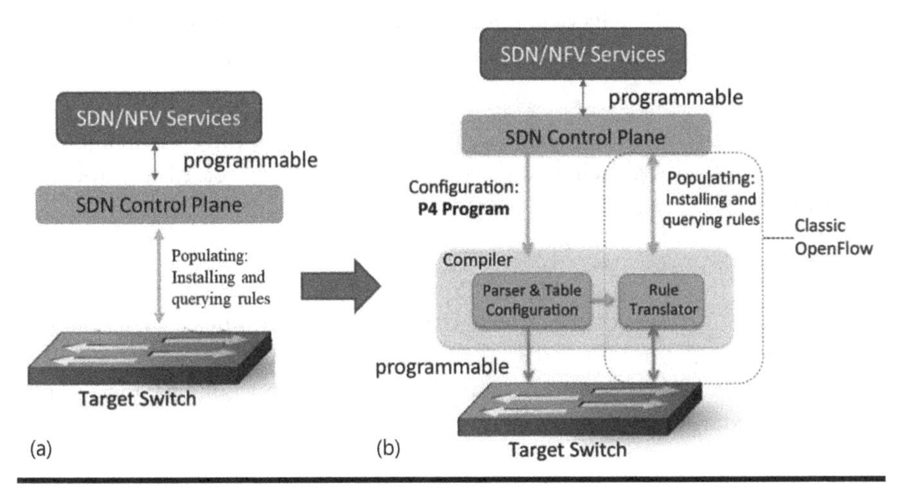

Figure 4.9 (a) Switch without P4 support, and (b) switch with P4 support.

4.4 P4 and CORD Becoming Popular and Their Impacts

With the availability of ONOS, P4 and CORD are becoming popular in the ON community. This creates significant impacts to the SDN ecosystem, which includes a high-speed network vendor (programming protocol-independent packet processors [P4] switch), Rack solution provider (M-CORD [Central Office Re-architected as a Datacenter] box) silicon provider (system-on-a-chip/integrated circuit [SoC/IC]), commodity provider (white box switch, server, storage I/O, blade), and network service operator (services).

4.4.1 P4 as an SDN Switch Programming Language

P4 [12–15] is a reconfigurable, multi-platform, protocol, and target-independent packet processing language that is used to facilitate dynamically programmable, extensible packet processing in the SDN data plane. In SDN, a switch uses a set of "match+action" flow tables to apply rules for packet processing, and P4 provides an efficient way to configure the packet processing pipelines to achieve this goal. Specifically, a packet consists of a packet header and payload, and the header includes several fields defined by the network protocol. A P4 program describes how packet headers are parsed and their fields are processed by using the flow tables where matched operations may modify the header fields of the packet or the content of metadata registers.

Figure 4.9 shows how P4 makes the SDN switch more flexible and programmable [12,13]. Figure 4.9a shows the SDN switch without the supporting P4, and Figure 4.9b with P4 support. P4 switch powered by Barefoot's Tofino chip can achieve 6.5Tb/s (65 × 100GE or 260 × 25GE) performance [16, 23].

We have conducted the P4 switch system performance evaluation by developing two applications related to traffic classification and load balancing:

1. Traffic classification for dynamic QoS control based on the P4 switch [24]
2. Dynamic load balancing and congestion avoidance based on the P4 Switch [25]

Figure 4.10a–c shows the topology of experiment, architecture, and the sample percentage of applications distribution, respectively, for the traffic classification for dynamic QoS control based on the P4 switch. Figure 4.10d and e illustrates the two topologies and the architecture, respectively, for dynamic load balancing and congestion avoidance based on the P4 switch.

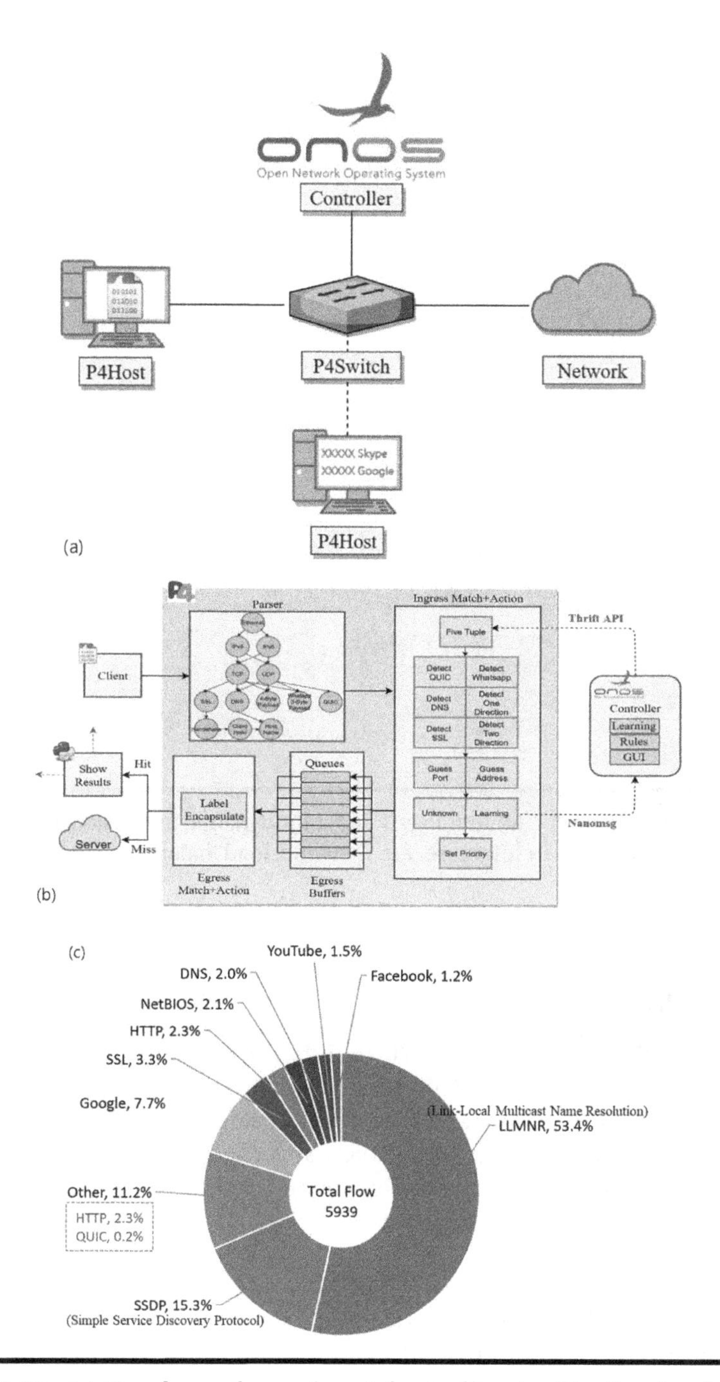

Figure 4.10 **(a) Topology of experiment for traffic classification for dynamic QoS control. (b) architecture of P4-based traffic classification for dynamic QoS control, (c) percentage of applications for a sample traffic flow.** ***(Continued)***

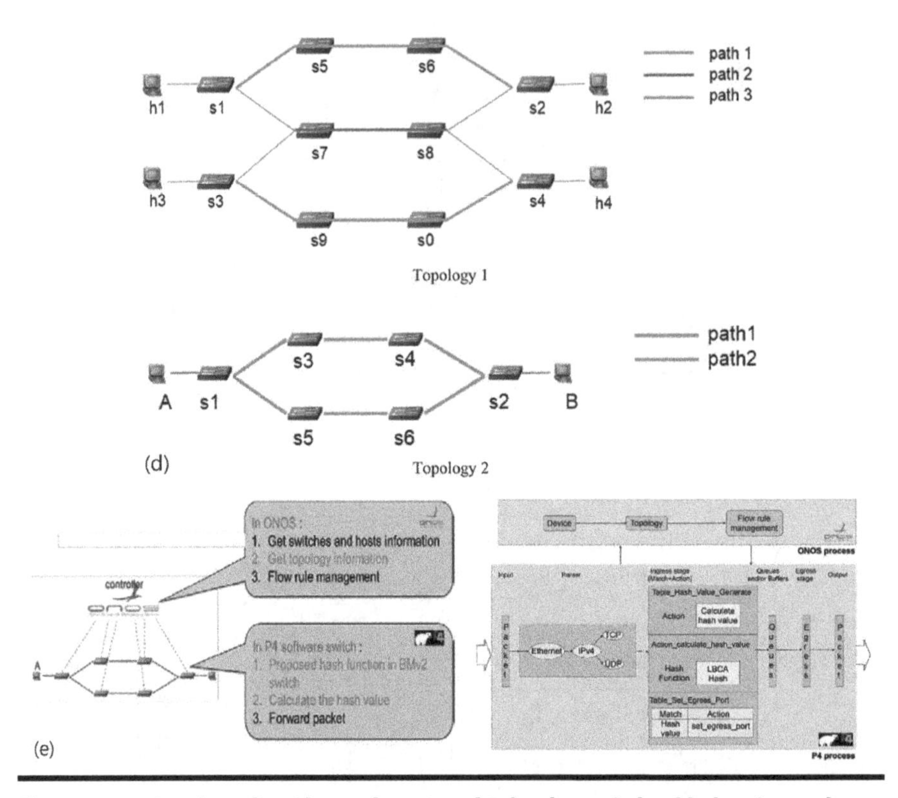

Figure 4.10 (Continued) (d) topology 1 and 2 for dynamic load balancing and congestion avoidance. (e) architecture for dynamic load balancing and congestion avoidance.

4.4.2 CORD Architecture

The basic CORD architecture [18,20] transforms telecommunications Central Office (CO) to Data Center (DC) (Figure 4.11). This transformation will change the CO with propriety hardware/software to DC with white-box commodity and open source software. Based on this CORD architecture, many possible applications may apply SDN/NFV and the cloud (local cloud and remote cloud) to telecommunications networks, including

1. Mobile/Wireless network—M-CORD
2. Residential/wireline network—R-CORD
3. Enterprise network—E-CORD

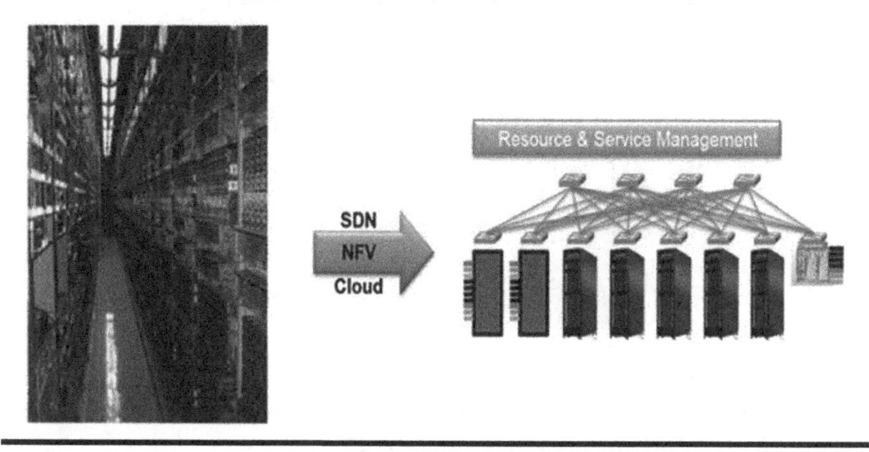

Figure 4.11 Basic CORD architecture. (From Snow, B., "Transforming Service Provider Networks with ONOS and CORD," Open Networking Lab, 2015; Snow, B. and Parulkar, G., CORD—Central Office Re-architected as a Datacenter, ON.Lab Technical Report, Collaboration with AT&T, March 2016.)

4.5 Challenges and Technical Issues

Through the joint R&D of SDN/NFV technology between Chunghwa Telecom Laboratories (CHT-TL) and NCTU, we have discovered and identified six technical and human resource issues [3,4,8]. In this section, we address how to resolve those issues.

4.5.1 *Performance for Network Based on Operating Network/Open Source versus Propriety*

The network performance comparison is an issue between the approach with ON and open source versus the approach with propriety hardware and software. However, if network scalability and programmability are the requirements, then the ON/open source approach or solution is a better choice. In addition, the network performance of the ON/open source approach continues to improve automatically while the propriety approach requires paying more to buy a new model with better capacity and capability. [21, 26].

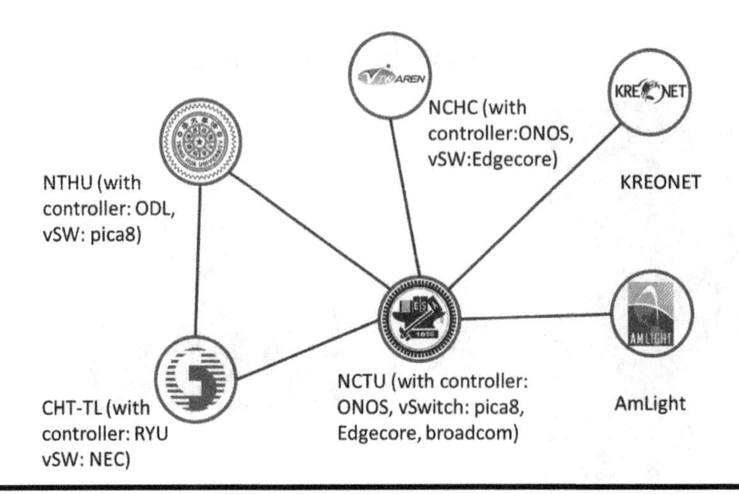

Figure 4.12 Network topology in North Taiwan.

4.5.2 End-to-End Interoperability

The choice of SDN popular controllers' OSs includes ONOS, ODL, and RYU. The available switches from various vendors are built on top of different SoC/IC solutions. How to test the end-to-end interoperability among SDN of North Taiwan universities and to connect to global are shown in Figure 4.12. Figure 4.12 indicates the network topology, controller OS of each site and various brands of switches of each site:

1. NCTU SDN controller runs ONOS as its OS, with switches by PICA 8, Edgecore, and Broadcom
2. CHT-TL (CHT: Taiwan's top telecom operator) controller runs RYU as its OS, with switches by NEC
3. National Tsing Hua University's controller runs ODL as its OS, with switches by PICA 8
4. Kreonet and AmLight are global links to South Korea and to Florida then to South America

The issue of interoperability is resolved through intensive exercises and experiments, following the successful global deployment led by the ONF Lab Team (formerly called ON.LAB).

4.5.3 Global Deployment

To deploy SDN global peering, the NCTU team worked together with the ONF Lab Team. Eight cities on five continents participated in this global SDN-IP peering. The result was a demonstration at the Open Network Summit (ONS) that went very well. Figures 4.13 and 4.14 provide two screens shown during the demonstration.

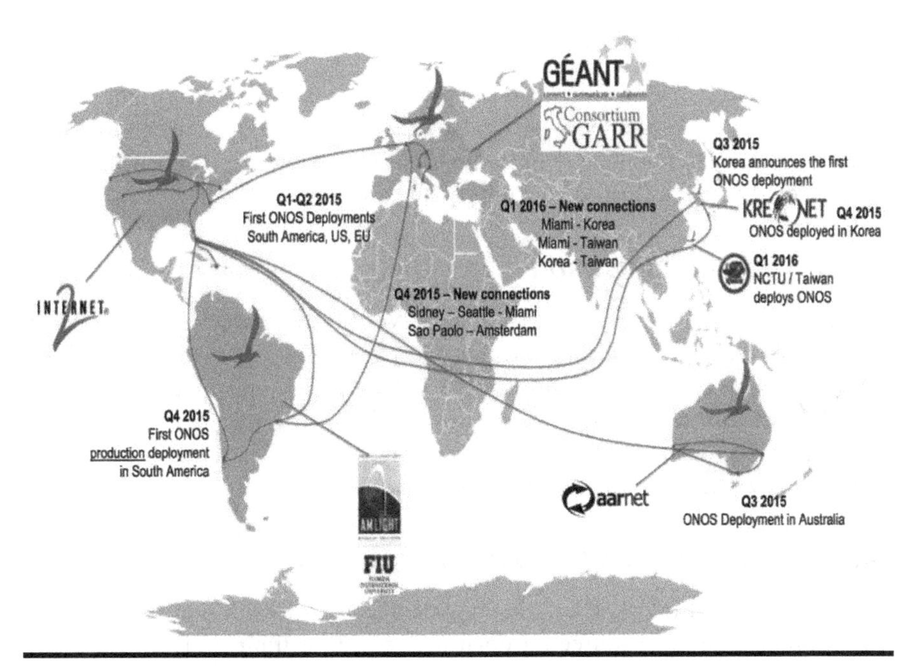

Figure 4.13 SDN-IP global peering deployment with five continents.

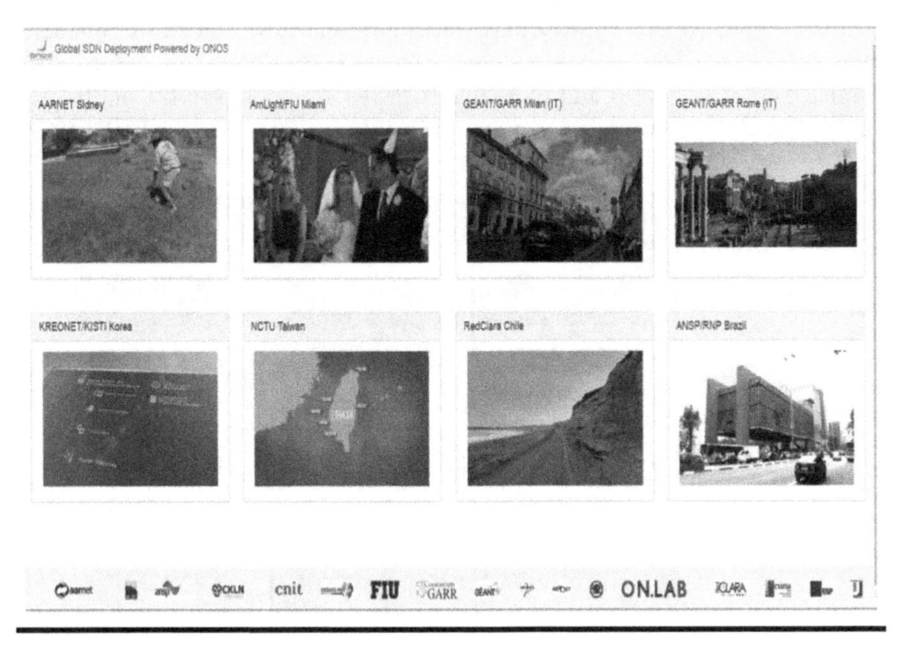

Figure 4.14 Video shown from eight cities and regions during SDN-IP global deployment.

4.5.4 Software Compatibility

Familiarity with the availability and functionality of the proper version of released open-source software and tools, particularly those related to open networking software, is essential to resolving the incompatibility issue. Version control and development methodology are also very important.

4.5.5 Vertical Integration

The challenge of integrating SDN and NFV was addressed by several papers [3, 4, 5]. By creating a vertically integrated stack as a SDN distribution, the open-source software and tools related to SDN will continue to be updated or revised, and the virtualization of network elements defined in NFV may change or be updated from time to time. The case illustrated within a black shape in Figure 4.15 shows a successful vertically integrated SDN distribution.

4.5.6 Human Resource—Shortage of Talent

The availability of talent who have the knowledge of P4, ONOS, CORD, and the familiarity of open source for open networking is limited. Technical training and practice are needed. In addition, motivated people must be sent, as interns and/or contributors, to collaborate with reputable, well-known organizations in this area. The NCTU experiences of working with ONF (Lab Team) in the areas of P4 and CORD have been quite successful. For this collaboration, NCTU has sent several graduate students to the ONF Lab Team (in the United States) as interns and participants in the ambassador roles and brigade team. In addition, NCTU's doctorate students have participated in EURECOM

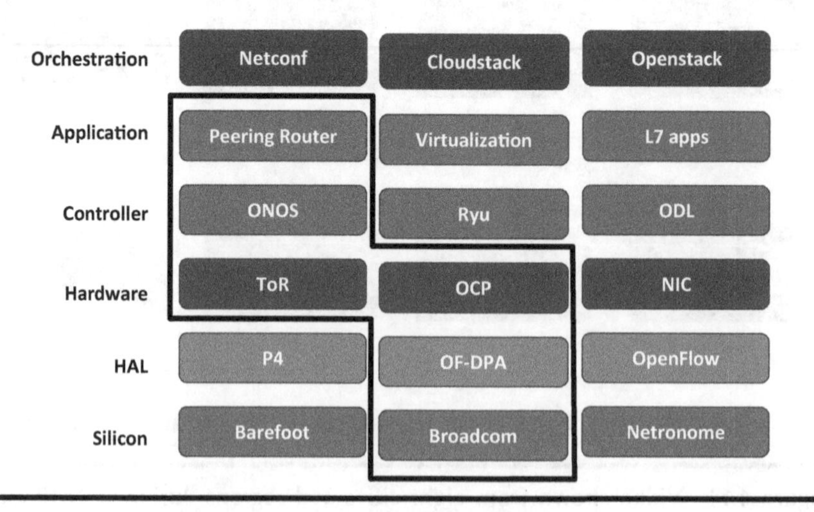

Figure 4.15 Vertically Integrated Stack as SDN Distribution.

(France) Open Air Interface (OAI) open networking project and they have contributed significantly in the area of virtual Radio Access Network (vRAN) performance improvement.

4.6 R&D Status of the P4 Switch and M-CORD at NCTU

4.6.1 Testbed of P4 Switches and Applications Development

At NCTU, we have done the performance measurements based on the testbed of P4 switches designed by Edgecore (Figure 4.16). P4 switches supplied by Inventec or Wistron NeWeb Corporation (WNC) can be used as alternative switches for the testbed for the two applications—Traffic Classification for Dynamic QoS Control and Dynamic Load Balancing and Congestion Avoidance—as described in Section 4.4.1 [24,25].

Today this P4 testbed is open to universities and research institutes (with proper agreements) who are interested in P4 switch-related research and technology exploration and P4 applications development.

4.6.2 M-CORD Architecture, Prototype, and Applications

Figure 4.17 shows a 4G OAI-based M-CORD Prototype at NCTU. This prototype consists of two components: (1) USRP X310 (RRU+BBU) as OAI eNB, and (2) CORE-in-a-box as OAI EPC with ONOS as SDN controller (MME) and P4 switches (S+P GW) + HSS.

The M-CORD-in-a-box and M-CORD prototype at NCTU shown in Figure 4.18 is prepared and ready for 5G UE-eNB as RAN/C-RAN.

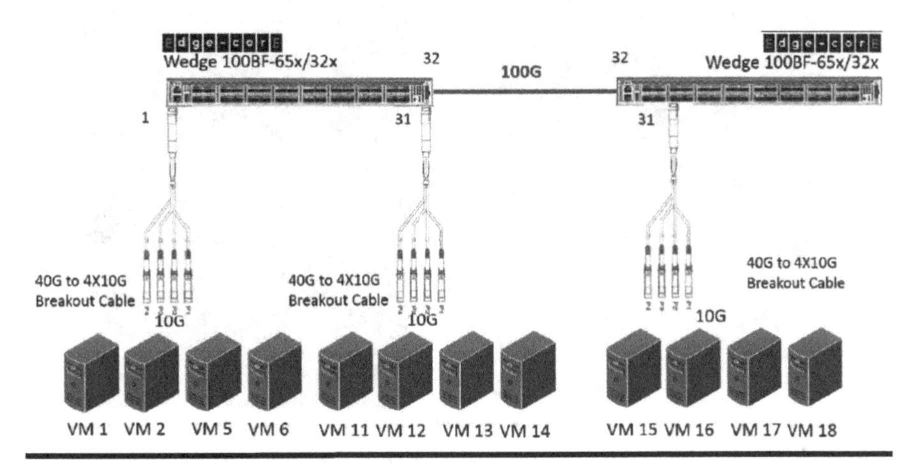

Figure 4.16 Topology of P4 testbed at NCTU.

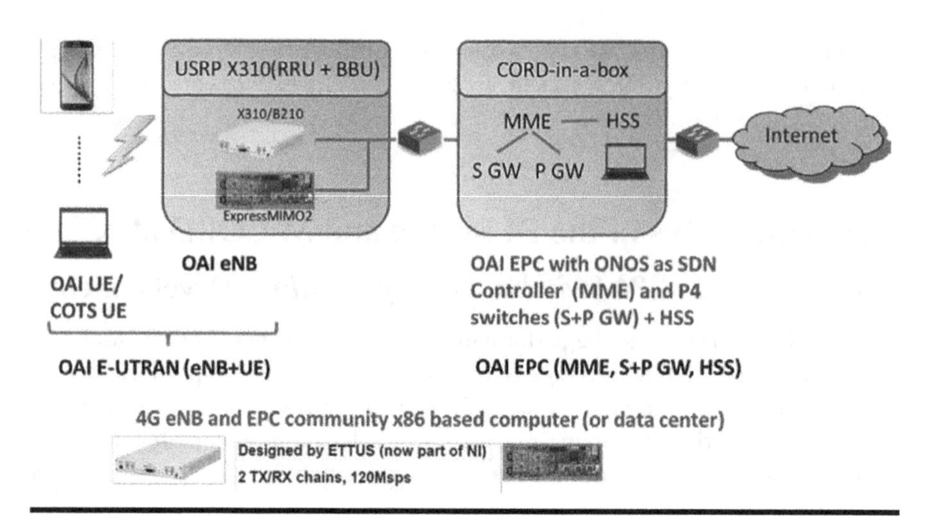

Figure 4.17 An OAI-based M-CORD prototype at NCTU.

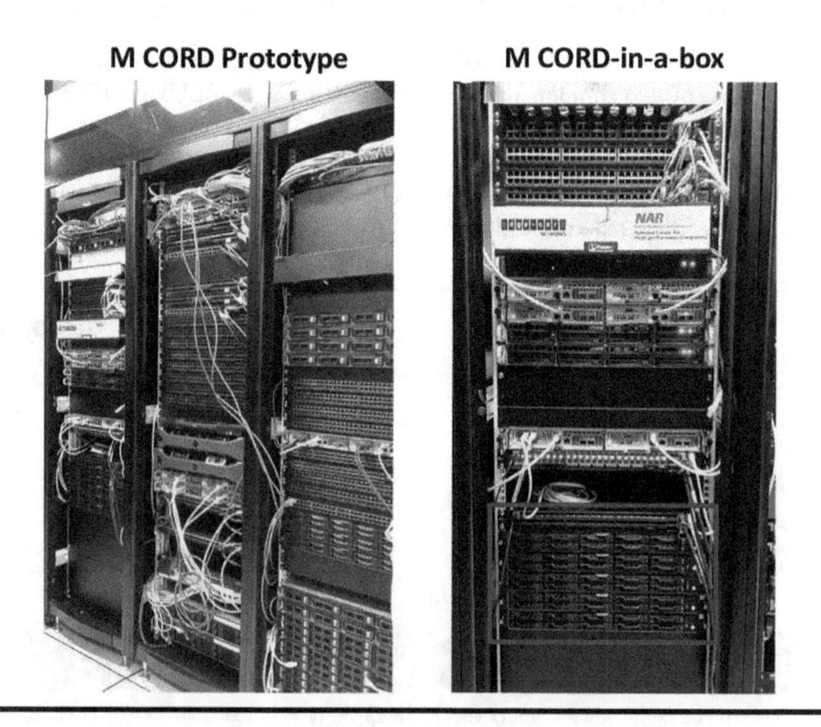

Figure 4.18 M-CORD-in-a-box and M-CORD prototype at NCTU.

4.7 IoT Technology Development and Industrial IoT at NCTU

4.7.1 IoT Technology Development at NCTU

The emerging of M2M technologies will provide all machines with network connectivity and achieve the goal of "network everything." Forrester, a research firm in the United States, forecasts that the value of the M2M industry will be 30 times the value of the Internet and will form the next trillion level of Information and Communication Technology (ICT) business.

In this area, we focus on the research of M2M Communications and the IoT including the following topics: (1) IoT/M2M communication protocols, (2) IoT/M2M service platform, (3) IoT/M2M sensor platform, (4) IoT/M2M security and privacy, (5) IoT/M2M cyber physical system, (6) IoT/M2M gateway, (7) IoT/M2M architecture and related international standards, and (8) IoT/M2M core network technology. As a new standard, NB-IoT is under development for applications that require high QoS, low latency, high reliability, and long range. Based on the study [36], NB-IoT is more suitable for future IoT applications.

4.7.2 IoT Applications Development at NCTU

Many IoT applications have been developed and their technical papers published. Here we list three applications:

1. CampusTalk: IoT devices and their interesting features on campus applications [32]
2. Flower Sermon: an interactive visual design using IoTTalk [33]
3. An Arduino Network application development platform based on LoTTalk [34]

Through the academia–industry collaboration program, currently, at NCTU we are developing smart farm, intelligent health care, and smart city and those R&D efforts will lead NCTU to work with the industries in the areas of the Industrial IoT (IIoT).

4.8 Conclusion

This chapter looked into the roles of 5G in the development of the future IoT. The 5G-based IoT will be better than the current 4G/LTE, not only in the areas of faster performance and lower latency, but also in having many more new/smart applications/services and the IIoT. To conduct softwarization and virtualization by applying SDN/NFV to 5G and IoT, we identified six challenges and ways to resolve them. Finally, we described the development efforts in SDN/NFV for open networking architecture, applications development based on P4 and CORD, and platforms and applications R&D of IoT at NCTU.

Acknowledgments

The work as supported by the Ministry of Science and Technology (MOST) of Taiwan and National Chiao Tung University under grants: 107-2221-E-009-056, 106-2221-E-009-008, and 105-2221-8-009-067.

References

1. Z. Qin et al., "A Software Defined Networking Architecture for the Internet of Things," *IEEE Network Operations and Management Symposium*, Krakow, Poland, May 5–9, 2014.
2. N. Bizanis et al., "SDN and Virtualization Solutions for the Internet of Things," *IEEE Access*, 4, 5591–5606, 2016.
3. C. Chappell, "The SDN/NFV Integration Challenge," *Light Reading*, October 2013.
4. J. Costa-Rquena et al., "SDN and NFV Integration in Generalized Mobile Network Architecture," *2015 European Conference on Network and Communications* (*EuCNC*), Paris, France, 2015.
5. J. Matias et al., "Toward an SDN-Enabled NFV Architecture," *IEEE Communications Magazine*, 53(4), 187–193, 2015.
6. E. Hossain and M. Hsan, "5G Cellular: Key Enabling Technologies and Research Challenges," *IEEE Instrumentation and Measurement Magazine*, 18(3), 11–21, 2015.
7. Y. Jararweh et al., "SDIoT: A Software-Defined Based Internet of Things Framework," *Journal of Ambient and Humanized Computing*, 6(4), 453–461, 2015.
8. Galis, "Challenges in 5G Network and Service Softwarization and Infrastructure," *Workshop on Network Softwarization*, Turin, Italy, September 21, 2015.
9. M. Corici et al., *Unleashing the Potential of Virtualization by the Right Toolkits and Open Testbeds*, Fraunhover Institute FOKUS, Berlin, Germany, 2015.
10. M. Liyanage et al., *Software Defined Mobile Networks beyond LTE Network Architecture*, Wiley, Chichester, UK, 2015.
11. C.J. Bermardo et al., "An Architecture for software Defined Wireless Networking," *IEEE Wireless Communications*, 21(3), 52–61, 2014.
12. P. Bosshart et al., "P4: Programming Protocol-Independent Packet Processors," *ACM SIGCOMM Computer Communication Review*, 44(3), 87–95, 2014.
13. A. Sivaraman et al., "DC.p4: Programming the Forwarding Plane of a Data-Center Switch," *SOSR 2015*, Santa Clara, CA, June 17–18, 2015.
14. S. Signorello et al., "NDN.p4: Programming Information-Centric Data-Planes," *2016 IEEE NetSoft Conference and Workshops* (*NetSoft*), Seoul, South Korea, June 6–10, 2016.
15. P. Bosshar et al., "Forwarding Metamorphosis: Fast Programmable Match-Action Processing in Hardware for SDN," *SIGCOMM'13*, Hong Kong, China, August 12–16, 2013.

16. Barefoot Networks, "The World's Fastest Most Programmable Networks," *Barefoot Networks*, June 15, 2016.
17. E.R. Sanchez, "Clash of titans in SDN: OpenDaylight vs ONOS," Techday MARID Spain, March 11, 2016.
18. B. Snow, "Transforming Service Provider Networks with ONOS and CORD," Open Networking Lab, 2015.
19. P. Berde et al., "Towards an Open, Distributed SDN OS," *hotSDN'14*, Chicago, IL, August 22, 2015.
20. B. Snow and G. Parulkar, "CORD—Central Office Re-architected as a Datacenter," ON.Lab Technical Report, Collaboration with AT&T, March 2016.
21. K. Prabhu, "Delivering a Software-based Network Infrastructure," AT&T Labs President and AT&T CTO, Speaker of EE of Columbia University Engineering, October 15, 2015.
22. L.V. Le et al., "SDN/NFV, Machine Learning, and Big Data Driven Network Slicing," *IEEE 1st 5G World Forum*, Santa Clara, CA, July 9–11, 2018.
23. World's fastest and most programmable switch series up to 6.5Tbps, Tofino by Barefoot Networks, June 15, 2016.
24. Y.-C. Chang, "Traffic Classification for Dynamic QoS Control Based on P4-Switch," MS thesis, National Chiao Tung University, Hsinchu, Taiwan, July 2017.
25. H.-T. Liu, "Dynamic Load Balancing and Congestion Avoidance based on P4 Switch," MS thesis, National Chiao Tung University, Hsinchu, Taiwan, July 2017.
26. L.V. Le et al., "Applying Big Data, Machine Learning, and SDN/NFV to 5G Traffic Clustering, Forecasting, and Management," *IEEE NetSoft Conference and Workshops (NetSoft)*, Montreal, Canada, June 25–29, 2018.
27. C.A. Akpakwu et al., "A Survey on 5G Network Internet of Things: Communications Technologies and Challenges," *IEEE Access*, 6, 3619–3647, 2018.
28. M. Manning, "5G as an innovative catalyst in IoT," *Technology*, TG, May 9, 2018.
29. S.K. Lee et al., "Future of IoT Networks: A Survey," *Applied Sciences*, 7(10), 1072, 2017. doi:10.3390/app7101072.
30. Y. Stein, "5G-A Glimpse Into the Future," DRC, December 2017. https://www.drc.ltd/5g-a-glimpse-into-the-future/.
31. P. Collela, "5G and IoT: Ushering in a new era," Ericsson, March, 2017. https://www.ericsson.com/en/about-us/india/authored-articles/5g-and-iot-ushering-in-a-new-era.
32. Y.-P. Lin et al., "CampusTalk: IoT Devices and Their Interesting Features on Campus Applications," *IEEE Access*, 6, 26036–26046, 2018.
33. C.Y. Hsiao et al., "Flower Sermon: An Interactive Visual Design Using IoTTalk," *Mobile Networks and Applications*, 2018.
34. Y.-W. Lin et al., "An Arduino Network Application Development Platform Based on LoTTalk," *IEEE System Journal*, 13(1), 468–476, 2018.
35. I.F. Akyildiz, "Internet of Things: Trends, Directions, Opportunities, Challenges," BWN Lab, School of ECE, Georgia Institute of Technology, IFA'2017. www.ece.gateck.edu/researh/labs/bwn.
36. R.S. Sinha et al., "A Survey on LPWA Technology: LoRa and NB-IoT," *ICT Express*, 3(1), 14–21, 2017.

Chapter 5

Achieving Scalability in the 5G-Enabled Internet of Things

Fuchun Joseph Lin and David de la Bastida

Contents

5.1 Introduction to Scalability of the Internet of Things Platform

The oneM2M is a global Internet of Things/machine-to-machine (IoT/M2M) platform that provides the service layer capacity for M2M communications. Not only can it manage heterogeneous IoT/M2M devices, but it also provides an integration platform among different IoT/M2M technologies. Nevertheless, scalability is still an issue not well addressed for the oneM2M platform. Our research focuses on the scalability of IoT platforms such as oneM2M.

Figure 5.1 depicts the oneM2M architecture, which consists of both infrastructure domain and field domain. In addition to IoT/M2M devices represented by non-oneM2M device, application-dedicated, and application service nodes (NoDN, ADN, and ASNl, respectively) (Figure 5.1), oneM2M also defines two other types of nodes: the middle node (MN) and the infrastructure node (IN). INs are IoT/M2M servers that normally reside in the cloud, whereas MNs are service nodes

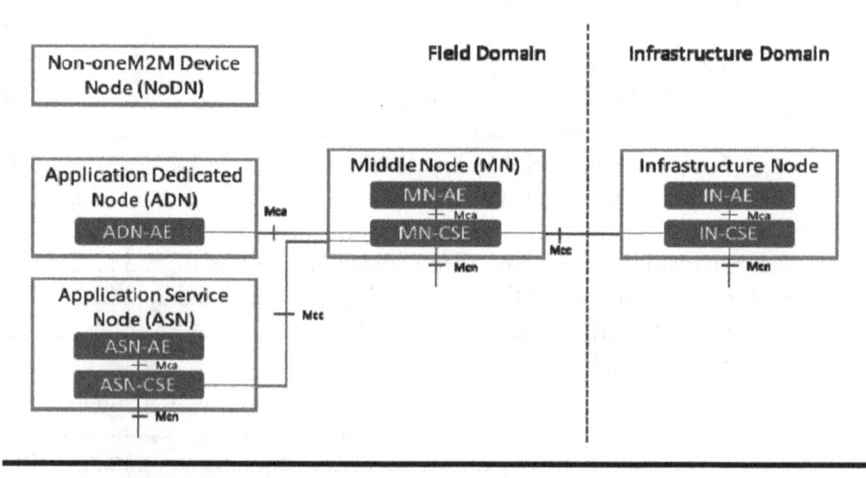

Figure 5.1 OneM2M architecture.

in the continuum from the cloud to the things. Both MNs and INs consist of two main functional entities: an application entity (AE) and a common service entity (CSE) [1]. An AE represents an application residing on the oneM2M node, and a CSE provides useful common service functions (CSFs).

The oneM2M defines several communication interfaces among its entities: Mca between AE and CSE; Mcc between CSEs; and Mcn between CSE and the underlying network service entity. The oneM2M supports the binding of these interfaces to the HTTP and the constrained application protocol (CoAP) and thus provides representational state transfer application programming interfaces (REST APIs) over all these interfaces. To store and share a large amount of IoT/M2M data, oneM2M also defines a hierarchical data structure called a resource tree, accessible via a CSE.

5.2 5G-Enabled Technologies

With the development of upcoming 5G technologies, we see some very useful tools that can be applied to tackle the IoT scalability problems. Notably, cloud computing, fog computing, and software-defined networking (SDN) technologies in 5G that can work together to enable the scalability of the IoT system. We introduce each of these enabling technologies below.

5.2.1 Cloud

Our research largely depends on OpenStack for cloud operations. OpenStack is an open-source infrastructure-as-a-service (IaaS) cloud operating system that offers many advanced features such as multiple hypervisors compatibility (Nova), advanced images management (Glance), embedded identity and security mechanisms (Keystone), virtualized internal network (Neutron), efficient message queue (RabbitMQ) and Representational State Transfer (REST) APIs for managing all those features. The OpenStack components interact with one another via APIs and their communication occurs via a message queue. OpenStack is based on the micro services paradigm where an application is decoupled into small pieces of code that are able to run independently, but when combined can build complex systems.

Our research is based on the 11th release of OpenStack called Kilo. Thus, the following explanations may vary from the latest release (Newton up to middle of 2018). OpenStack Kilo marks a turning point for the open source project with contributions from nearly 1,500 developers and 169 organizations worldwide. As the core of platform matures, its focus turns to interoperability in the market, raising the bar for driver compatibility, and extending the platform to fit workloads with bare metal and containers.

5.2.1.1 Compute (Nova)

Kilo offers new API versioning management with version 2.1 and micro-versions to provide reliable, strongly validated API definitions. This makes it easier to write long-lived applications against compute functionality. Major operational improvements include live upgrades when a database schema change is required in addition to better support for changing the resources of a running Virtual Machine (VM).

5.2.1.2 Networking (Neutron)

The load-balancing-as-a-service API is now in its second version. Additional features support network function virtualization (NFV), such as port security for Open vSwitch, Virtual Local Area Network (VLAN) transparency and Maximum Transmission Unit (MTU) API extensions. Additional architectural updates will improve scale for future releases.

5.2.1.3 Storage (Cinder)

Major updates to testing and validation requirements for backend storage systems across 70 options ensure consistency across storage choices as well as continuous testing of functionality for all included drivers. In addition, users can now attach a volume to multiple compute instances to enable new high-availability and migration use cases.

5.2.1.4 OpenStack Message Queue

OpenStack projects use an open standard for messaging middleware known as advanced message queuing protocol (AMQP). This messaging middleware enables the OpenStack services that run on multiple servers to talk to each other. OpenStack Oslo remote procedure call (RPC) supports three implementations of AMQP: RabbitMQ, Qpid, and ZeroMQ. OpenStack Oslo RPC uses RabbitMQ by default.

5.2.1.5 OpenStack Micro Services

Micro services are a software development architectural style that structures an application as a collection of loosely coupled services. In a micro services architecture, services are fine-grained and the protocols are lightweight. The benefit of decomposing an application into different smaller services is that it improves modularity and makes the application easier to understand, develop, and test and more resilient to architecture erosion. It also parallelizes development by enabling small autonomous teams to develop, deploy and scale their respective services independently. Furthermore, it allows the architecture of an individual service to emerge through continuous refactoring.

5.2.2 Fog

Our research is inspired by fog computing architecture as defined by the OpenFog Consortium. The OpenFog Consortium was launched in 2015 and aimed to promote the concept of fog computing and define a set of system-level architectural framework. OpenFog introduced a major component in fog computing called fog node, which provides the computing, networking, storage and acceleration capacity for the end-devices. In the OpenFog reference architecture [2], it has highlighted eight essential pillars: security; scalability; openness; autonomy; reliability, availability, and serviceability (RAS); agility; hierarchy; and programmability to support the OpenFog architecture. In our research, we focus on the scalability pillar of the fog architecture. In order to provide a highly scalable fog architecture, the fog network not only has to be capable of deploying different number of fog nodes according to different workloads, but it also has to be able to balance the workload among these nodes. In addition, for heterogeneous services the fog network also needs to provide a mechanism to discover required fog nodes, and direct the data flow to the correct nodes for service processing.

Figure 5.2 shows the cloud-to-thing continuum. Given the proximity to the things, the fog has the better real-time processing capacity than the cloud. Furthermore, the raw data generated by things can be preprocessed hierarchically by fog nodes. Only the information that needs further processing goes to the next level. Consequently, fog can not only save the access bandwidth to the core network but release the burden of the cloud.

There are many common features between the middle nodes in oneM2M and the fog nodes in OpenFog. In addition to supporting the hierarchical processing

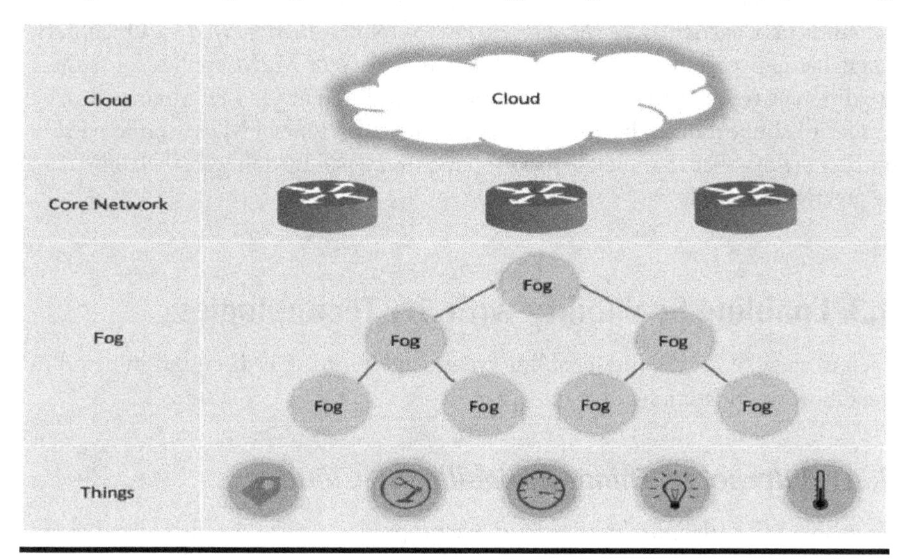

Figure 5.2 Cloud-to-Thing continuum.

model, oneM2M middle nodes resemble a localized IN that can offload the processing load of the IN normally residing in the cloud by providing compute, storage and network closer to the IoT/M2M devices. Consequently, middle nodes can assume the role of fog nodes. Our research targets at integrating oneM2M and fog into a coherent architecture.

5.2.3 Network Slicing of Software-Defined Networking

Network slicing in our research is enabled by SDN technologies in OpenStack. In the SDN paradigm, the lower layer comprises commodity network forwarding devices (SDN switches); the middle layer is a network operating system, also known as SDN Controller which runs the core services needed to manage the forwarding devices in the lower level; and the third, or top, layer consists of SDN applications that interact with the forwarding devices through the SDN controller.

SDN switches are controlled by the SDN controller through one southbound-interface protocol like OpenFlow, OVSDB, NETCONF, SNMP, among others, with OpenFlow being the most popular at the moment. The SDN controller run different core services such as topology, inventory, statistics, and host tracking services. There are several open source and proprietary SDN controllers such as Ryu, ONOS, OpenDayLight (ODL), and OpenContrail. SDN applications communicate with the SDN controller through a northbound interface, where RESTful communications are the main method employed so far.

Network slicing technology allows a single physical network to be partitioned into multiple independent virtual networks, where each virtual network can be configured with specific quality of service (QoS) capabilities in order to satisfy the QoS demands of a particular M2M application. SDN can be deployed in OpenStack to enable network slicing. Due to the heterogeneity of M2M applications, there are different types of M2M traffic patterns with different QoS requirements. As a result, it is important to be able to direct a particular type of M2M traffic pattern onto a network slice that matches its QoS requirements. By doing so, scalability can be greatly improved.

5.3 Enabling Scalability with 5G Technologies

Each of these 5G scalability-enabling technologies for IoT is discussed in detail in the following subsections.

5.3.1 Internet of Things Scalability by Cloud

To enable IoT scalability by the cloud, we propose to utilize the virtualized cloud environment enabled by OpenStack. OpenStack is the de facto cloud operating system. It runs on almost any Linux-compatible hardware, with a minimum of requirements. Its versatility and flexibility make it the preferred platform for both research and production environments. OpenStack is composed of modules for virtualization (Nova), networking (Neutron), identity and security (Keystone), images administration (Glance), system measurements and statistics (Ceilometer), orchestration (Heat), storage (Cinder and Swift), and others. These modules interact among them via plugins and allow user management via REST APIs.

5.3.1.1 Architecture

Our approach adopts the micro services paradigm [3] and the advanced queueing services from OpenStack while designing an IoT-specific master node and load balancing queue. Our proposed system, as shown in Figure 5.3, includes the following components that communicate with one another using OpenStack Neutron:

5.3.1.1.1 Master Node

The master node is aware of the CPU state of the platform nodes in order to determine when to do scaling up or scaling down. It utilizes the OpenStack Nova API to turn on and off virtual machines. The master node continuously receives the CPU status from each platform node. It then utilizes a linear function to forecast the CPU status of every platform node; this value is the input for our scalability algorithms.

5.3.1.1.2 Load Balancing Queue

Our load balancing queue is designed based on RabbitMQ [4], which is the messaging exchange method in OpenStack. RabbitMQ is an implementation of the AMQP. We adopt the "basic.qos" method in RabbitMQ with the "prefetch_count = 1" setting. This setting tells our load balancing queue not to dispatch a new message to a platform node until it has processed and acknowledged the

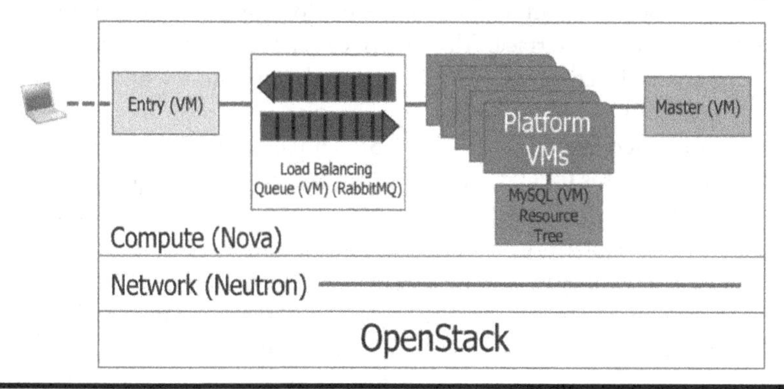

Figure 5.3 Our proposed cloud OpenStack-based architecture.

previous one. If there are no platform nodes available, the load balancing queue will store the messages until at least one platform node is available.

The queue has two channels: one for distributing the incoming requests uniformly among the platform nodes, the other for returning the request responses back to the client. It communicates with the platform nodes via a load balancing queue–platform node interface. This interface is implemented based on the micro services paradigm and follows the RPC style [5]. On the load balancing queue side, a request is forwarded to an available platform node. Then on the platform node side, a piece of software called the "load balancing queue client" is deployed with each oneM2M IN-CSE instance to translate the received message (via the request channel) into a specific platform function (e.g., a HTTP POST request into the IN-CSE) and finally sends back the result to the load balancing queue (via the response channel).

The load balancing queue then translates and forwards the response back to the outside client.

5.3.1.1.3 Entry Node

The entry node is the connection point of our system with external clients. It is a RESTful server waiting for requests. The incoming requests are forwarded to the load balancing queue. The entry node communicates with clients: It passes clients' input messages to the request channel of the load balancing queue while receiving response messages from the response channel of the load balancing queue and returning them to clients.

5.3.1.1.4 Platform Nodes

Platform nodes consist of a fixed number of homogeneous virtual machines running on top of OpenStack Nova, where each virtual machine has been provisioned with an oneM2M IN-CSE instance.

The master node communicates with the platform nodes to perform scalability via a master–platform interface. This interface has two purposes: to inform the master node about the current CPU state of the platform nodes, and to execute the scaling up and scaling down procedures.

We first modified the oneM2M IN-CSE source code in order to add a CPU notification plugin. This CPU notification plugin periodically inquiries the current CPU loading of a platform node and forwards it to the master node. This procedure occurs on each active platform node. Secondly, we utilize OpenStack Nova to perform scaling up or scaling down according to the rules of the scalability algorithms.

OpenStack Nova manages the life cycle of the platform nodes. When a virtual machine needs to be spawned, the master node will call the corresponding OpenStack Nova APIs to trigger its creation. Similarly, when a virtual machine needs to be shut down, the master node will call the corresponding Nova APIs.

5.3.1.1.5 MySQL Node

The MySQL node is the data repository and resource tree of our system. It is shared among all platform nodes. We have utilized MySQL Release 5 as the target data repository.

5.3.1.2 Key Enablers

In this section, we discuss both scaling up and scaling down algorithms for cloud-based IoT systems. Scaling up happens when the system is overloaded, whereas scaling down happens when the system is underloaded.

5.3.1.2.1 Scaling Up Algorithm

The scaling up algorithm, as shown in Table 5.1, compares the forecasted CPU status of active platform nodes, versus two thresholds: overloaded and scale up. The overloaded threshold tests a single platform node, whereas the scale up threshold is applied to the whole pool of active platform nodes. Our algorithm ensures that no new platform nodes are booted if the load can be handled by current active platform nodes. It also ensures that only one platform node is booted at a time.

5.3.1.2.2 Scaling Down Algorithm

If the scaling up algorithm does not match, the scaling down algorithm, as shown in Table 5.2, is evaluated right after it. It also utilizes two thresholds: idle and scale down. First, the idle threshold is tested against each active platform node. If met, a platform node will be designated as the candidate for scale down. Secondly, the scale down threshold is tested against the average CPU usage of all active platform nodes. If met, the master node will ensure that no pending process are left on the virtual machine designated for scale down, and then shut it down.

Table 5.1 Scaling Up Algorithm

```
At time tn, receive CPU status from Platform node i
forecast CPU status at time tn+5
if forecast > Overloaded Threshold {
   if all platform nodes' avg CPU > Scale Up Threshold
   and booted Platform nodes < Total Platform nodes
   and booting Platform nodes = 0 {
     Scale up 1 Platform node
   }
}
```

Table 5.2 Scaling Down Algorithm

```
At time tn receive update from Platform node i and forecast
if forecast < Idle Threshold {
  if all Platform nodes' avg CPU < Scale Down Threshold
  and not shutting down Platform nodes
  and active Platform nodes > 2 {
     designate Platform node i for termination
  }
}
At time ti+1 receive update from Platform node i
If Platform node i has no pending work {
Scale down Platform node i
}
```

On our proposed system, we have set the scaling up threshold at 50% of CPU utilization, and the scaling down threshold at 10%.

5.3.1.3 Evaluation

In order to evaluate our system, we compared its performance in terms of response time, power consumption and computational cost against that of three native OpenStack scalability methods: Load Balancer as a Service (LBAAS), Heat/Ceilometer, and combination of LBAAS and Heat/Ceilometer, using four types of IoT/M2M traffic patterns generated by the traffic generator tool called Jmeter [6].

Each IoT/M2M traffic pattern was tested in isolation from others for each of our four designs. It is clear that in a more realistic deployment these different IoT/M2M traffic flows could appear together, flowing simultaneously to the same destination server. Such a consideration of mixed IoT/M2M traffic patterns will be addressed in Section 5.3.3.

5.3.1.3.1 Experimental Setup

We implemented the four designs using four desktop machines with Intel Xeon 2.7 GHz processors (four cores), 16 GB of RAM and 1 TB of hard disk. We used OpenStack Kilo running on top of Ubuntu Server 14.04.

As depicted in Figure 5.4, our setup of OpenStack Kilo consisted of one physical machine for controller, one physical machine for network, and two physical machines for compute. We used an additional physical machine to run Jmeter and act as the client. The OpenStack deployment and the Jmeter client were located in two different buildings in our campus, separated by at least 500 m, and interconnected by a private segment of our campus LAN. We assume there was no other traffic flowing on our private segment of the network during the execution of our experiments.

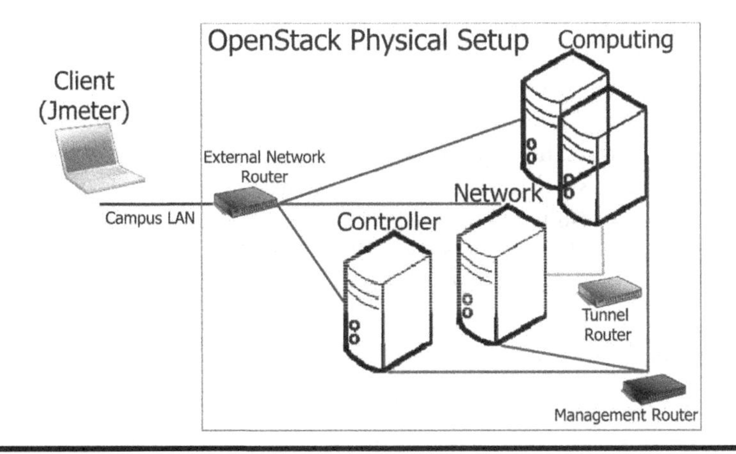

Figure 5.4 Our OpenStack testbed.

The master node was built using NodeJS and Express framework, and Python 2.7 was used for the implementation of the load balancer node and the entry node. Our platform nodes are instances of oneM2M IN-CSE with MySQL as its resource tree data repository. We set up five platform nodes, each running on a virtual machine with 1 CPU, 2 GB of RAM, and 20 GB of HDD.

In all of our tests, we used the following configuration via the Jmeter Ultimate Thread Group Plugin: during the first 60 s, 10 users were set to send data (warming up period); then, 50 users would be sending data during 90 s (high traffic period); finally, 10 users would send data again for 60 s (cool down period). The total duration of a test is 210 s. We have performed a fair number of repetitions on each test and taken the average of all the executions as the values reported in this paper. For all the four systems, we established two and five as the minimum and maximum number of platform nodes running at any time, respectively.

The resource tree contains an AE resource created for each type of IoT/M2M application. Each AE resource contains a container called "data," whose ContentInstances would be cleaned each time before running a new test. The data stored in the ContentInstances are JSON objects having attributes related to each type of IoT/M2M application. The attributes and values of JSON objects are built using Jmeter.

5.3.1.3.2 Experimental Results

We conducted experiments in order to compare the performance of our proposed OpenStack scalability system with OpenStack native LBAAS, Heat/Ceilometer functions, and a combination of both [8], in terms of their response time, power consumption and computational cost.

5.3.1.3.3 Response Time Results

For the response time tests, we expect each system will scale up according to the increasing workload. Our proposed system utilizes the CPU state of the platform nodes to trigger the scaling up function; LBAAS utilizes the number of concurrent connections; Heat as well as the combined system utilize a CPU Ceilometer alarm. The response time results of smart meter, Bluetooth tags, eHealth and video traffic in milliseconds are shown in Figures 5.5 through 5.8, respectively. The response time of our proposed system is the fastest in most cases.

5.3.1.3.4 Power Consumption Results

For the power consumption tests, though we expected more energy would be consumed for those systems with faster response time because more resources are utilized, the results are actually more dependent on the nature of applications than on the scalability of systems. The power consumption results of smart meter, Bluetooth tags, eHealth and video applications in watts are shown respectively in Figures 5.9 through 5.12.

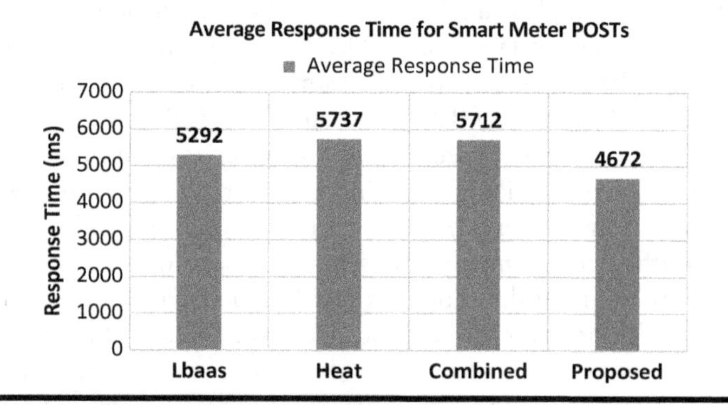

Figure 5.5 Results for smart meter response time.

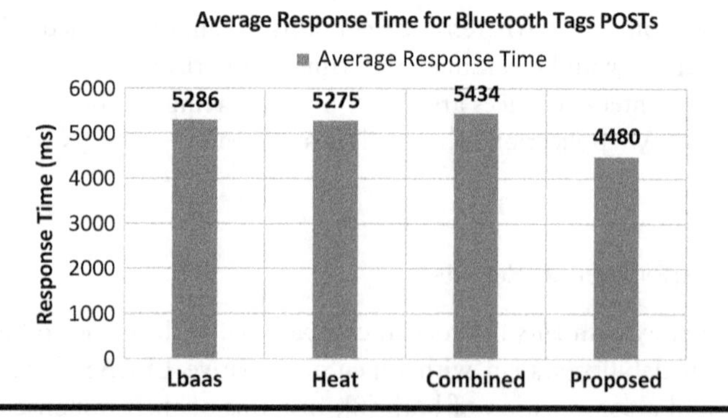

Figure 5.6 Results for Bluetooth tags response time.

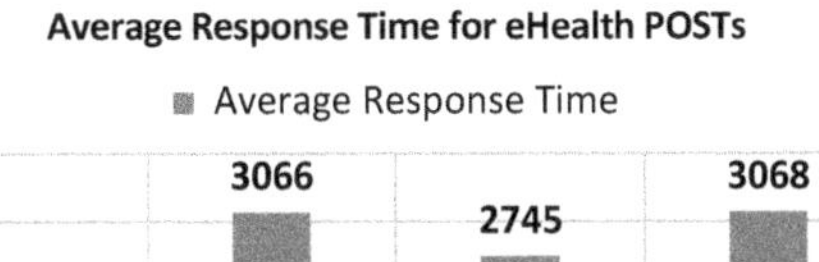

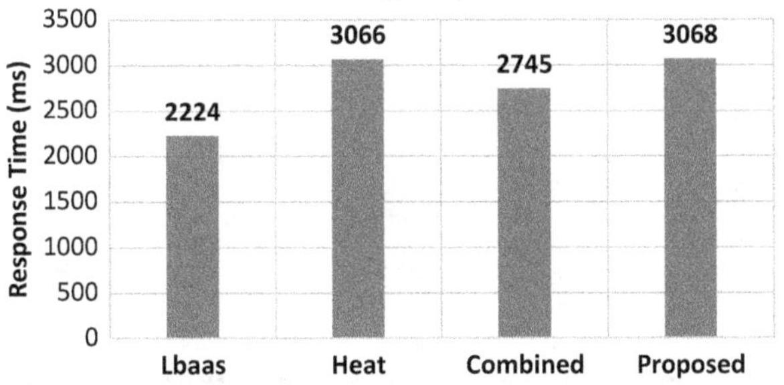

Figure 5.7 Results for eHealth response time.

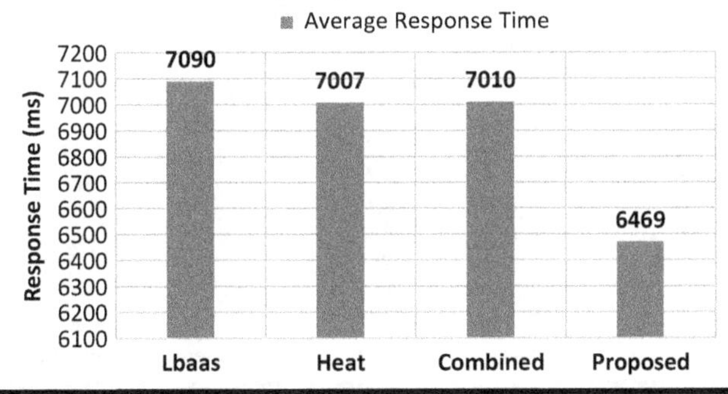

Figure 5.8 Results for video response time.

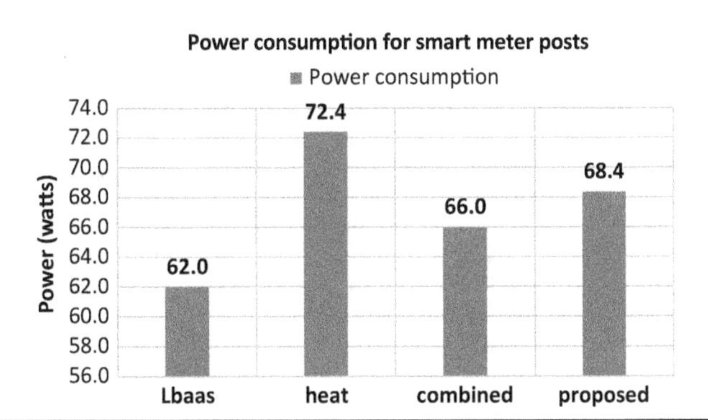

Figure 5.9 Results for smart meter power consumption.

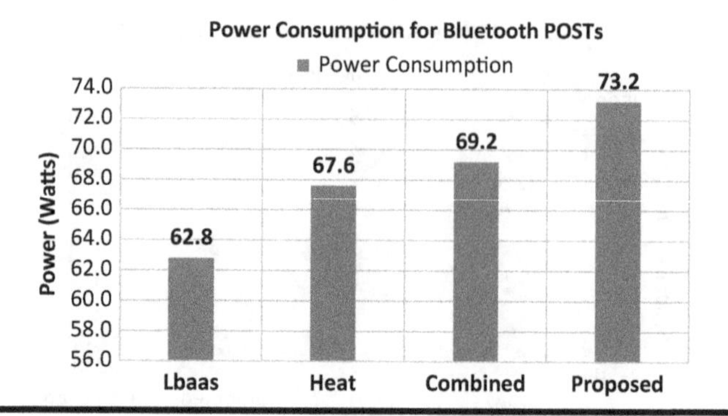

Figure 5.10 Results for Bluetooth tags power consumption.

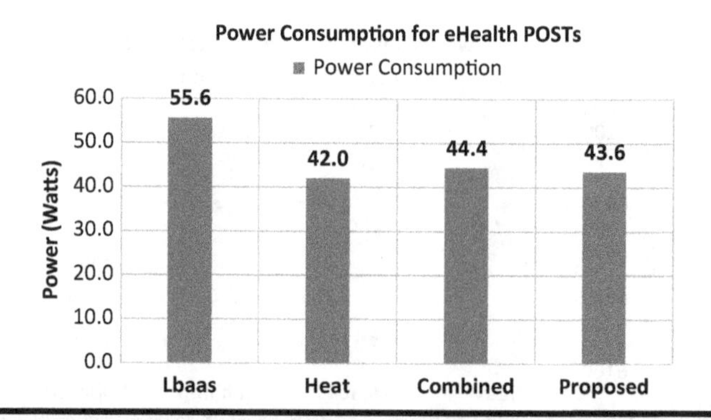

Figure 5.11 Results for eHealth power consumption.

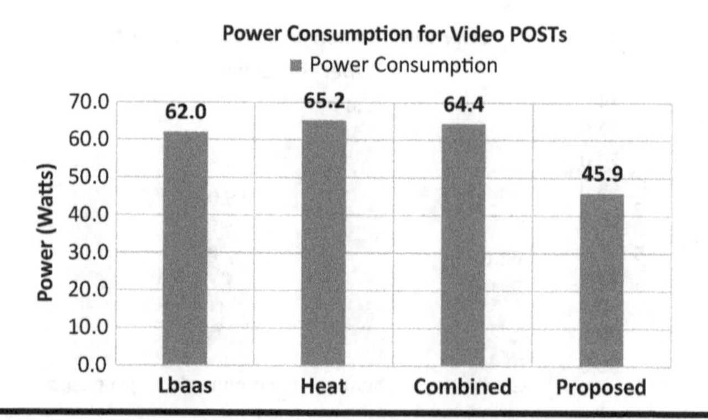

Figure 5.12 Results for video power consumption.

5.3.1.3.5 Computational Cost Results

To compare computational costs, we measured CPU and memory utilization of the software components that enable scalability and load balancing on each platform while excluding the cost of the OpenStack core itself. The computational cost of the platform nodes is not taken into consideration because these nodes are the processes enabled by the scalability software components. Only the computational costs of those system components that enable scalability are measured.

As shown in Figures 5.13 and 5.14, respectively, our proposed system is the one using the lowest amount of memory and CPU. These results are to be expected because LBAAS, Heat and the Combined require several sophisticated OpenStack processes such as heat-api-cfn, heat-engine, ceilometer-alarm, ceilometer-agent, ceilometer-api, and ceilometer-collector, load-balancer and haproxy for their execution, whereas our proposed system does not rely on any of those components.

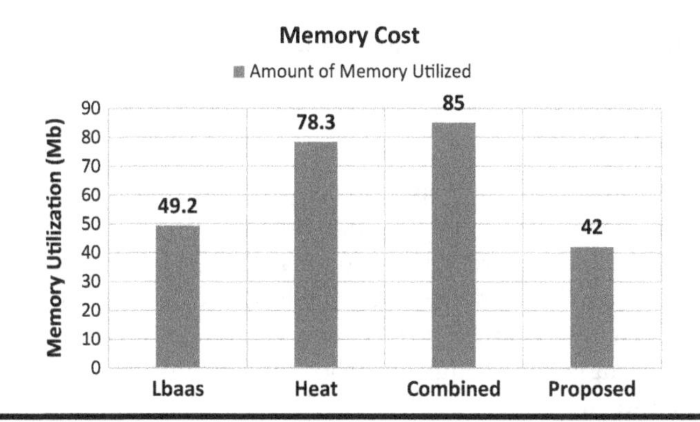

Figure 5.13 Results for memory cost.

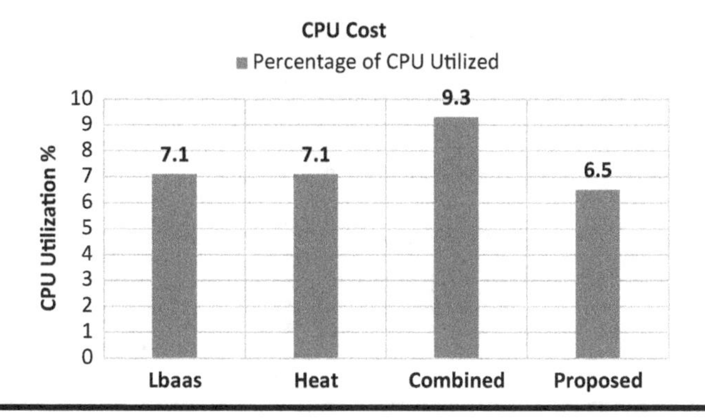

Figure 5.14 Results for CPU cost.

5.3.2 IoT Scalability by Fog

In Section 5.3.1, we described how to support the scalability of an IoT/M2M platform in the cloud with OpenStack [7,8] where the resources in the cloud can be scaled up or down according to the loading of IoT traffic. However, dealing with IoT traffic solely in the cloud will not completely solve the scalability problems. To improve the scalability of the IoT/M2M platform, IoT traffic must be managed before it reaches the cloud. This means that the fog network, as a continuum between the cloud and the things, must provide the scalable capacity; that is, it must be able to flexibly and dynamically scale in/out the serving instances across fog nodes to match with incoming traffic.

One use case ideal for applying fog scalability is Industry 4.0, under which an enormous number of sensors and actuators have been deployed on the factory floor. With these connected sensor and actuators, manufacturers can easily develop Industrial IoT (IIoT) applications on their platforms [9]. Most data collected from factory sensors are still sent to the cloud for analytics processing. Nevertheless, such initial deployment has revealed the need to improve real-time responsiveness and system scalability of such a cloud-centric design.

5.3.2.1 Architecture

As depicted in Figure 5.15, we propose to construct the middle nodes of oneM2M as a hierarchical fog architecture. In our proposed fog architecture, there are two types of fog node: fog worker and fog manager. The former provides virtual resources to run oneM2M middle nodes (both MN-CSEs and MN-AEs) as containers. The latter is responsible for supervising the loading conditions of these resources, including orchestrating and scheduling containers and assigning tasks across different fog workers.

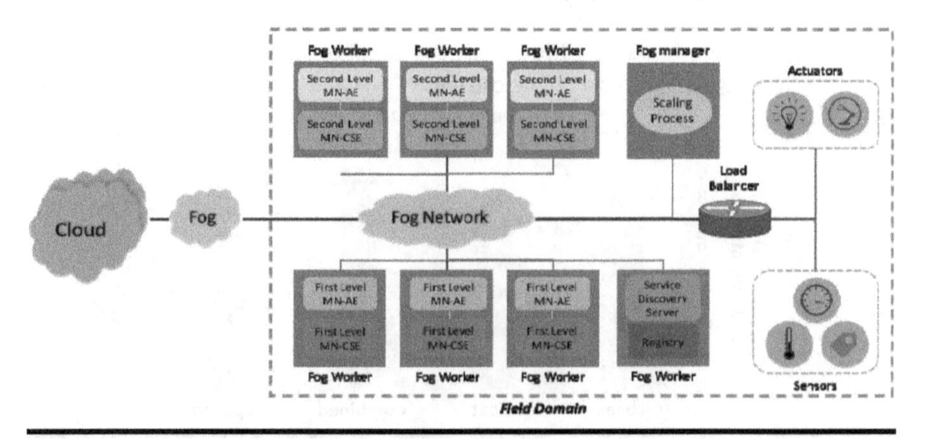

Figure 5.15 Proposed fog architecture.

As there are sensors and actuators in the field domain, we separate sensor-related tasks from actuator-related tasks in our proposed architecture. The former are served by the first-level fog workers, whereas the latter are served by the second-level fog workers. These fog workers will run the following services in the containers:

First-level MN-CSE: This level of MN-CSEs establishes the resource tree corresponding to the underlying sensor devices and registers to the second-level MN-CSE. The second-level MN-AE would subscribe to certain resources in First-Level MN-CSE in order to be notified about the events of interest.

First-level MN-AE: This AE is responsible for monitoring the sensor input. Thus, it will constantly perform resource discovery and subscribe to the newly discovered resources of interest at the first-level MN-CSE. It will keep monitoring the sensor data in the first-level MN-CSE's resource tree. It thus can detect abnormal events and notify the second-level MN-AE.

Second-level MN-CSE: This MN-CSE establishes the resource tree corresponding to the underlying actuator devices. It stores the information of the actuator devices in the field domain.

Second-level MN-AE: This AE is responsible for controlling the actuator devices. It first registers to the second-level MN-CSE then will constantly perform resource discovery to discover and subscribe to the actuator-related resources of interest. Furthermore, it will also subscribe to the abnormal events in first-level MN-CSE. Whenever an abnormal event notification is received, this AE will immediately send the commands to the second-level MN-CSE to control the corresponding actuator.

Registry: The registry is a private repository for storing service images. Whenever fog nodes need to launch services, they will fetch the images from the registry and spawn them as serving instances.

Service discovery server: This is a server provisioned for service discovery. It will periodically discover the services running in the fog network and assist the newly spawned oneM2M instances (MN-CSEs and MN-AEs) to choose a proper MN-CSE to register. Moreover, to prevent too many MN-AEs or MN-CSEs from registering to the same MN-CSE, the service discovery server would maintain a priority list in which an MN-CSE is rated with lower priority if it is registered by more number of registered entities. Every time an MN-AE or an MN-CSE instance is created, it will send a request to the service discovery server asking for a specific kind of MN-CSE service. The service discovery server will respond with the IP address of the highest priority MN-CSE.

Within the fog manager, a scaling process will periodically check each fog node's CPU utilization and decide whether to scale out or scale in the serving instances of fog worker. In addition, a load balancer is deployed between IoT/M2M devices and the fog cluster to distribute all the incoming requests from the IoT/M2M devices to the first-level MN-AE/MN-CSE according to a chosen load-balancing policy.

5.3.2.2 Key Enablers

We leverage virtualization enabled by the container technology [10] to create high scalable fog-based IoT/M2M systems. Compared to virtual machines, containers provide lightweight virtual environment and are thus more suitable to achieve a fast response in the fog network. We introduce two approaches to extend the scalability of IoT/M2M systems in the fog.

Static pool of serving instances: This is the traditional approach where a fixed number of oneM2M middle node container instances is deployed among the fog nodes before the service starts to work. The load balancer then will distribute the traffic to all the serving instances.

Dynamic scaling mechanism: In this approach, the fog network have a scaling server. This server will check each fog node's status such as its CPU and memory usage and automatically scale in/out oneM2M middle node container instances according to the incoming traffic.

5.3.2.3 Evaluation

We implemented our testbed in six laptops with Intel Core i5-4200H 2.8 GHz processor (two cores), 8 GB of RAM and 1 TB of hard disk. Each physical machine runs the Ubuntu 14.04 as its host operating system.

In our testbed, the IoT/M2M platform used is OpenMTC Rel. 4 [11] that is a oneM2M-based IoT platform developed by Fraunhofer FOKUS. We also adopt Docker [12] as the container-based virtualization technology. Starting with Docker version 1.12, swarm mode has been included in the Docker engine. Swarm mode is a native cluster management tool of Docker.

In Swarm mode, we can assign some of nodes that contains Docker Engine to be fog manager nodes. With fog managers, we can group our fog workers to be a fog cluster and use an overlay network called fog network to connect these fog nodes. Each fog node can run the oneM2M CSE instance as a container.

When different oneM2M instances (CSE or AE) are deployed in containers as services, each service will be assigned a virtual Internet Protocol Address (VIP). When a query is sent to the fog node, the Docker Engine on each fog node has an internal Domain Name System (DNS) server to resolve the incoming query and forward the query to the target container. Moreover, we use the HAProxy [13] as our external load balancer because, in our scenario, we use HTTP as the communication protocol between the IoT devices and the fog nodes. HAProxy supports TCP/HTTP load balancing, which is suitable for balancing HTTP requests.

In order to evaluate the performance of our proposed architecture, we compare it to the static approach with a fixed pool of serving instances. We manipulate the transfer rate of each thread of our IIoT application into 100 requests per second, 25 requests per second and 10 requests per second, which represent high traffic load, moderate traffic load and low traffic load, respectively. The application will generate 10-min traffic to the fog cluster during the test.

Our testbed consists of a fog manager, a load balancer and four fog workers. The scaling process of our proposed architecture will deploy the oneM2M service instances to these fog workers. In our scenario, most of data processing happens at the front end for handling the sensor data; so our testing is primarily for the scalability of the first-level oneM2M services.

In the test, we set 80% CPU utilization as the scale-out threshold. This means that once the CPU utilization of a fog node goes over 80%, the scaling server would scale out the serving instances by one. We set the minimum size of first-level oneM2M as one, and the maximum size as four. Therefore, the number of the first-level oneM2M instances that are deployed in the fog cluster will fall in the range of between one and four.

In the beginning of the static system, we deploy a pool of three first-level oneM2M middle node service instances (both MN-CSE and MN-AE) to the fog cluster. On the other hand, in our proposed system, we deploy only one first-level oneM2M instance to the fog cluster initially and let the system dynamically deploy more if needed.

Our testing results are shown in Table 5.3. We list the number of active containers after 10-min traffic generation, and report the average power consumption, the average CPU usage per fog node and the average response time for each request.

Table 5.3 Comparison of Test Results

Metrics for Comparison Between Static and Dynamic Systems		*Simulator with Transfer Rate of 100 Requests per Second*	*Simulator with Transfer Rate of 25 Requests per Second*
Number of Active Containers	Static system	3	3
	Proposed dynamic system	4	3
Average Power Consumption per Fog Node	Static system	30.72 (Watt)	16.69 (Watt)
	Proposed dynamic system	25.62 (Watt)	15.44 (Watt)
Average CPU Usage per Fog Node	Static system	100 (%)	50.02 (%)
	Proposed dynamic system	78.35 (%)	48.33 (%)
Average Response Time per Request	Static system	37.25 (ms)	14.62 (ms)
	Proposed dynamic system	19.44 (ms)	16.57 (ms)

In our proposed architecture, the high traffic load (100 requests/s) makes the number of first-level oneM2M instances scaled to four, while the moderate traffic load (25 requests/s) makes the number of first-level oneM2M instances scaled to three. Finally, the low traffic load (10 requests/s—not shown here due to the space limit) only makes the number of first-level oneM2M instances scaled to two. The result shows that our approach is able to deploy the proper numbers of serving instances for different workloads. In the static system, the number of serving instances always remains at three.

To measure power consumption, we estimate each fog node's power consumption using the powerstat command, which is a tool for showing power consumption statistics. In the cases of low traffic load (10 requests/s) and moderate traffic load (25 requests/s), the difference of power consumption between the static and the proposed dynamic system is not significant because there is no overloaded oneM2M middle node instances among the fog cluster. However, as the traffic load reaches 100 requests/s, the power saved by the dynamic approach is much more than that of the static approach. This is because the scaling server in the dynamic system is able to monitor each fog node's workload and scale in/out the MN instances automatically when needed. On the other hand, the static system always runs three MN instances and when the traffic load exceeds its capacity, it becomes overloaded and consumes a large amount of energy.

For average CPU usage, our approach provides a flexible mechanism that ensures each fog node's CPU utilization is below the scale-out threshold of 80%, while the static system will be overloaded once the workload exceeds the capacity of a fixed number of serving instances. Hence, overall, our proposed architecture has better CPU utilization than that of the static system.

For average response time, our proposed approach also has faster response time than the static system. In the case of the high traffic load, the response time of our proposed system is 1.9 times faster than that of the static system. Without any autoscaling mechanism, the static system would be slowed down once the traffic load goes beyond its capacity.

In summary, our proposed approach gives the system good elasticity under different traffic loads and does not create too much overhead with an additional scaling server. The scaling server automates the scale-in and scale-out procedure. Compared to the static approach, our proposed approach allows the system capacity to dynamically grow and shrink as needed.

5.3.3 Internet of Things Scalability by Network Slicing

Network slicing can help us take care of different QoS requirements of heterogeneous M2M traffic. However, there are several technical challenges associated with this idea such as how to identify different traffic patterns, how to enable network slices with distinct QoS, and how to decide which traffic pattern goes to which network slice. In this research, our goal is to design an algorithm for optimal matching between M2M traffic patterns and SDN network slices based on mandatory QoS requirements of traffic patterns. Moreover, we also intend to demonstrate the usability of our algorithm in an SDN-enabled OpenStack testbed.

5.3.3.1 Architecture

In this research, we utilize the ODL controller [14], the Open vSwitch (OVS) [15], and the OpenFlow protocol [16] in coordination via the ODL RESTful API. ODL is a modular open-source SDN controller and platform that allows programmability and customization of SDN-based networks. The core of ODL is the model-driven service abstraction layer (MD-SAL). In ODL, underlying network devices and network applications are all represented as objects or models whose interactions are processed within the SAL.

The SAL is a data exchange and adaptation mechanism between the YANG models representing network devices and applications. The YANG models provide generalized descriptions of a device or application's capabilities without requiring either to know the specific implementation details of the other. Within the SAL, models are simply defined by their respective roles in a given interaction. A "producer" model implements an API and provides the API's data; a "consumer" model uses the API and consumes the API's data. While "northbound" and "southbound" provide a network engineer's view of the SAL, "consumer" and "producer" are more accurate descriptions of interactions within the SAL. For example, a protocol plugin and its associated model can be either a producer of information about the underlying network or a consumer of application instructions it receives via the SAL. ODL includes support for commonly used SDN protocols including OpenFlow, OVSDB, NETCONF, and BGP. ODL uses OSGi and Maven to build project packages based on Karaf (a Java core abstraction of ODL) that constitute ODL components and features.

Open vSwitch is a production quality, multilayer virtual switch licensed under the open-source Apache 2.0 license. It is designed to enable massive network automation through programmatic extension while still supporting standard management interfaces and protocols (e.g., NetFlow, sFlow, IPFIX, RSPAN, CLI, LACP, 802.1ag). In addition, it is designed to support distribution across multiple physical servers similar to VMware's vNetwork distributed vswitch or Cisco's Nexus 1000 V. It also supports forwarding layer abstraction to ease porting to new software and hardware platforms

OpenFlow is considered one of the first SDN standards. It originally defined the communication protocol in SDN environments that enables the SDN controller to directly interact with the forwarding plane of network devices such as switches and routers—both physical and virtual (hypervisor-based) ones—so it can better adapt to changing business requirements. By using the OpenFlow protocol rules, an SDN controller pushes down changes to the switch/router flow-table allowing network administrators to partition traffic, control flows for optimal performance, and start testing new configurations and applications. OpenFlow can be used with network traffic involving a variety of protocols and network services. Note that at the Media Access Control/link layer, only Ethernet is supported. Thus, OpenFlow as currently defined cannot control Layer 2 traffic over wireless networks.

In an SDN architecture, the northbound APIs are usually RESTful APIs used for communications between the SDN controller and the network applications. The northbound APIs can be used to facilitate innovation and enable efficient orchestration and automation of the network to align with the needs of different applications via SDN network programmability. Many different sets of northbound APIs are emerging. Currently, more than 20 different SDN controllers are available, all featuring different northbound APIs. ODL utilizes RESTCONF API, which consists of a series of MD-SAL's RESTful interfaces designed based on the RESTCONF protocol [17].

An SDN is enabled in OpenStack via Neutron; Neutron provides virtual networking services, management of software-defined internal subnets, multi-tenant networking isolation, integration of internal and external physical subnets, networking security services, and so on. The OpenStack Neutron API is the foundation for OpenStack networking services. It allows other physical and virtual networking technologies that support the Neutron API to be easily integrated, thus opening the door for compatible SDN controllers to be part of OpenStack. In OpenStack, the end nodes can be instantiated by Nova (typically via Kernel-Based Virtual Machine [KVM]), while via Neutron these nodes can be linked by OVS [18]. So far OpenStack is compatible with a number of SDN controllers [19] such as ONOS, Ryu, and ODL.

5.3.3.2 Key Enablers

To match an M2M traffic pattern optimally with an SDN network slice based on their key features of reliability, delay, and bandwidth, we need to first define what these features are.

Reliability, $R(t)$, [20] is defined as the probability of a network operating without packet loss until a time threshold, t, and can be denoted by $R(t) = P(\text{TF} \geq t)$ where TF means Time to Failure. Reliability is a top requirement in M2M, particularly in critical applications such as eHealth where patients trust their lives to devices and applications, expecting no failure during the utilization of such services.

Delay [21], D, is defined as the latency for bits of data to be transmitted over the network end-to-end and can be denoted as D = bits/ratio, where delay is calculated by the number of bits transmitted over a transmission ratio (i.e., bits/time). Delay is usually measured in milliseconds. It plays a fundamental role in both critical and massive M2M applications. For example, in real-time tracking systems it is important to reflect the location of a specific target with as little delay as possible.

Bandwidth [22] B is defined as the quantity of bits transmitted per unit of time and can be denoted as B = bits/time, where "bits" and "time" indicate the quantity of bits transmitted and the amount of time taken for transmission, respectively. Bandwidth is usually measured in bits per second. It is another crucial metric for all types of M2M applications, not only due to the increasing number of devices connected to the cloud, but also because of a large variety of applications such as video-on-demand and virtual reality requiring a large bandwidth. In general, reliability and bandwidth are considered the higher the better, while for delay it is the lower the better.

We assume there are N given network slices and N different patterns of M2M traffic. Our challenge is to (1) develop an optimal matching algorithm to map N traffic patterns to N network slices and (2) to discover whether there exists no slice that can match a traffic pattern. We consider a static environment where a set of N M2M traffic patterns $T = \{t_1, t_2, \dots, t_n\}$ and a set of N SDN network slices $S = \{s_1, s_2, \dots, s_n\}$ are given. Our task is to find the optimal matching between the set of traffic patterns and the set of network slices. To tackle such a problem, we first need to be able to formally describe traffic patterns and network slices.

Each M2M traffic pattern can be described in a tuple of tuples to indicate its reliability, delay, and bandwidth requirements as $t_i = <<R_i, \hat{m}R_i>, <D_i, \hat{m}D_i>, <B_i, \hat{m}B_i>>$ where each of reliability (R), delay (D), and bandwidth (B) requirements is specified in a 2-tuple in sequence: the first element of the tuple indicates the required minimum reliability, maximum delay, and minimum bandwidth; the second element of the tuple (denoted by $\hat{m}$) indicates whether this is a mandatory (if $\hat{m} = 1$) or optional requirement (if $\hat{m} = 0$) for the M2M traffic.

An M2M network slice can also be described in a tuple as $s_i = <R_i, D_i, B_i>$ where the guaranteed levels of reliability, delay and bandwidth that can be delivered by the network slice are specified. For network slices there is no need to specify whether a QoS requirement is mandatory.

We modeled this matching problem as an assignment problem [23], where a set of M2M traffic patterns with some mandatory QoS requirements has to be matched with a set of SDN networks slices that offer particular QoS capabilities, with the condition that one traffic pattern can be assigned to only one network slice. We define G = (T, S, E), where G is a bipartite weighted graph composed of two disjointed set of vertices, T and S, representing traffic patterns and network slices, respectively, and a set of edges, E, representing the cost of connecting the vertices.

In Figure 5.16, we present an example of weighted graph G, where the nodes on the left side represent M2M traffic patterns, the nodes on the right side

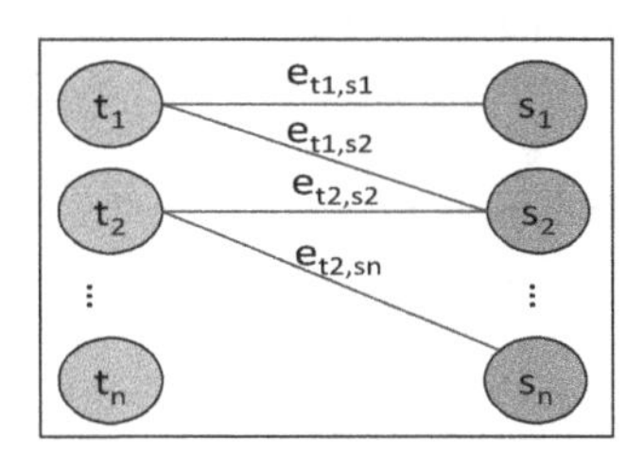

Figure 5.16 Weighted graph.

represent SDN network slices, and the edges represent the cost of connecting these two types of nodes. Notice that a traffic pattern could be connected to more than one network slice (e.g., t_1, t_2), and in some cases a single network slice is connected to multiple traffic patterns (e.g., s_2). Also notice that the last traffic pattern, (t_n), is left unconnected because it finds no match with any network slice (i.e., there exists one of its mandatory QoS requirements that cannot be fulfilled by any network slice).

A network slice is considered feasible for a traffic pattern if it can meet the minimum QoS requirements mandated by the traffic pattern. When we construct the bipartite weighted graph G, the edges will only connect traffic patterns with feasible network slices. In the bipartite weighted graph G, the cost of each edge is defined by a Euclidean distance, E [24] to indicate how close a network slice meets the QoS requirements of a traffic pattern. Intuitively, the closer the distance, the better is the match between the traffic and the slice. Our objective is to find the best match, thus the optimal assignment between traffic patterns and network slices.

We use the Hungarian algorithm [25] to achieve this objective. The Hungarian algorithm utilizes a cost matrix C, which can be derived from the bipartite weighted graph G, where each element in the matrix represents the cost, E, of assigning the ith traffic pattern to the jth network slice. With the Hungarian algorithm, an optimal solution can be found by calculating the minimal cost of matching N traffic patterns with N network slices.

5.3.3.2.1 Traffic-Slice Mapping Algorithm

The following are the main steps of our proposed algorithm:

1. Calculate cost matrix.
2. Calculate optimal matching pairs.
3. Generate the set of matching tuples.
4. Find the set of unmatched network slices.
5. Enforce the matching tuples via the SDN controller.

Table 5.4 presents the first step of our algorithm, calculating the cost matrix $C: T \times S \rightarrow E$, with the given sets of N M2M traffic patterns, T, and N SDN network slices, S, as input. The entries in C represent the cost (Euclidean distance, E) of using each network slice to serve each traffic pattern based on their QoS capabilities and requirements. For the infeasible network slice (i.e., a slice that does not fulfill the mandatory QoS for a traffic pattern) we assign a very high value (e.g., 999999) in the cost matrix.

In the second step, we use the Hungarian algorithm with C as input to find a set of network slices that can serve the QoS needs of the traffic patterns with the minimal cost (i.e., optimal matching). The Hungarian algorithm outputs a set

Table 5.4 Calculate Cost Matrix

```
1   Foreach ti in T
2      Foreach sj in S
3         If isFeasibleSlice(ti,sj) then
4            C[i,j]=EuclideanDistance(ti,sj)
5         Else
6            C[i,j]=999999
7   Return C
8   Function isFeasibleSlice(t,s)
9      If (t.MR==1 and s.R<t.R) or
10        (t.MD==1 and s.D>t.D) or
11        (t.MB==1 and s.B<t.B) Then
12        Return True
13     Else
14        Return False
15  Function EuclideanDistance(t,s)
16     Return √((s.R − t.R)² + (s.D − t.D)² + (s.B − t.B)²)
```

of optimal matching pairs, H, of [row, column] indexes that locate the optimal matching items in the original cost matrix C. In our particular case, a row index represents a traffic pattern and a column index indicates a network slice in T and S, respectively.

The Hungarian algorithm consists of four steps: Steps 1 and 2 are executed once, while Steps 3 and 4 are repeated until an optimal matching is found.

Step 1: For each row, find the lowest value and subtract it from each element in that row.

Step 2: For each column, find the lowest value and subtract it from each element in that column.

Step 3: Cover all zeros in the resulting matrix using a minimum number of horizontal and vertical lines. If n lines are used, then an optimal matching exists among the zeros. The algorithm stops. If less than n lines are used, continue with Step 4.

Step 4: Find the smallest value that is not covered by a line in Step 3. Subtract that value from all uncovered elements, and add that value to all elements that are covered twice.

The third step consists of establishing the actual matching by inspecting H, C, T, and S. In this step we generate the set of matching tuples defined as $M = \{m_1, m_2, \ldots, m_n\}$. Whenever a value 999999 is found in C (i.e., an infeasible network slice), as indicated by any of the [row, column] index pairs in H, we create a new entry in M as follows $m_i = <T_{row}, -1>$, otherwise the entry is created as $m_i = <T_{row}, S_{column}>$.

The fourth step also consists of finding the set of unmatched network slices defined as $\hat{S} = \{\hat{s}_1, \hat{s}_1, \ldots, \hat{s}_k\} \mid k \leq n$, where an unmatched network slice is defined as $\hat{s}_i = <R_i, D_i, B_i>$. This can be done by simple inspection of M and S. Finally, the set of matching tuples M is passed to the SDN controller whose task is to enforce the appropriate OpenFlow rules in the corresponding OVS switches. The complexity of our matching algorithm is O (N^3).

Next, we illustrate the operation of our algorithm by showing an example.

Consider the set of traffic patterns $T = \{t_1, t_2, t_3, t_4\}$ in Table 5.5 and the set of network slices $S = \{s_1, s_2, s_3, s_4\}$ in Table 5.6 as the input for running our matching algorithm.

In Table 5.7 we can see the cost matrix C obtained after calculating the Euclidean distance, E, between each traffic and each slice based on their mandatory metrics

Table 5.5 Set of M2M Traffic Patterns (A)

Id	*Traffic*	*R (%)*	$\hat{m}_R$	*D (ms)*	$\hat{m}_D$	*B (Mbps)*	$\hat{m}_B$
t_1	Smart meter	90	0	100	1	50	0
t_2	BLE tags	90	0	100	1	70	0
t_3	eHealth	100	1	20	0	60	0
t_4	Video	90	0	50	0	80	1

Table 5.6 Set of SDN Network Slices (A)

Network Slice	*Reliability (%)*	*Delay (ms)*	*Bandwidth (Mbps)*
s_1	100	10	80
s_2	90	20	70
s_3	80	30	80
s_4	70	40	60

Table 5.7 Cost Matrix (A)

	Column Index	*1*	*2*	*3*	*4*
Row Index	*Id*	s_1	s_2	s_3	s_4
1	t_1	90	80	70	60
2	t_2	90	80	70	60
3	t_3	0	999999	999999	999999
4	t_4	0	999999	0	999999

and feasibility. For example, for $[t_1, s_1] = \sqrt{(100-10)^2} = 90$ is feasible since t_1 asks for a mandatory delay equals or lower than 100 ms, while s_1 offers 10 ms.

Notice how $[t_3,s_2]$, $[t_3,s_3]$, $[t_3,s_4]$, $[t_4,s_2]$, and $[t_4,s_4]$ in Table 5.7 have a value of 999999, which is the representation of an infeasible network slice. This happens because t_3 has a mandatory requirement for reliability of 100%, but the capabilities of $s_2 = 90$, $s_3 = 80$, and $s_4 = 70$ are not sufficient; whereas t_4 has a mandatory request for bandwidth of at least 80 Mbps, but $s_2 = 70$ and $s_4 = 60$ are not enough. Also, notice the row and column indexes and their equivalences with the corresponding traffic patterns and network slices. For example $[t_1,s_1]$ can be accessed using [row index = 1, column index = 1].

In Table 5.8, we show the set of optimal matching pairs H that is the result of executing the Hungarian algorithm with C as input. All pairs in H in conjunction represent the optimal solution with the minimum cost found by the Hungarian algorithm. Each pair in H has a row and a column index in the cost matrix C.

In Table 5.9, we show M, which contains the matching tuples of traffic patterns and network slices. Notice how all the traffic patterns have found a feasible network slice in this example. Because there is no unmatched traffic pattern, $\hat{S}$ is thus empty and not shown.

Finally, the SDN controller will issue the corresponding OpenFlow rules to realize the optimal matching tuples in M.

Table 5.8 Rows and Column Indexes and Their Cost (A)

Row Index	*Column Index*	*Cost in C*
1	4	60
2	2	80
3	1	0
4	3	0

Table 5.9 Matching Tuples (A)

IoT Application	*Traffic*	*Slice*	*Status*
Smart Meter	t_1	s_4	Matched
BLE Tags	t_2	s_2	Matched
eHealth	t_3	s_1	Matched
Video	t_4	s_3	Matched

5.3.3.3 Evaluation

In this section, we introduce our testing environment based on OpenStack and OpenDaylight. In order to verify that the M2M system scalability by network slicing is feasible, we implemented the matching algorithm defined in Section 5.3.3.2.1 our testbed. Furthermore, we evaluate the performance of the system using the test data of the complete matching example in Section 5.3.3.2.1. We measure our system performance in terms of its average response time, energy consumption, and memory and CPU cost, then compare it with the one without network slicing under the same four M2M traffic patterns discussed in Section 5.3.3.2.1.

5.3.3.3.1 Setting the Testing Environment

As depicted in Figure 5.17, we installed OpenStack Kilo on four physical nodes with Ubuntu 14.04 as the host operating system. Each physical node has a 3.2 GHz

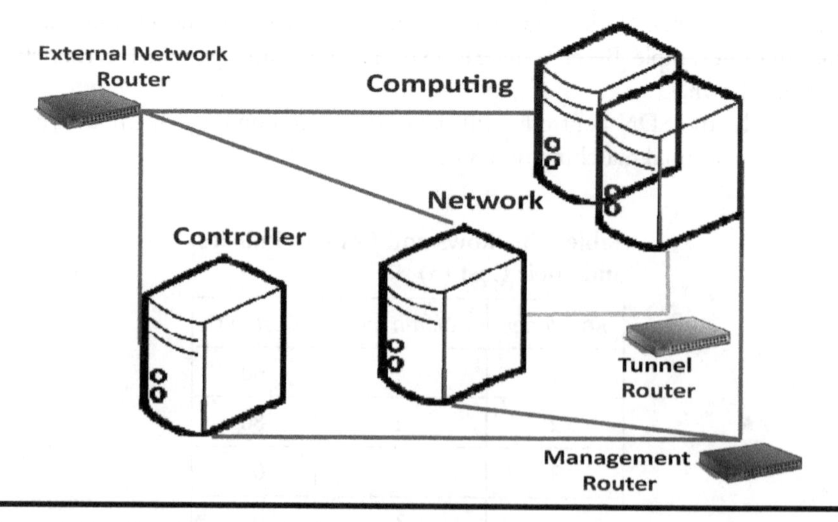

Figure 5.17 Our OpenStack physical setup.

CPU (four cores), 16 GB of RAM and 1 TB of hard disk. We used one physical node for controller, one for network, and two for compute. Moreover, we used three 100 Mbps 8-port switches for management, tunnel, and external networks, respectively. On the controller node, we installed ODL Beryllium SR2 and configured OpenStack Neutron, along with OVS required for the network and compute nodes to work with ODL.

Figure 5.18 shows the logical topology of our testbed. By default our OpenStack testbed had three OVS (i.e., normally OpenStack Neutron creates one OVS per each physical node, except controller), but it was not enough for creating the four network slices needed in our experiment. For this reason, we manually configured an extra OVS in the network node, allowing us to create at least four network slices. We connected the additional OVS with the OVS in the two compute nodes via Virtual Extensible Local Area Network (VXLAN) links, and an OVS patch link for the two OVS in the same network node. In addition, we configured the extra OVS with the appropriate pipeline table of OpenFlow rules to set up correct communications among all virtual switches.

As illustrated in Figure 5.19, each network slice consists of a client running an instance of our traffic generator and a server running an instance of oneM2M Infrastructure Node-Common Service Entity (IN-CSE), connected by a feasible SDN network slice. The clients are virtual machines instantiated via OpenStack Nova in the Compute 1 node, with 2 GB of RAM, 1 CPU, and 20 GB of storage, and while the servers have similar characteristics, they are all instantiated in the Compute 2 node.

The expected reliability, delay, and bandwidth requirements of a network slice is realized in our testing environment using tc qdisc6. For example, if we want to realize the network slice s_1 = <100,10,80> of Table 5.6 in the network interface called

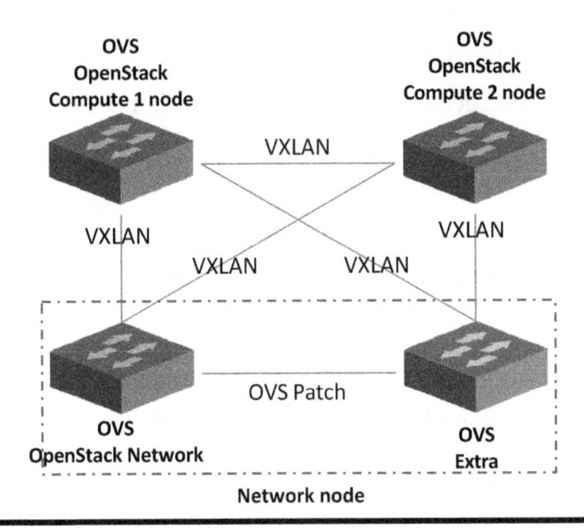

Figure 5.18 Network logical topology.

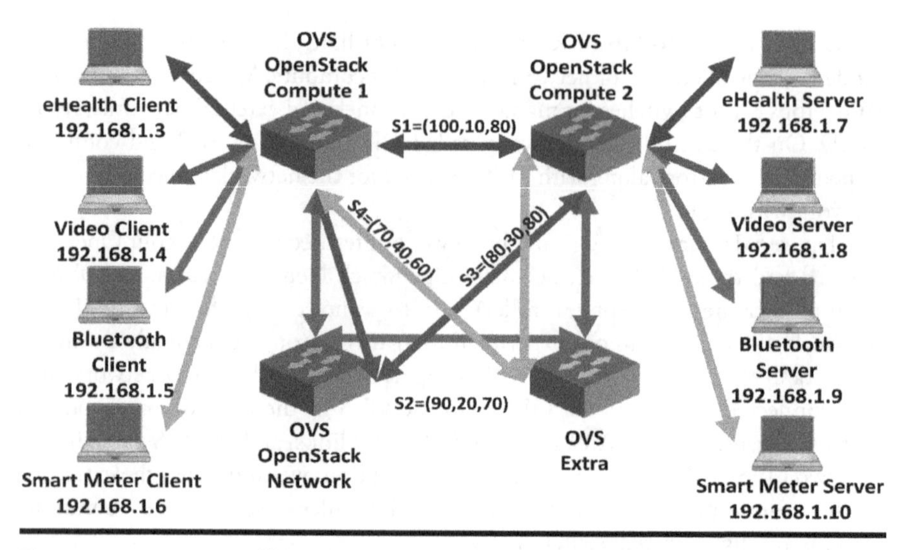

Figure 5.19 M2M traffic patterns and SDN network slices logical view.

port_C1_C2 that connects OVS OpenStack Compute 1 with OVS OpenStack Compute 2 in the Compute 1 node (Figure 5.19), the following are the tc qdisc commands that will enable such a configuration:

tc qdisc add dev port_C1_C2 root handle 1:0 tbf rate 80mbit buffer 1600 limit 3000
tc qdisc add dev port_C1_C2 parent 1:0 handle 10: netem delay 10ms loss 0%

5.3.3.3.2 Traffic Generator

We developed a traffic generator in Python to simulate four M2M traffic patterns: smart meter, Bluetooth tags, eHealth, and video. Our generator allows us to configure payload size and frequency of requests for each traffic pattern. As shown in Table 5.5,

1. Smart meter requires low bandwidth, can tolerate long delay, but needs high reliability.
2. Bluetooth tag simulates a Bluetooth low-energy (BLE) tag that can be attached to objects for tracking purposes. Such an application requires low bandwidth, high delay tolerance and high reliability.
3. eHealth simulates blood glucose measurements. Its bandwidth is low, but it needs the lowest delay and highest reliability.
4. For video traffic, we assume the server will store only video metadata. It requires high bandwidth, long delay tolerance but allows low reliability.

Using the complete matching example of Section 5.3.1 as an example, the controller will issue the following OpenFlow 1.3 commands in both Compute 1 and Compute 2 nodes to configure the network slice s_1 in Table 5.6:

[Compute 1] ovs-ofctl -O OpenFlow13 add-flow br-int "priority=60005, tun_id=0 × 10011, table=110, ip, nw_src=192.168.1.3,nw_dst=192.168.1.7, actions=output:3"

[Compute 2] ovs-ofctl -O OpenFlow13 add-flow br-int "priority=60005, tun_id=0 × 10011, table=110, ip, nw_src=192.168.1.7,nw_dst=192.168.1.3, actions=output:2"

In this example we are assuming that the OVS in the Compute 1 and the Compute 2 nodes are called br-int. The ports that interconnect br-int in Compute 1 with br-int in Compute 2 are Port 2 and Port 3, respectively. The VXLAN ID for this particular network is 0 × 10011, and the OpenFlow table ID in which these rules are stored is 110. Openflow Table 110 contains OpenFlow matching and forwarding rules for Layer 3, so that IP addresses are used as matching criteria. Notice the parameters "nw_src" and "nw_dst" are used to specify IP source and IP destination addresses, respectively, on each of the OpenFlow rules. Also notice how the OpenFlow rules have to be created in both directions, that is, from client to server and vice versa. A similar approach is utilized for the remaining network slices s_2, s_3, and s_4.

5.3.3.3.3 Results of Performance Evaluation

We evaluate and compare two systems: one with network slicing support, and one without. For the latter, it is equivalent to supporting only slice s_1 because it is the default slice when no SDN controller is installed in OpenStack and it connects Compute 1 with Compute 2 directly. For the former, the ODL controller is utilized to set the appropriate OpenFlow rules in all corresponding OVSs, based on the traffic-slice matching pairs.

The tests were conducted by simultaneously running a traffic generator on each client to generate four patterns of IoT traffic. Each experimental test lasted for 1 min and this was repeated for a sufficient number of times in order to calculate the average of response time, power utilization and CPU/memory cost.

5.3.3.3.4 Response Time Results

The results of average response time is shown in Figure 5.20. Across all traffic patterns, there is a big improvement in response time for the system with network slicing support. The improvement is 52.7%, 56.5%, 52.2% and 55.4% for smart meter, eHealth, Bluetooth tags and video, respectively.

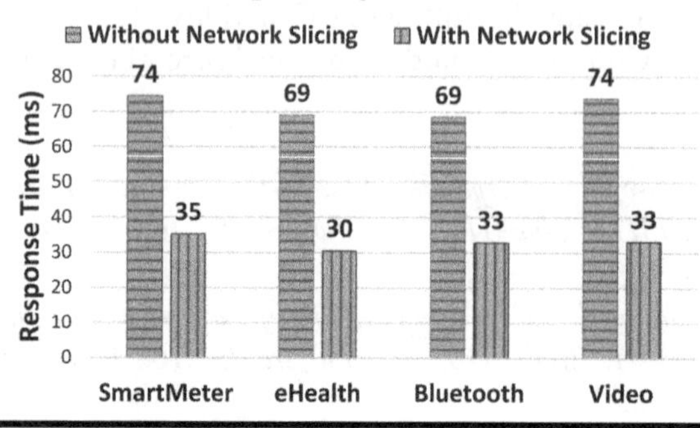

Figure 5.20 Average response time.

5.3.3.3.5 Power Utilization Results

Figure 5.21 shows the results of average power utilization per request. We calculate the power consumption and derive the amount of power required for each individual request by dividing the total power used during the whole period of time by the total number of requests. We have calculated the power consumption using the following formula (1):

$$\textit{Power Consumption} = TDP*CPU\% + K*Memory\%$$

where TDP is the microprocessor's Thermal Design Power, a reference measurement of CPU running in normal conditions and given by the manufacturer. For our particular testbed, TDP is 80 W. In addition, *K* is the common power consumption of memory modules, typically 3 W for DDR3 memory cards. CPU% and Memory% are the average CPU and memory measurements, respectively, on each host system. Figure 5.21 shows that there is 16.5%, 9%, 12.5% and 12.7% less power consumption for smart meter, eHealth, Bluetooth tags and video, respectively. This is due

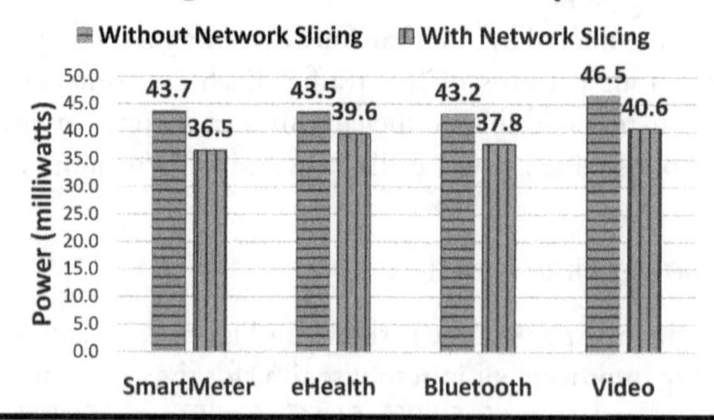

Figure 5.21 Average power consumption.

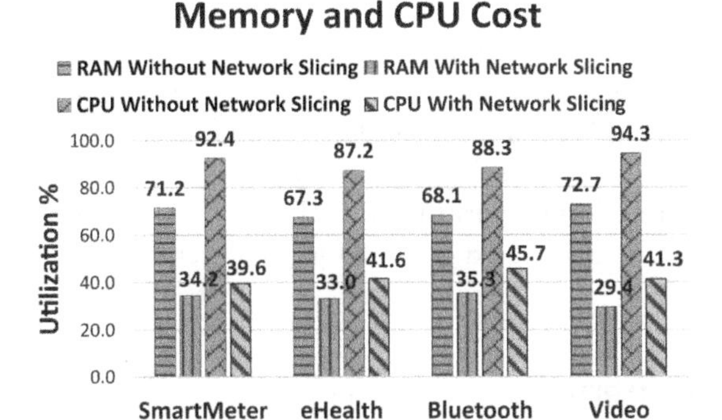

Figure 5.22 Memory and CPU cost.

to the fact that the system is able to handle a larger number of requests during the same period of time with network slicing enabled, thus a lower power consumption per request can be achieved.

5.3.3.3.6 Computational Cost Results

Figure 5.22 shows the results of average memory and CPU utilization. These results measured the average amount of computational resources that the M2M platforms utilized to fulfill a number of requests during a period of time. The system with network slicing again outperforms the one without slicing support.

The above improvement of average response time, energy consumption per request and memory/CPU cost indicates that via network slicing we can achieve a more scalable M2M system.

5.4 Conclusion and Future Work

In this chapter, we discuss the utilization of several 5G-enabled tools to tackle the IoT scalability problems. These scalability enablers in 5G include cloud computing, fog computing and network slicing technologies. We show how each of them can enable the scalability of the IoT systems.

To enable IoT scalability by the cloud, we propose to utilize the virtualized cloud environment enabled by OpenStack. We developed a scaling up algorithm that constantly compares the forecasted CPU status of active platform nodes, versus two thresholds: overloaded and scale up. Our algorithm ensures that no new platform nodes are booted if the load can be handled by current active platform nodes. It also ensures only booting one platform node at a time.

To enable IoT scalability by fog, we propose to manage the IoT traffic in fog before it reaches the cloud. This means that the fog network, as a continuum between the cloud and the things, must provide the scalable capacity, that is, it must be able to flexibly and dynamically scale in/out the serving instances across fog nodes to match with the incoming traffic.

To enable IoT scalability by network slicing, we propose to utilize the ODL controller, OVS, and OpenFlow protocol in coordination via ODL RESTful API to create network slices of specific QoS for different IoT traffic flows. We also propose an optimal matching algorithm (the Hungarian algorithm) for IoT traffic patterns and network slices based on their QoS requirements.

Acknowledgment

The research reported in this chapter is sponsored by Ministry of Science and Technology (MOST) of Taiwan Government under Project Number MOST 106-2221-E-009 -055 -MY3.

References

1. oneM2M, Functional Architecture, oneM2M Technical Specification TS-0001-V.2.10.0, August 30, 2016.
2. OpenFog Consortium, OpenFog Reference Architecture for Fog Computing, 2017.
3. J. Lin, L. Lin and S. Huang, Migrating web applications to clouds with microservice architectures, in *International Conference on Applied System Innovation (ICASI)*, Okinawa, Japan, May, 2016.
4. A. Jain and R. Kumar, A multi stage load balancing technique for cloud environment, in *International Conference on Information Communication and Embedded Systems (ICICES)*, Chennai, India, 2016.
5. W. Sze and A. Srivastava, Hardening OpenStack Cloud Platforms against Compute Node Compromises, in *11th ACM on Asia Conference on Computer and Communications Security*, ACM, New York,, 2016.
6. M. Sharma, V. Iyer, S. Subramanian and A. Shetty, A comparative study on load testing tools, *International Journal of Innovative Research in Computer and Communication Engineering*, 4(2): 1906–1912, 2016.
7. E. Cerritos, F. J. Lin and D. De la Bastida, High scalability for cloud-based IoT/M2M systems, in *ICC'16*, Kuala Lumpur, Malaysia, May, 2016.
8. D. De la Bastida and F. J. Lin, OpenStack-based Highly Scalable IoT/M2M Platforms, in *IEEE International Conference on Internet of Things (iThings)*, Exeter, UK, June, 2017.
9. M. S. D. Brito, S. Hoque, R. Steinke and A. Willner, Towards programmable fog nodes in smart factories, *IEEE 1st International Workshops on Foundations and Applications of Self* Systems*, Augsburg, Germany, 2016, pp. 236–241.
10. D. Merkel, Docker: Lightweight Linux containers for consistent development and deployment, *Linux Journal*, 2014(239): 2, 2014.
11. OpenMTC, OpenMTC 2016, Document, http://www.openmtc.org. Accessed on April 2018.
12. Docker, Docker, https://www.docker.com. Accessed on April 2018.

13. HAProxy, The reliable, high performance TCP/HTTP load balancer, http://www.haproxy.org. Accessed on April 2018.
14. OpenDaylight, OpenDaylight, https://www.opendaylight.org. Accessed on April 2018.
15. Open vSwitch, Production quality, multilayer open virtual switch https://www.openvswitch.org. Accessed on April 2018.
16. OpenFlow, OpenFlow, https://www.opennetworking.org/sdn-resources/openflow. Accessed on April 2018.
17. RESTCONF, RESTCONF protocol, https://datatracker.ietf.org/doc/draft-bierman-netconf-restconf. Accessed on April 2018.
18. S. R. Sivaramakrishnan, J. Mikovic, P. G. Kannan, C. M. Choon and K. Sklower, Enabling SDN experimentation in network testbeds, in SDN-NFVSec '17 *Proceedings of the ACM International Workshop on Security in Software Defined Networks & Network Function Virtualization*, 2017.
19. T. Bakhshi, State of the art and recent research advances in software defined networking, *Wireless Communications and Mobile Computing*, 2017, 2017.
20. N. Maalel, E. Natalizio, A. Bouabdallah, P. Roux and M. Kellil, Reliability for emergency applications in Internet of Things, in *Distributed Computing in Sensor Systems IEEE International Conference on*, May 20–23, 2013.
21. Y.-Y. Shih, W.-H. Chung, A.-C. Pang, T.-C. Chiu and H.-Y. Wei, Enabling low-latency applications in fog-radio access networks, *IEEE Network*, 31(1): 52–58, 2017.
22. Z. Ma, Q. Zhao and J. Huang, Optimizing bandwidth allocation for heterogeneous traffic in IoT, *Peer-to-Peer Networking and Applications*, 10(3): 610–621, 2017.
23. R. Burkard, M. Dell'Amico and S. Martello, *Assignment Problems*, Society for Industrial and Applied Mathematics, 2012. doi:10.1137/1.9781611972238.
24. P. Hu, H. Ning, T. Qiu, Y. Xu, VIII. Luo and A. K. Sangaiah, A unified face identification and resolution scheme using cloud computing in Internet of Things, *Future Generation Computer Systems*, 2017.
25. J. Munkres, Algorithms for assignment and transportation problems, *Journal of the Society for Industrial and Applied Mathematics*, 5(1): 32–38, 1957.

5G ACCESS NETWORK FOR THE INTERNET OF THINGS

Chapter 6

5G Small Cells: The Harbinger of the Internet of Things and Connected Living

Abhishek Roy, Navrati Saxena, and Sukhdeep Singh

Contents

6.1 Introduction

The seminal work of British entrepreneur Kevin Ashton unveiled the new concept of the Internet of Things (IoT) [1]. The IoT refers to physical objects, or things, equipped with sensors, software and network connectivity for automatic information exchange. Interestingly, in the domain of the IoT, "things" refers to a wide variety of devices including monitoring cameras, health-care devices, biochips, and smart cars. The rapid increase in the usage of smart phones and the ratification of new communication standards are gradually taking us toward the vision of the IoT. We expect that this momentum of the IoT will continue to increase its impact on industry and household applications, thereby offering a better quality of life and new industrial opportunities in the near future. Such a massive connectivity, involving a myriad of devices, mandates the use of some sophisticated gateway for efficient Internet connectivity. Naturally, the network designer can opt for different candidate access technologies for this gateway design. Although wireless fidelity (Wi-Fi) could be an obvious choice for many applications, it suffers from increasing packet collisions with an increasing number of access uplinks. Moreover, traditional Wi-Fi is not suitable for quality of service (QoS) support and consumes a fair amount of power. Thus, some or the other form of flexible wireless access technology seems to be necessary for a massive roll out of IoT. Note that a set of emerging applications like smart cities, traffic monitoring and smart grids (SGs) also offer new markets to the legacy wireless operators for enhancing their revenues. As a result, de facto wide, the Third Generation Partnership Project (3GPP) has included narrowband IoT (NB-IoT) in its standards. However, the inherent human-centric fourth-generation (4G) long-term evolution (LTE)/LTE Advanced (LTE-Adv.) cellular networks are not efficient enough to support this massive IoT connectivity.

First, the IoT is expected to include a large number of devices with automated connectivity. Unfortunately, current commercial 4G systems mostly support only around 1,000 connected users or devices per cell [2]. With increasing numbers of users, the LTE Radio Resource Control (RRC) block triggers session admission control to reject the new connections [2]. Moreover, the communications from a myriad of IoT devices might also sporadically impose huge traffic bursts on wireless cellular networks. With an exponential increase in wireless data usage, it is quite unlikely that 4G wireless networks will be able to sustain the traffic demand. On top of all this, there is the "curse of heterogeneity," which involves a variety of devices with different applications, data size and QoS requirements. These challenges are continuously fueling new research investigations in different areas of wireless communications for IoT. 5G wireless networks, under the aegis of 3GPP New Radio (NR), are gradually exploiting NR technologies for providing better coverage, higher data rates, dense connectivity and superior IoT experience at a reasonable cost. However, ultra-dense 5G wireless networks also suffer from high energy consumption. The emerging concepts of cloud radio access networks (C-RAN) and virtualized radio access networks (V-RAN) are offering some solutions for energy efficiency, environmental sustainability and reduction in a network's OPerating EXpenditures (OPEX). We believe that in future, V-RAN will evolve toward supporting massive IoT-enabled devices and applications. This motivates us to discuss some of our own 5G-enabled solutions for providing efficient connectivity to IoT devices.

We begin with a brief discussion on the major related works in IoT and also highlight the generic and some country-specific IoT developments. Subsequently, we explain our own design and prototyping of 5G-enabled IoT gateways. The gateways explore efficient uplink IoT traffic classification and optimal uplink data (traffic) compression strategies to reduce the uplink traffic burden and increase the utilization of scarce orthogonal frequency division multiple access (OFDMA) resources. Next, we point out how optimal 5G small cell planning and emerging features like evolved Multimedia Broadcast and Multicast Services (eMBMS) could be used to improve demand response (DR) programs of SGs. Our solutions are verified by prototype development and laboratory experiments as well as simulation results on the actual wireless data.

6.2 Related Work

The IoT is a result of convergence of three visions and three technologies, which are "Internet oriented," "things oriented," and "semantic oriented" and wireless sensor networks (WSN), smart objects, and radio frequency identification (RFID), respectively [1,3]. Complex knowledge, intelligence and a wide variety of wired and wireless communication technologies are three crucial requirements of IoT design. Table 6.1 depicts the five important layers present in the generic architecture of the IoT:

1. Perception layer or the device layer
2. Network layer
3. Middleware layer
4. Application layer
5. Business layer

Research has proved that next generation 5G mobile wireless will play an important role in deployment of IoT, constituting billions of interconnected smart sensors and objects that will represent the real world. It is also expected to support mission-critical use cases, which in turn requires automation of dynamic processes and real-time

Table 6.1 Layers in the Generic Architecture of IoT

Layer	*Description*
Perception Layer	Physical objects (sensors, RFIDs)
Network Layer	3G, Wi-Fi, ZigBee
Middleware Layer	Information processing database
Application Layer	Smart applications and management
Business Layer	Graphs, models

messages across a wide variety of operational fields such as vehicle-to-vehicle, vehicle-to-infrastructure, control systems, and high-speed use cases [4]. 5G wireless, with its aim to connect massive smart devices using the benefits of mobile cellular networks is expected to cater to growing demands of machine type communications. For the same, enhancements are being made in 3GPP releases 14 and 15 for cellular IoT (first phase of 5G standards) in NB-IoT and machine-to-machine (M2M) systems [5]. 3GPP's initiative to standardize and enhance the existing 4G networks from the Key Performance Indicator (KPI) perspective to ease the design and deployment of 5G networks from scratch has the potential to accommodate IoT use cases and minimize CAPital EXpenditures (CAPEX), and OPEX of new networks. Upcoming release of 3GPP and existing releases 14 and 15 are expected to lay down key performance enhancements and features for NB-IoT and M2M systems for a wide range of critical and massive IoT applications. Some of the important ones are as follows [4]:

- General Enhancements to MTC
- Enhancements of NB-IoT
- NB-IoT RF requirement for coexistence with Code Division Multiple Access (CDMA) networks
- Release-14 extensions for the Cellular IoT (CIoT)
- New band support for Release-14NB-IoT
- New services and markets technology enablers

Various countries have identified the major capabilities of the IoT. Environment monitoring, remote monitoring, location sensing, and ad hoc networking are a few of the important capabilities. In order to develop their economies, many countries are adapting 5G and innovating it together with the IoT. Korea is one such country and it is ranked among the top three countries for "suitability of preparedness for IoT opportunities" [6]. At present, from the IoT perspective, the Korean government envisions converging ICT across various industries like smart farming, SGs, smart logistics, smart health care, smart city, smart homes, and smart vehicles. Figure 6.1 presents the functional elements and layered model of IoT architecture from Korea's perspective [2]. The aforementioned model is open, layered and service oriented. Some of the core features of the architecture include

- Underlying connectivity protocols
- Connectivity and network management
- Resource and service management
- Semantics and knowledge management
- Security and privacy
- Application and services

The Ministry of Science Information and Communications Technology (ICT) and Future Planning (MSIP) has prepared the plan for building the IoT testbed and

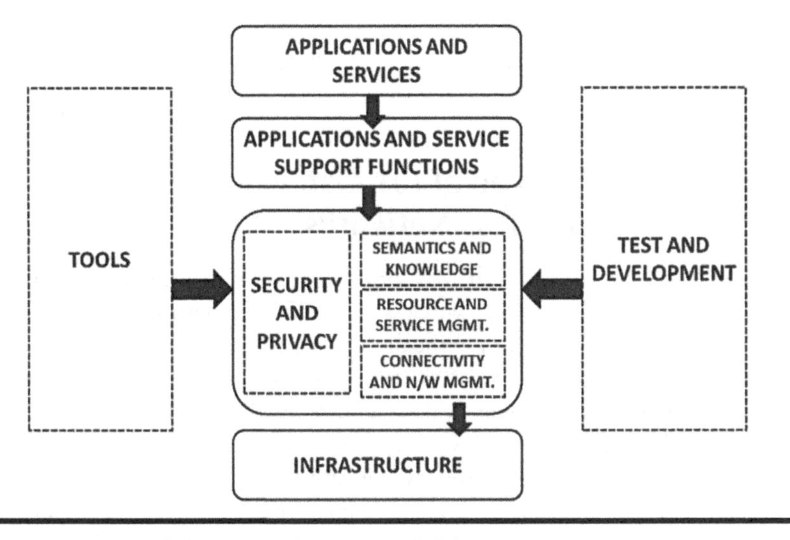

Figure 6.1 IoT architecture reference model for Korea.

has laid down the IoT Promotion Act. The development of the testbed has already started and is expected to be complete by 2020. Following are four important strategies for building an IoT-specific testbed:

1. Develop various technologies and systems that are operational on pilot basis
2. Develop an open service platform
3. Build an IoT, cloud, Big Data, mobile (ICBM) platform
4. Establish the IoT Promotional Act covering applications such as health care and SGs

The United States, one of the giants with respect to the global connectivity index, has done much research and development in the era of the Industrial IoT (IIoT) [7]. Their aim has been to gather data insights and digitize their assets with the help of industries such as aviation, health care, and so on. This in turn is predicted to improve their longevity, productivity and save lives. The next leader in IoT is Sweden. with more than 80% of cashless transactions [7]. They are adopting IoT-based wearables and contactless cards for shopping and trading in order to create a safe society, especially for small business and startups. Japan, being one of the leading global contenders in the case of Internet usage, is working closely with mobile network providers to provide hyper-connected travel solutions for its citizens. Sixty Bluetooth devices have been installed in the Tokyo city station to determine the exact location of commuters with the help of an app in user's smartphones while making use of augmented reality (AR) [7]. IoT is playing a crucial role in Japan for city management.

In the next section, we discuss our research effort in the era of 5G IoT wherein we designed the 5G wireless network to enhance the functionality of IoT networks.

6.3 A New Prototype of IoT Gateway Over 5G Wireless Networks

With its huge opportunities, the IoT promises to improve the quality of life. It raises many challenges in today's wireless networks, which are resource constrained given that today's network is specifically deployed for human-to-human interaction. It fails to satisfy diverse traffic requirements and enormous connectivity. On the other hand, emerging 5G networks tend to exploit features like massive multiple input–multiple output (MIMO), C-RAN, and millimeter Wave (mmWave). These features have the potential to overcome the current challenges of IoT and thus ease the rollout of IoT. Exploring the previously mentioned features, we developed [8] a model for 5G C-RAN to communicate with remote radio heads (RRHs) with the help of 5G-enabled IoT gateways. We present the model by exploiting the major features of 5G communications that are (1) Massive MIMO, (2) mmWave, and (3) C-RAN. Our 5G-enabled IoT gateways are equipped with efficient coherent compression schemes to facilitate efficient resource utilization for uplink transmissions. 5G-enabled gateways dispense exhaustive information visualization, data analysis, and data aggregation to connect day-to-day wireless devices with the 5G wireless Internet. We propose novel uplink data classification, buffering, and encoding techniques at proposed IoT gateways. The uplink data is first classified as delay-tolerant or delay-sensitive data. Thereafter, delay-sensitive data is passed on to the next destination over 5G radio interface, whereas delay-tolerant data is buffered in the IoT gateway's uplink Media Access Control (MAC) buffers until (1) the arrival of more delay-sensitive data, (2) the expiration of the periodic uplink timer threshold, or (3) the expiration of the maximum data size threshold. After buffering, the data is encoded using the Lempel-Ziv-Storer-Szymanski (LZSS) scheme and then sent to 5G C-RAN's RRH. A corresponding decoding process is carried over a C-RAN that needs (1) the reading of dictionary offset and (2) the copying a specified number of characters.

6.3.1 Simulation Setup and Results Discussion

To validate the proposed C-RAN prototype, we set up a real-time testbed consisting of two virtual base station clusters (VBSCs) that are executed over two IBM X350 servers (connected with each other using gigabit backhaul interfaces) containing 40 VBSs (20 for each). Each VBS instance was further connected to Ethernet Remote Radio Head (eRRH) with the help of 10 GBps optical front haul cables. Furthermore, we collected 1 month's actual IoT traffic data from thousands of IoT devices (http://fiesta-iot.eu/fiesta-experiments/ [October 22, 2016]). With the help of the IXIA traffic generator, real-time IoT traffic was generated, emulated and fed into proposed IoT gateways. We had the help of dense urban 5G channel model [9] and RF parameters [10] (31 dBm RRH transmission power, 5 dBm idle power, 20 dB penetration loss, 0.8 shadowing deviation, 10 dBm IoT gateway maximum power in 27.925 GHz frequency band, and 200 MHz bandwidth). With the help of the testbed, we [8] obtained some promising results. Following are the key gains achieved with the help of the proposed setup:

1. Due to efficient data classification and encoding techniques, there was 80% reduction in overall uplink data access (from 120 to 25).
2. Results, in Figure 6.2, point out that 5G-enabled gateways can accommodate up to 110 IoT gateways without proposed data classification and encoding techniques, that is 9 times higher than with the legacy LTE. On the other hand, proposed gateways empowered with LZSS encoding techniques can support more than 210 IoT gateways.
3. The round-trip latency of 5G MAC is reduced to 5 and 10 ms from 50 ms for without and with data compression, respectively.
4. Figure 6.3 shows that the proposed 5G C-RAN module curtailed the power consumption by up to 60%, as compared with legacy 5G wireless.
5. Compared with existing 5G wireless, the proposed prototype provides 50% improvement in the energy efficiency, as shown in Figure 6.4.

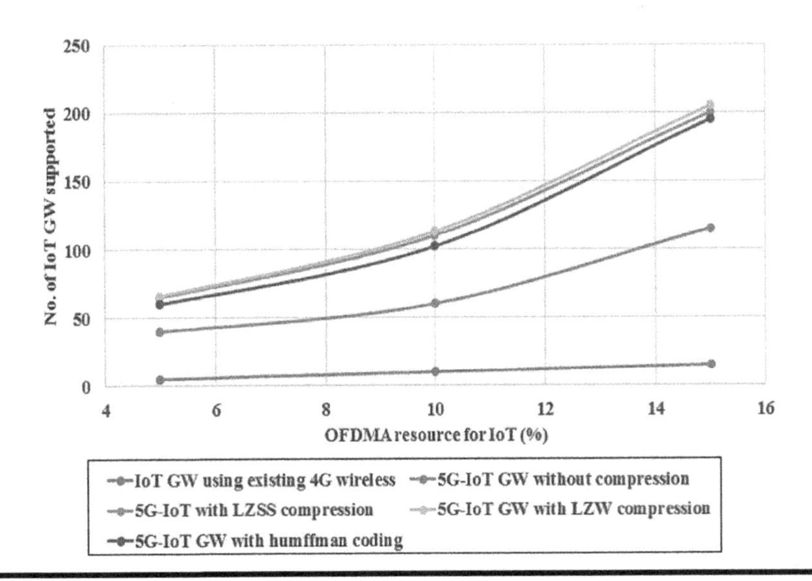

Figure 6.2 Device capacity of cells. Huffman coding (also known as Huffman encoding) is an algorithm for compression of data and it forms the basic idea behind file compression.

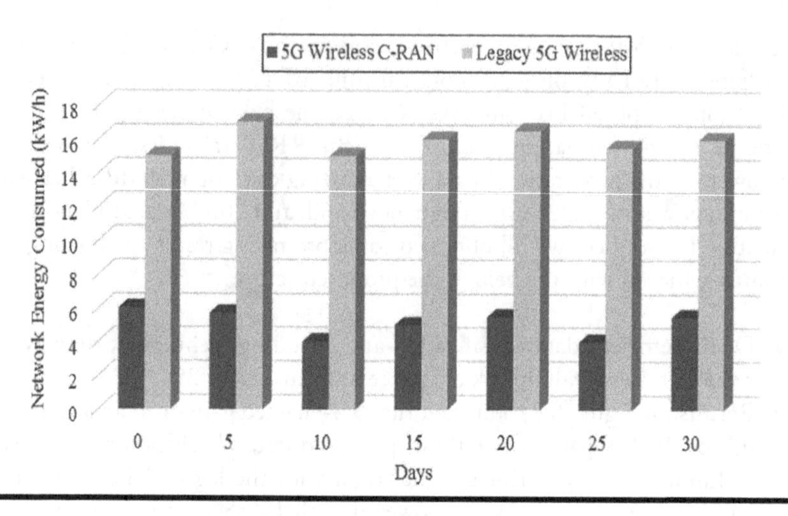

Figure 6.3 Improvement in energy consumption.

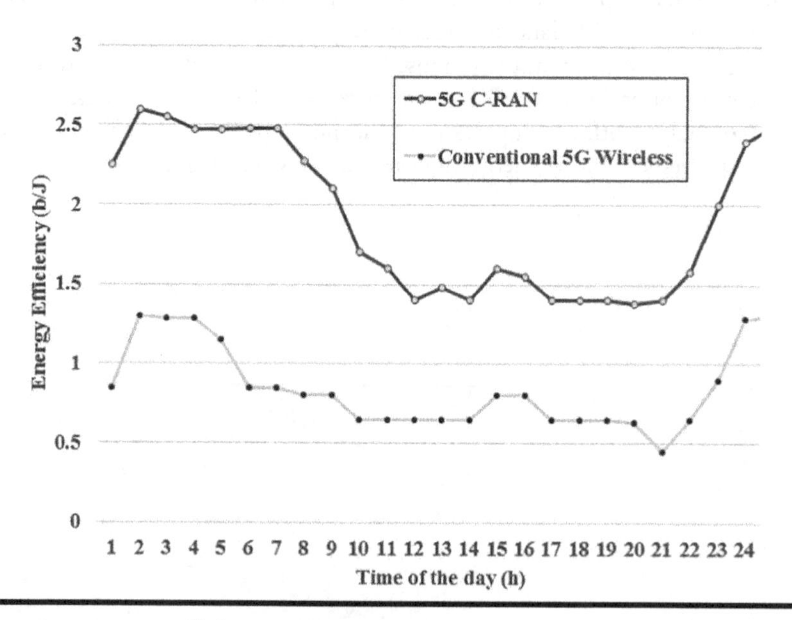

Figure 6.4 Energy efficiency using C-RAN.

Briefly, the proposed 5G C-RAN model, 5G-enabled gateways and live air laboratory experiments promised massive improvement in IoT device capacity in the network with the help of efficient uplink data encoding techniques along with optimal virtualization of wireless resources. Next, we present the proposal of improving DR programs of SGs with the help of optimal 5G small cell planning and eMBMS.

6.4 Efficient Smart Grids Demand Response Using Optimal 5G Small Cell Planning and eMBMS

The other noticeable research work in the era of 5G has been carried out specifically for SGs, wherein we have introduced the concept of planning of 5G small cells efficiently using 5G-evolved eMBMS to optimize DRs in SGs [11]. SGs contain communication technologies that are seamlessly integrated with energy grids. This provides the sustainable and systematic creation and management of electricity. The information and electricity are shared between consumer devices and energy grids bi-directionally. SGs encourage consumers to curtail the electricity consumption in peak hours with the help of DR programs by facilitating incentives to the customers. The dynamic pricing with the help of DR programs help the consumers to reduce their consumption and flatten their demands, which in turn reduces their electricity bills [12].

Figure 6.5 depicts the architecture of SG communication's DR program. In the program, Base stations (BSs) serve as aggregators between electricity consumers and the SG utility. The DR engine, which is part of the SG utility companies, helps to collect the energy data and forecast the load. The residence of electricity consumers have smart meters that use BSs to provide power-related information to the utility. The utility optimizes the load with the help of real-time data collected from smart meters. This is done with the help of time-dependent pricing (TDP) [12]. A DR engine installed inside the utility runs DR programs on the energy usage patterns of the electricity consumers, and the results are sent to aggregators.

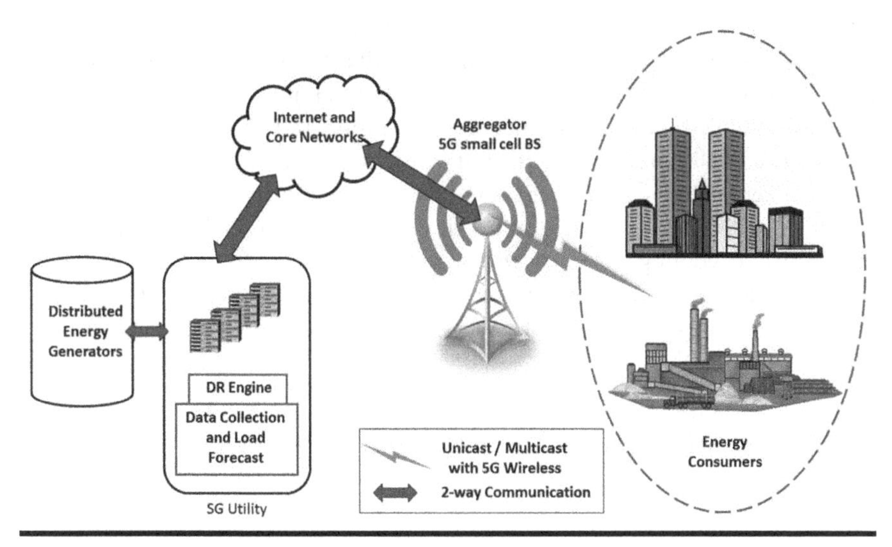

Figure 6.5 SG communications with DR programs.

Consumers are given energy usage recommendations based on the data received by the aggregators. The information that flows between the electricity consumers and the utilities requires effective communication systems. We assert that DR programs' results should be used by BSs to provide selective information to the group of electricity consumers [11]. BSs should perform the selection of electricity consumers with the help of response factor, current demand of energy and consumption level of consumers to minimize the use of electricity during peak hours. Recent surveys put forward the argument of using effective communication technologies for SGs with (1) a 6% error rate or real-time metering and sensing packets [9], (2) a network latency of up to 12–20 ms [9], (3) advance broadband access networks [9], and (4) a secure, faster medium for communication [13]. 3GPP and 5GPP identified the use of 5G communication technology suitable for SG because legacy 3G or LTE networks are incapable of providing support for (1) critical tasks, (2) flexible usage of spectrum, and (3) massive IoT. Furthermore, separation of control and data plane solution is identified as effective solution for provisioning reliable and robust SG traffic for 5G communication [14]. RRHs of 5G small cells with increased spatial multiplexing and reduced interference due to directional narrow mmWave beams can be utilized for dense areas like Korea, Japan, and the United States for provisioning of optimal performance [15,16]. Moreover, with the help of effective OFDMA radio resources, multicast transmissions are being considered best suitable for DR programs involving the exchange of information among multiple electricity consumers [17]. This in turn motivated us to explore avenues of 5G eMBMS [18] for designing an effective communication system between electricity consumers and aggregators. Following are the proposed important developments for building an optimal DR in SGs: We have introduced the concept of optimal planning of 5G small cell RRH in order to increase the efficiency of SG DR programs. We formulated planning problem of 5G small cell RRH. We did it by choosing the location of small cell RRH with the help of directional antennas in such a way that the small cell RRH number needed for dispensing coverage to all the DR customer of SGs is minimal. By doing so, we take care that service quality parameters like cell–edge and average SINR are maintained. Two different solutions—dynamic programming and greedy heuristics—can be utilized by DR customers to curtail their energy cost. This is done with the help of effective scheduling of 5G OFDMA radio resources. Last, with the help of an estimation of the 5G cell performance, we performed analysis of multicast capacity to extend our solution for including 5G Multicast Broadcast Single Frequency Network (MBSFN).

6.4.1 Simulation Setup and Results Discussion

We developed 5G eMBMS Optimized Network Engineering Tools (OPNET) OPNET simulator and perform simulations over actual energy data. Table 6.2 represents the simulation parameters.

Table 6.2 Major Simulation Parameters

5G Network-Related Parameters		*SG and Device-Related Parameters*	
Frequency	27.925 GHz	Number of SGs	10
Channel Band	520 MHz	Antenna gain	4 dBi
Channel Model	Dense urban	Device power	1 mW
Cell Radius	200 m	Area	7 km × 7 km
Antenna Gain	9 dBi	Packet size	128 B
Penetration	20 dB	Backhaul and core network delay	5 ms
Path Loss	3.8	Small cells	100
Cell's Tx power	1 Watt	Device noise	9 dB
Shadowing	8 dB	—	—

Figure 6.6 shows the average of 5G network latency dynamics by varying the network load. It delineates that at the highest cell load of 0.6, the greedy heuristic has a latency of ≤10 ms, whereas MBSFN exhibits a latency of ≤9 ms. Unicast services, greedy solutions and point-to-multipoint single-cell eMBMS scheduling with DP exhibits a latency of 21, 17, and 15 ms, respectively.

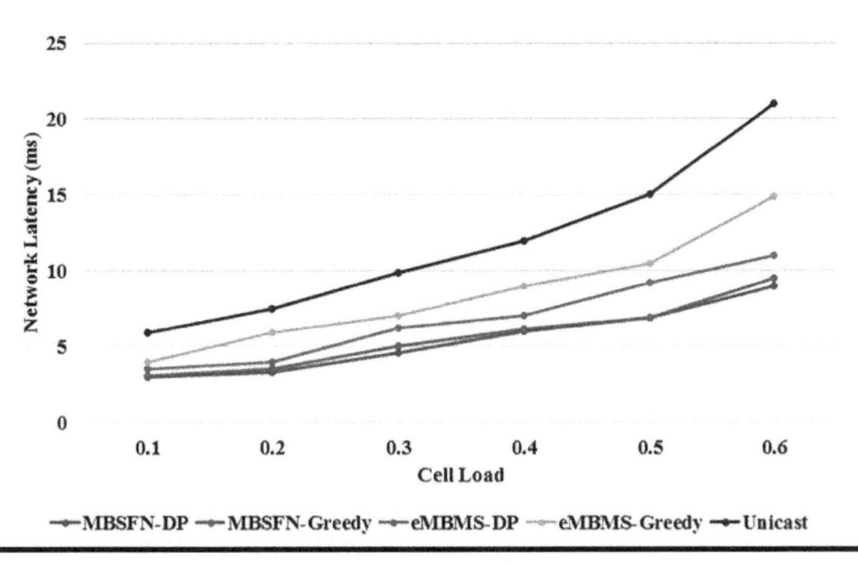

Figure 6.6 Latency dynamics with network load.

Figure 6.7 shows the effectiveness of our proposed schemes that can curtail the peak energy load significantly. We are able to achieve a 35% reduction in maximum peak load. This in turn helps to preclude a power outage during a peak load.

Figure 6.8 shows that discounts/incentives varied from 17% to 0% during off-peak and peak hours, respectively. A similar pattern was seen daily with respect to incentives during the entire simulation period.

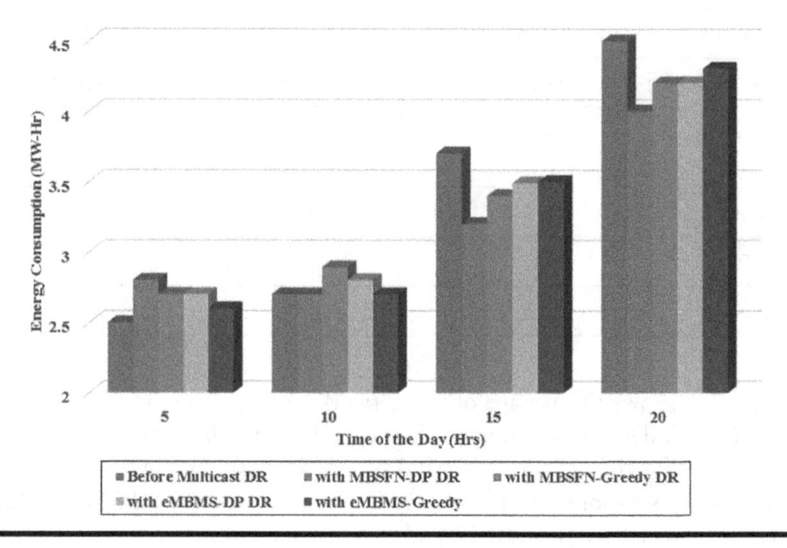

Figure 6.7 Energy consumption dynamics with SG demand response.

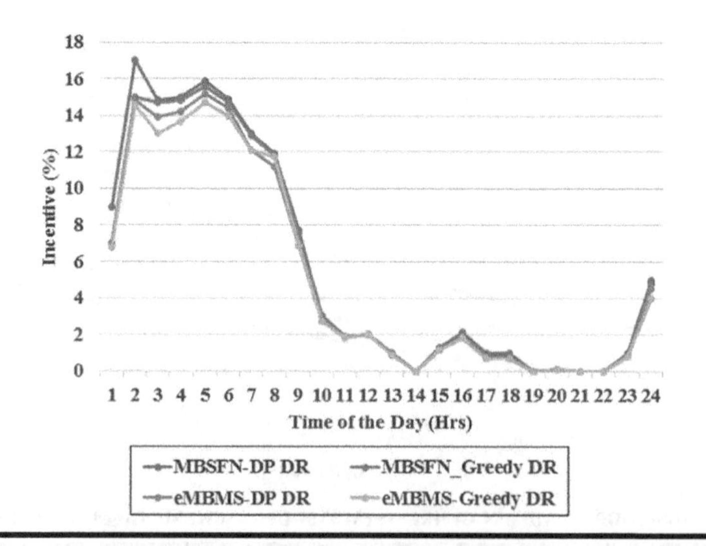

Figure 6.8 Average daily reward dynamics.

6.5 5G IoT Open Issues and Solutions

In this section, we discuss the open issues and their respective solutions for 5G-based IoT networks. Subsequently, we depict the research needs of IoT for 2020 and beyond. A summary of open issues is provided in Figure 6.9.

1. Architectural issues: An open, interoperable and scalable architecture is required for 5G IoT. It should be capable of supporting the context awareness of various events of smart devices [19,20] because communication (ad hoc, wireless or wired) among devices can take place anytime and anywhere. Therefore, a common architecture cannot be followed for all types of

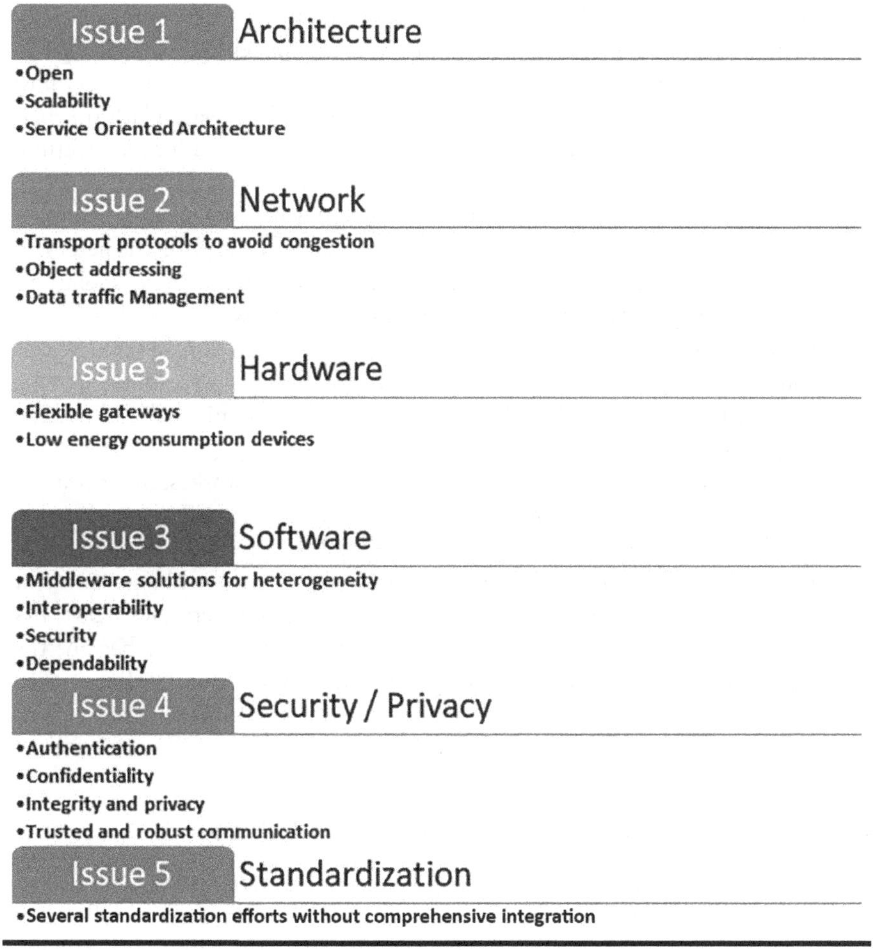

Issue 1 Architecture
- Open
- Scalability
- Service Oriented Architecture

Issue 2 Network
- Transport protocols to avoid congestion
- Object addressing
- Data traffic Management

Issue 3 Hardware
- Flexible gateways
- Low energy consumption devices

Issue 3 Software
- Middleware solutions for heterogeneity
- Interoperability
- Security
- Dependability

Issue 4 Security / Privacy
- Authentication
- Confidentiality
- Integrity and privacy
- Trusted and robust communication

Issue 5 Standardization
- Several standardization efforts without comprehensive integration

Figure 6.9 5G IoT open issues and solution.

events or all types of IoT devices. Hence, service-oriented architecture (SOA) is proposed for addressing the aforementioned biggest issue of 5G IoT.

2. Network issues: An extensive number of devices are used in the IoT [21], including RFID, WLAN and cellular communication, each having their own respective features. When they are used under one umbrella, issues like compatibility, interoperability and security have to be considered, which may in turn hinder the development of IoT. In addition, huge content is created by the large number of heterogeneous devices in the network. Obviously, such content/data has to be retrievable from any position and by any authorized user. Therefore, efficient addressing policies have to be taken into consideration when deploying a 5G IoT network.
3. Hardware and software issues: Intelligent hardware is a key to effective development of diverse applications in an IoT network. Hardware should be small in size, be available at low cost, have low weight and be energy efficient [22]. These diverse hardware requirements increase the challenges of an IoT significantly from the hardware perspective. On the other hand, the next foremost challenge is constructing a unified 5G-enabled IoT software platform. It is a middleware and demands an SOA solution, as mentioned for architectural issues [23].
4. Security and privacy issues: Compared with legacy networks, the deployment of IoT has comparatively more privacy and security issues [24]. The heterogeneous communication and dynamic activities of IoT devices demand solutions that are separate from traditional communication networks. Authentication and data integrity are the major issues. To some extent, lightweight encryption technologies and various encryption/decryption algorithms can be employed as solutions. Digital forgetting is another technique for addressing the privacy and security issues.
5. Standardization issues: For commercializing the IoT network, standardization is another critical issue for having a common platform for IoT network across the globe by multiple parties. For example, in 2012, the Telecommunication Technology Association (TTA) was launched by the Korean government and included 267 companies such as Cisco, Samsung Electronics, IBM, LG Electronics, Intel, SKT, Oracle, KT, 3M, and LGU+ [6]. Korea became the chair of IoT standardization working group of ISO in 2014 [25]. Figure 6.10 shows a schematic diagram of these open research challenges.

6.6 Conclusion and Future Research Issues

Emerging 5G NR is expected to explore a set of new, emerging features such as mmWave and V-RAN for providing huge connectivity for the massive roll out of IoT. Exploring these new features, we discuss our new 5G-enabled IoT-gateway design and highlight ways to explore 5G small cell planning and eMBMS for

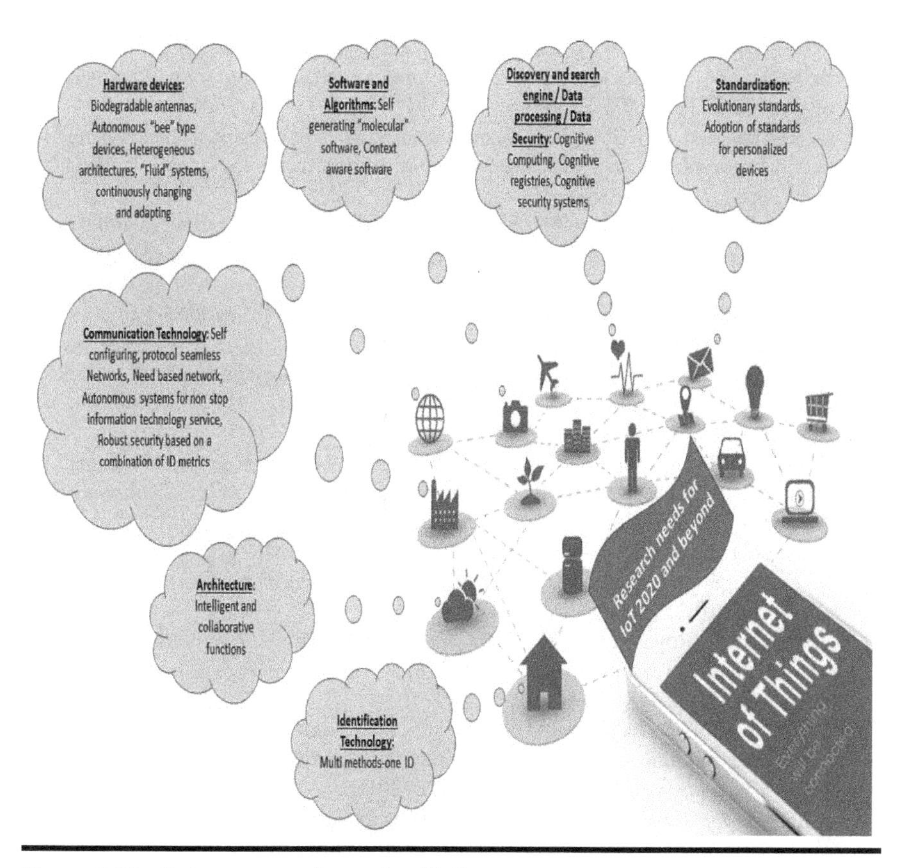

Figure 6.10 Future research needs of IoT.

improved DR programs in SGs. Our 5G C-RAN prototype, laboratory experiments and simulation results point out that huge improvement in IoT-device capacity and DR programs of SGs are possible by using sophisticated uplink data compression, efficient wireless resource virtualization, optimized 5G small cell planning and efficient 5G eMBMS scheduling. We hope our work will help in resolving some of the future research issues mentioned in Figure 6.10.

References

1. A. Al-Fuqaha, M. Guizani, M. Mohammadi, M. Aledhari and M. Ayyash, "Internet of things: A survey on enabling technologies, protocols and applications," *IEEE Communications Surveys & Tutorials*, vol. 17, no. 4, pp. 2347–2376, 2015.
2. H. Holma, A. Toskala and J. Reunanen, "LTE small cell optimization: 3GPP evolution to release 13," *John Wiley & Sons Publications Ltd.*, pp. 1–462, 2015.
3. L. Atzori, I. Antonio and M. Giacomo. "The internet of things: A survey." *Elsevier Computer Networks*, vol. 54, no. 15, pp. 2787–2805, 2010.

4. 5G Americas, "LTE and 5G technologies enabling the internet of things," 5G Amer., Bellevue, WA, White Paper, Dec. 2016. Online. Available: htpp://www.5gamericas.org/les/3514/8121/4832/ Enabling__IoT_WP_12.8.16_FINAL.pdf.
5. G. A. Akpakwu, B. J. Silva, G. P. Hancke and A. M. Abu-Mahfouz, "A survey on 5G networks for the internet of things: Communication technologies and challenges," *IEEE Access*, vol. 6, pp. 3619–3647, 2018.
6. S. Muralidharan, A. Roy and N. Saxena, "An exhaustive review on internet of things from Korea's perspective," *Springer Wireless Personal Communications*, vol. 90, no. 3, pp 1463–1486, 2016.
7. Online available at: https://www.yourreadybusiness.co.uk/5-iot-nations-championing-connected-world/.
8. N. Saxena, A. Roy, B. J. R. Sahu, H. Kim, "Efficient IoT gateway over 5G wireless: A new design with prototype and implementation results," *IEEE Communications Magazine*, vol. 55, no. 2, pp 97–105, 2017.
9. Y. Yan, Y. Qian, H. Sharif and D. Tipper, "A survey on smart grid communication infrastructures: Motivations requirements and challenges," *IEEE Communication Surveys & Tutorials*, vol. 15, no. 1, pp. 5–20, 2013.
10. R. Ma, H.-H. Chen, Y.-R. Huang and W. Meng, "Smart grid communication: Its challenges and opportunities," *IEEE Transactions on Smart Grids*, vol. 4, no. 1, pp. 36–46, 2013.
11. N. Saxena, A. Roy and H. Kim, "Efficient 5G small cell planning with eMBMS for optimal demand response in smart grids," *IEEE Transactions on Industrial Informatics*, vol. 13, no. 3, pp. 1471–1481, 2017.
12. J. Vardakas, N. Zorba and C. Verikoukis, "Survey on demand response programs in smart grids: Pricing methods and optimization algorithms," *IEEE Communication Surveys Tutorials*, vol. 17, no. 1, pp. 152–178, 2015.
13. J. Rifkin, *The Third Industrial Revolution: How Lateral Power is Transforming Energy the Economy and the World*, New York, NY: Palgrave McMillan, 2011.
14. G. Wunder, "5GNOW: Non-orthogonal asynchronous waveforms for future mobile applications," *IEEE Communications Magazine*, vol. 52, no. 2, pp. 97–105, 2014.
15. T. S. Rappaport, F. Gutierrez, E. Ben-Dor, J. N. Murdock, Y. Qiao, J. I. Tamir, "Broadband millimeter wave propagation measurements and models using adaptive beam antennas for outdoor urban cellular communications," *IEEE Transactions on Antennas Propagation*, vol. 61, no. 4, pp. 1850–1859, 2013.
16. J. G. Andrews, S. Buzzi, W. Choi, S. V. Hanly, A. Lozano, A. C. Soong and J. C. Zhang, "What will 5G be?" *IEEE Journal on Selected Areas in Communications*, vol. 32, no. 6, pp. 1065–1082, 2014.
17. "UMTS-multimedia broadcast/multicast service (MBMS)," Antipolis, France, 2014.
18. R. Afolabi, A. Dadlani and K. Kim, "Multicast scheduling and resource allocation algorithms for OFDMA-based systems: A survey," *IEEE Communication Surveys & Tutorials*, vol. 15, no. 1, pp. 240–254, 2013.
19. Y. Chen, F.Han, Y. H. Yang, H. Ma, Y. Han, C. Jiang, H. Q. Lai, D. Claffey, Z. Safar and K. R. Liu, "Time-reversal wireless paradigm for green internet of things: An overview," *IEEE Internet of Things*, vol. 1, no. 1, pp. 81–98, 2014.
20. C. Perera, A. Zaslavsky, P. Christen and D. Georgakopoulos, "Context aware computing for the internet of things: A survey," *IEEE Communications Surveys and Tutorials*, vol. 16, no. 1, pp. 414–454, 2013.

21. O. Vermesan et al., "Internet of things strategic research roadmap." *Internet of Things-Global Technological and Societal Trends*, 9–52, 2011.
22. G. Marrocco, C. Occhiuzzi and F. Amato, "Sensor-oriented passive RFID," *Proceedings of TIWDC*, 2009.
23. K. Finkenzeller, *RFID Handbook*, John Wiley & Sons, 2003.
24. H. Ning, H. Liu and L. T. Yang, "Cyberentity security in the Internet of Things," *Computer*, vol. 46, no. 4, pp. 4653, 2013.
25. Z. Sheng, S. Yang, Y. Yu, A. Vasilakos, J. Mccann and K. Leung, "A survey on the ietf protocol suite for the internet of things: Standards, challenges, and opportunities," *IEEE Wireless Communications*, vol. 20, no. 6, pp. 91–98, 2013.

Chapter 7

Mobile Edge Computing for the 5G Internet of Things

Haojun Huang, Wang Miao, Geyong Min, and Chunbo Luo

Contents

7.1 Introduction

7.1.1 5G Networks

Fifth generation (5G) mobile networks—characterized by ultra-high transmission throughput, ultra-high connection density, and ultra-high mobility—have been proposed as the next telecommunication standard beyond the current 4G/International Mobile Telecommunications-Advanced (IMT-Advanced) standards. It will revolutionize future ubiquitous and pervasive networking, and provide high quality-of-experience (QoE) services to billions of users globally. 5G networks are designed with the aim of satisfying the performance requirements of three kinds of applications, services and business trends:

- First, 5G technologies are expected to provide high-speed wireless communication services to mobile users. According to the International Telecommunication Union (ITU) IMT-2020 specification [1], 5G wireless communication networks should provide up to 20 Gbps to support various broadband-hungry applications, including high-definition videos, virtual reality, and 3D online gaming [2].
- Second, 5G technologies are required to offer ultra-reliable and ultra-low latency services to support the applications in critical communication scenarios. For instance, 5G has been regarded as a key pillar technology for realizing the Industry 4.0 by the German government [3] and autonomous robots and driving by the Japanese Government [4].
- Third, 5G technology needs to support a huge amount of network connections raised by Internet-of-Things (IoT)–related applications. According to the Statista Report, the number of devices connecting to the network are forecast to reach up to 75 billion by the end of 2025 [5]. Therefore, 5G should be designed, deployed and optimized to meet the large-scale connections of IoT devices in the near future.

These three aspects of performance requirements interact closely with each other. For instance, the successful deployment of autonomous driving requires the 5G network to provide a large amount of broadband, with low transmission latency and a huge amount of simultaneous communication connections. In this chapter, our focus is on the scenario of 5G networks to support IoT applications and services. With the increasing deployment of IoT devices and applications, a large amount of the data will be generated during the service provisioning. For system optimization and service quality guarantee, the data generated must be sent to remote cloud-based data center for processing, making latency-sensitive IoT applications vulnerable to performance degradation. To address this issue, cloud resources are migrated to the edge of the network to support local data processing. With this background, this chapter presents the architecture, challenges and key technologies of complementarily integrating mobile edge computing with 5G-enabled IoT services. Future work is discussed in order to guide researchers with the similar research interests to make contributions in this area.

7.1.2 Mobile Edge Computing

Mobile edge computing (MEC), which originated from a computing service platform, is a new network architecture designed to reduce the distance between the end users and computing resources, enabling unprecedented benefits for data processing, service provisioning and resource optimization. MEC has been widely adopted by industry companies, such as IBM and Nokia Siemens Networks, to provide services to mobile users [6]. The basic idea is to migrate the cloud computing resources from remote data centers to the edge of the mobile access network to improve utilization of computing and storage resources.

Bringing computation resources closer to mobile users is characterized by close range, low latency and location awareness. This realizes the localization of mobile service provisioning, improves their business capabilities, reduces latency and optimizes network performance [7]. MEC has very broad applications in today's mobile communications. for example, computational offloading, content delivery, mobile Big Data analytics, edge video caching, collaborative computing, connected cars, heath care, and smart grids.

7.1.3 Mobile Edge Computing in 5G

In the future 5G application scenario, 5G networks are required to quickly and efficiently respond to complex and variable services by leveraging cloud-based resource pools, fully meeting their performance requirements [8]. In this context, flexibility is a key feature for 5G networks to expand network resources and improve service provisioning. However, exploiting the existing mobile network architecture to provide the flexibility required by 5G scenario is a challenging task. MEC has been considered as a promising technology for 5G network to revolutionize the manner in which resources are managed, provided and optimized [6,9,10]. MEC enables the mobile operators to transform their operation modes, reduce OPerational EXpenses (OPEX) and enhance revenue capability.

Compared with the traditional centralized data center, MEC is deployed at the edge of 5G networks. The resources in MEC could be dynamically changed and optimized based on the associated services and applications. The computation resources of mobile devices could also be considered in the MEC architecture in order to conduct related tasks, entirely improving the system performance. The convergence of the 5G network and MEC creates a new business ecosystem for network operators, service providers and end users.

7.2 Convergent Network Architecture of 5G and Mobile Edge Computing

The focus of this subsection is to investigate the integration of 5G and MEC to support IoT applications and services. To obtain this architecture, the existing 5G and MEC architectures must be comprehensively analyzed and complementarily merged. The new network architecture should satisfy the system performance requirements of IoT applications and services. Therefore, we will first present the basic architecture of 5G and MEC. Then, the complementary aspects of the two architectures will be investigated with the aim of proposing a new architecture: the MEC-enabled 5G architecture.

7.2.1 5G Network Architecture

5G networks mainly consist of three functional planes: the access plane, the control plane, and the forwarding plane.

- The access plane is largely responsible for utilizing multisite collaboration, multiconnection mechanism and multi-system convergence technology to construct a flexible access network.
- The control plane is built based on reconfigurable network functions, providing on-demand access, mobility and session management, refined resource management and comprehensive capability optimization.
- The forwarding plane is part of a distributed data forwarding networks, responsible for managing dynamic anchor settings and providing packet processing and transmission.

Within 5G architecture, a modular-based pattern is used by the control plane to design and build the dedicated logical network for specific applications such as remote machine control. Various network functions are selected and chained to meet the application needs. The control plane is the core component for the 5G network architecture to design network service, schedule the underlying resources, and utilize the access and forwarding planes to provide the end-to-end services.

5G network architecture consists of three layers as shown in Figure 7.1: the management and orchestration layer, the network control layer, and the network resource layer.

In order to further improve the service flexibility, 5G adopts a three-level network architecture consisting of the distributed unit (DU), the centralized unit (CU) and the core networks 5G Core network (5GC). For 5G networks, baseband unit (BBU) function will be reconstructed into two functional entities, DU and CU, which together form a next generation NodeB (gNB). Within a gNB, one CU can be connected to one or more DUs. Different function-splitting schemes between the CU and the DU are used in 5G to adapt to different communication scenarios and communication requirements [11]. CU devices are responsible for managing the edge application deployment, providing high-level protocol stack functions, and supporting partial core network function, whereas DU devices mainly deal with physical layer functions.

5G is not a brand-new network architecture, but is evolving from 4G networks. Figure 7.2 shows how to map the network functions from 4G to 5G during the network evolution.

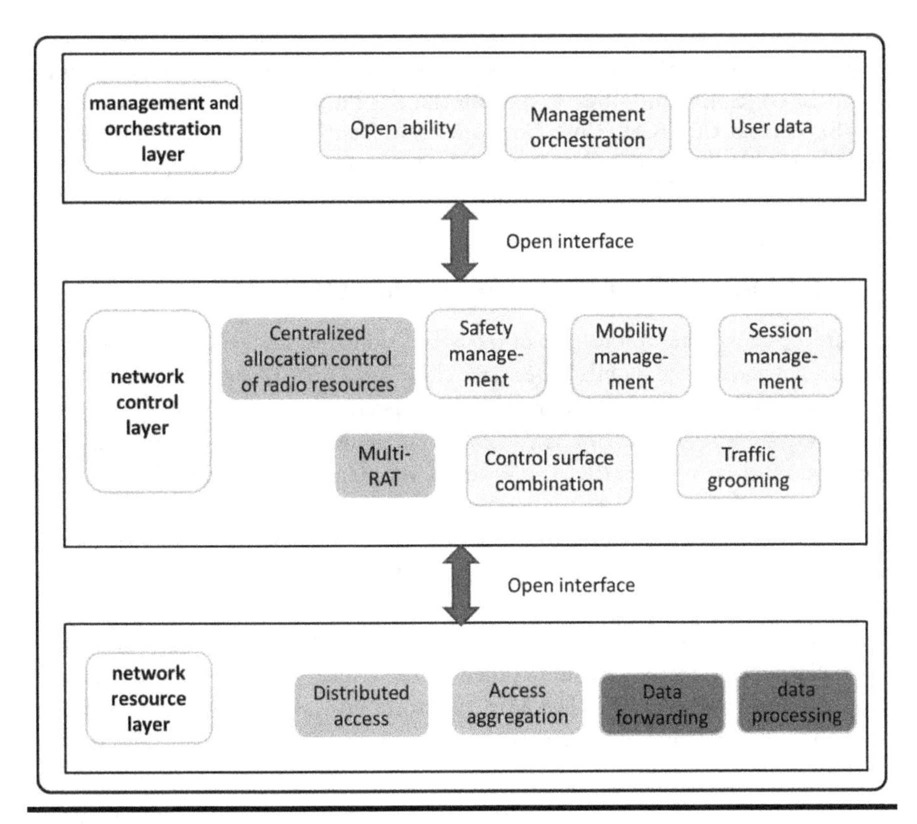

Figure 7.1 5G network function view. RAT: Radio Access Networks.

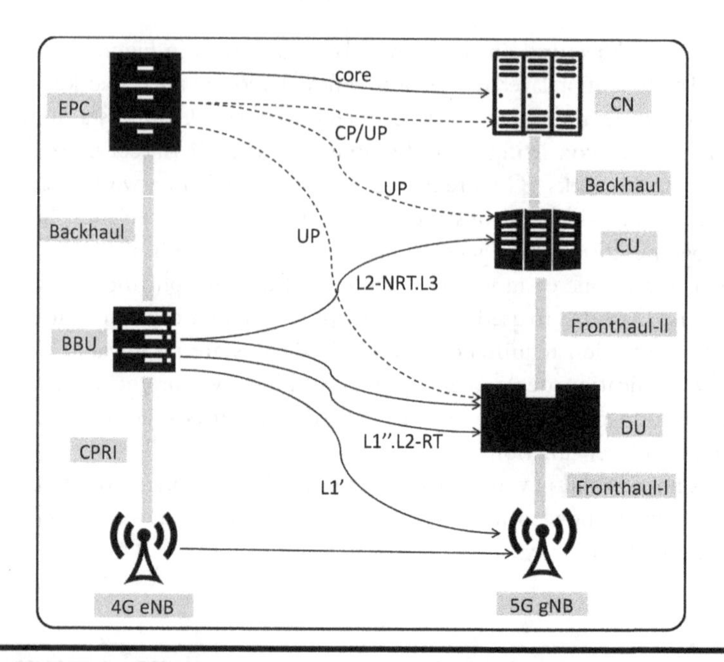

Figure 7.2 The evolution from 4G to 5G (Evolved Packet Core (EPC), Baseband Unit (BBU), Common Public Radio Interface (CPRI), Evolved Node Base-station (eNB), Central Unit (CU), Core Network (CN), Distributed Unit (DU), Neighbour Relation Table (NRT), Relation Table (NT), Layer 1/2/3 (L1/2/3).)

7.2.2 Mobile Edge Computing Architecture

Figure 7.3 shows the architecture of the MEC, which is a three-level network architecture, including the mobile edge system level structure, the mobile edge host level structure, and MEC user equipment (UE) respectively.

The system-level structure is composed of the MEC system-level management and the third-party users [9]. The system-level management is mainly responsible for policy design, upper-level service analysis and guiding the mobile edge host to provide MEC services. The host-level structure includes host-level management software and host infrastructure. The MEC host consists of the MEC platform, MEC application and virtualization infrastructure. Host-level management software is in charge of the management and operation of mobile edge platform and virtualized infrastructure, including the MEC platform manager and the virtualization infrastructure manager. The MEC platform and MEC applications share and exchange the information that they stored during the whole service provisioning, including information related to device discovery, network topology, and network load. The MEC platform has the detailed information of the end UE, including location, bandwidth, and network traffic and provides the MEC application

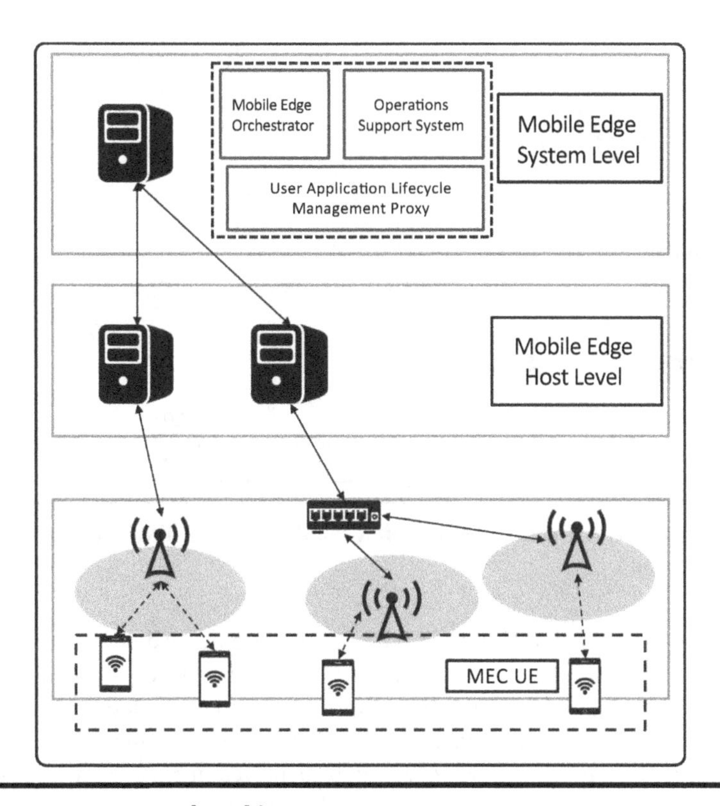

Figure 7.3 MEC network architecture.

with its operating environment during the service provisioning. In order to create an open ecosystem, the business rules of MEC services and the domain name resolution system (DNS) proxy/server of the MEC platform manager and application managers are also included in the system-level and host-level structures of the MEC architecture [8].

7.2.3 Integrated 5G Mobile Edge Computing Architecture

After introducing the network architecture, practical deployment and detailed components of 5G and MEC technologies, this subsection aims to combine these two architectures to improve the service provisioning capabilities of 5G networks for IoT applications. As we all know that the 5G networks are proposed to provide a huge amount of network broadband and support large-scale device connection. In 5G networks, the wide network broadband is achieved by exploiting the millimeter communication, multiple input–multiple output (MIMO) and small cells technologies. For supporting the large scale of network connections, 5G uses the tremendous computation and storage resources from the remote data center and

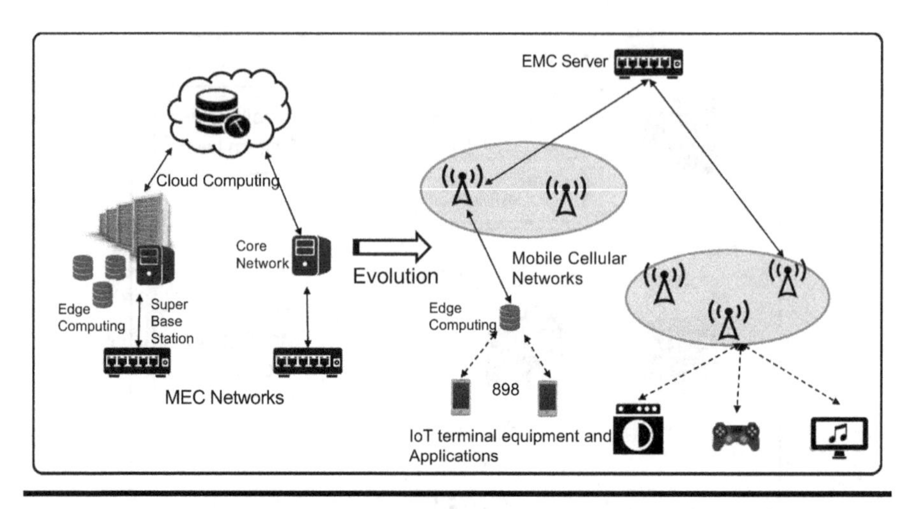

Figure 7.4 A 5G IoT infrastructure that combines MEC. EMC stands for Mobile Edge Computing.

utilizes network function virtualization and software-defined networking technologies to virtualize the network resources for achieving an end-to-end optimized system for service provisioning. However, one issue that 5G network suffers from is high latency, which cannot meet the requirements of the emerging IoT applications [12,13]. To solve this issue, MEC can be deployed in 5G gNB to eliminate the latency in the core network transmission, enhancing the service provisioning capability of 5G network for small-scale and ultra-low-latency services and application scenarios. As shown in Figure 7.4, the future 5G mobile communication network will be a heterogeneous communication network that includes both the centralized base station (BS) and multiple distributed BSs [10]. For integrating the MEC and 5G networks, Figure 7.4 shows how a multilevel computing network provides edge computing and cloud computing functions. Within this architecture, MEC computing resources are allocated in LTE Evolved Node Base-station (eNB), 5G gNB, super 5G BS, and the edge of core networks to provide the computing and storage resources for end users. The new network architecture provides 5G technology with more computation and storage resources, significantly improving the system performance and service provisioning capabilities.

7.3 Technical Considerations in the 5G Internet of Things

For building a high-performance IoT system, the idea of edge computing could be used to enhance the efficiency for data transmission and processing, coupled with the upper-layer network architecture. For practically supporting IoT applications, different applications have different requirements for network transmission, such as bandwidth, latency, and jitter [14]. For instance, automatic driving has been considered as a key application for IoT technology. It is unrealistic to exploit the remote cloud resources to process the requests from fast-moving cars. Once the remote cloud fails to respond to a request in time, problems such as fatal accidents, may occur on the road, resulting in serious economic and society losses. Future research work is required to investigate new data processing technologies to address these technical problems. With this background, this section introduces device-to-device (D2D) communications, cross-layer communications, computation offloading and customized network services in the 5G IoT scenario.

7.3.1 Device-to-Device Communications

Cellular network–based D2D communication, as shown in Figure 7.5 allows user data to be transmitted directly between terminals without being transited through the network [12]. The working mechanism of D2D communication is significantly different from that traditional communication system (Figure 7.5).

Similar to the machine-to-machine (M2M) concept in the IoT, the aim of D2D communication is to reduce the load of BSs through creating direct communications among UEs over a range of distances. Prior to the advent of D2D technology, similar communication technologies emerged, such as Bluetooth (short-range time division duplex communication), wireless fidelity (Wi-Fi) Direct (faster transmission speed and longer transmission distance), and Qualcomm's Flash LinQ technology (greatly improved the transmission distance of Wi-Fi). Among these three technologies, Bluetooth has been widely deployed and is supported by various wireless communication devices such as mobile phone, laptops and IoT devices.

The main difference between D2D communication and ISM band-based short-range communication technology such as Bluetooth and wireless local area network (WLAN) is that D2D uses the licensed frequency band for information transmission. Therefore, the interference environment is controllable, and the data transmission can be guaranteed. Bluetooth requires the end user to manually establish

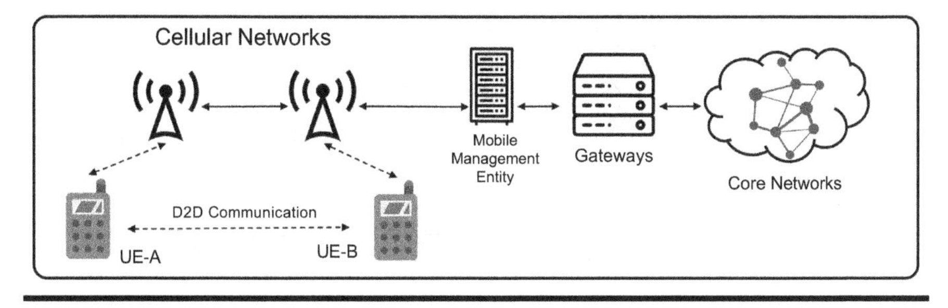

Figure 7.5 D2D communications.

communication. The WLAN needs to perform user-defined settings for the access point (AP) before communications are established. D2D communication does not require the above processes, thereby providing better quality of experience (QoE). In addition, D2D can support a large number of information interactions, provides higher transmission data rata and enable a better quality of service (QoS) guarantee for service provisioning.

7.3.2 Cross-Layer Communications

At the edge of the MEC-enabled 5G IoT system, a large amount of IoT sensors are deployed and linked into the system; they are built into mobile phones, televisions, washing machines, electric lights and so on. These devices are heterogeneous, and intelligent IoT applications require these devices to communicate with each other. Generally speaking, these devices are connected to the network with different protocols, including Wi-Fi and Bluetooth. In order to realize the communications among different devices, the network gateway is used to enable the communication between two networks. A gateway is a computer system or device that acts as a conversion, or translator, between two systems that use different communication protocols, data formats, and network architectures. As the scale of the IoT applications continues to increase, the number of gateways that need to be purchased and deployed will increase dramatically, resulting in an excessive cost of IoT deployment and application.

Considering that Wi-Fi, Bluetooth and ZigBee all work at the band of 2.4 GHz, it is reasonable to incorporate these three technologies under a unified communication frame as shown in Figure 7.6. For instance, to realize the interconnection among different devices with different protocols, it is possible to transmit the data signals generated by these three devices into a single-sided proprietary channel. This design provides the benefits of enabling the direct communication between heterogeneous senders and receivers, allowing heterogeneous devices to receive broadcast simultaneously from a sender by using overlapping frequencies (e.g., Bluetooth to Wi-Fi and ZigBee) and supporting a sender with a wider bandwidth (e.g., Wi-Fi)

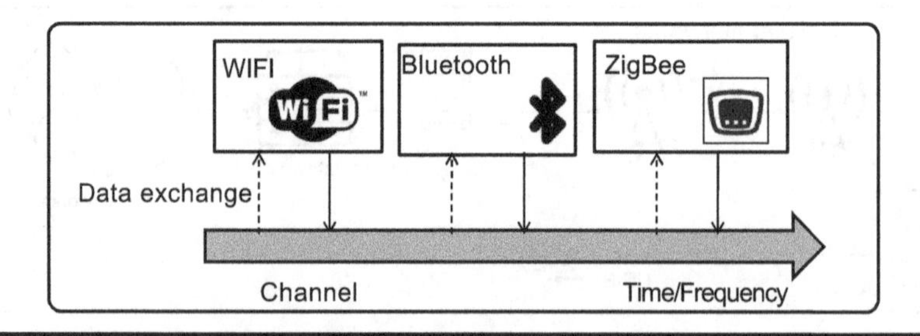

Figure 7.6 Cross-technology coordination.

to reach multiple narrower-band receivers (e.g., Wi-Fi to multi-channel ZigBee) [15]. In addition, this solution does not impose any modifications on the hardware devices in use or create any traffic within the network and the whole design is transparent to the upper-level wireless communication systems. As shown in Figure 7.6, this design solves the problem of cross-technology coordination, enabling heterogeneous devices to be easily interconnected.

7.3.3 Computation Offloading

The terminals of the IoT applications are deployed in the form of sensors. These devices can be homogeneous or heterogeneous. For fulfilling the various tasks of IoT applications, devices need to conduct various computation processes and access the network to upload the data and download the messages. However, with the explosive growth of the mobile devices and IoT applications, the computation capabilities, storage systems and network resources are approaching their limits, and hardly meet the performance requirements for IoT applications currently running on these devices [16].

It is worth noting that the limitations of mobile terminal resources are related to the mobility feature. Compared with static devices such as laptops, mobile devices are designed with lower computation, less storage and narrower broadband to meet the requirements of portability and mobility. In the design of mobile terminals, in order to satisfy the requirements of portability and mobility, the processing capability of the terminal device, the network connection and the like are transferred [17]. In this case, computing offloading is an effective method to extend the resources for mobile devices. During computation offloading, the intensive computing tasks of the IoT application are sent to a remote high-performance data center or to nearby mobile devices to be executed, and the execution results are downloaded back to the mobile devices for further processing, significantly enhancing the computation capability of mobile devices.

Nowadays, after 20 years of computation evolution, the computation offloading technologies can be classified into three stages—distributed, pervasive, and cloud computing—as shown in Table 7.1. In the evolution of computing migration, some computing migration systems, such as MAUI, Cloudlet, and CloneCloud, have emerged to implement the idea of computation offloading. Designed with different research objectives, these systems have their own advantages and disadvantages and were developed based on the architectures of surrogate, cloudlet-based and cloud-based systems, respectively. The authors in [18] demonstrated the detailed implementations for these three systems for computation offloading.

Although there have been some interesting research results reported in the literature, several major problems in computing migration remain and are listed as follows. First, there is a lack of a unified resource organization and management strategy to realize computation offloading [19,20]. Second, there is a lack of incentives and reputation systems necessary for mobile devices to cooperate with each other. Finally, the results of the research on the stability, security, data consistency and privacy of

Table 7.1 The Evolution of Computation Offloading

Stages	*Time*	*Research Mobile Terminal*	*Research Application*	*Research Content*	*Implementation*
Distributed Computing	1995–2000	Laptop performance/ energy consumption	Computationally intensive application	Feasibility	Manual division
Pervasive Computing	2001–2008	Performance of laptops and smartphone/ energy consumption	Resource-intensive application	System architecture and implementation	Base on surrogate
Cloud Computing	2009	Mobile device enhancement	Rich mobile app, augmented reality	Cloud-based computing migration system	Base on cloudnet/ cloud/mobile devices

computation offloading have not been fully investigated. In order to realize a successful deployment of computation offloading in the practical network environment, more research efforts are expected to address the aforementioned problems.

7.3.4 Personalized Network Service

Different IoT applications have different requirements that the network should satisfy. Therefore, the network should be managed and configured to cater to the upper-level applications, making personalized network services an important aspect in network evolution. For instance, 4G networks were mainly designed and deployed to provide services for mobile phones and the network architecture and protocols are optimized for mobile devices and services. However, in the 5G era, the devices connected to the mobile networks will be significantly increased and more importantly, the types of the devices, the performance requirements of applications and patterns of service provisioning are also changed a lot. This makes it important for the network to conduct data processing, understand the quality of service provisioning, and optimize the network resource to satisfy the transmission requirements.

In the MEC-enabled 5G IoT scenario, the sensors connected to the edge of the IoT generate data day and night. These data are transmitted from the end device, through the network links and devices, and reach the intermediate nodes. The intermediate nodes determine whether the data should be sent to the remote cloud node for processing or if the data can be processed directly in the edge computing nodes. Therefore, for MEC-enabled 5G IoT system, the function of filtering the data should be designed for saving bandwidth and reducing the processing latency. If the filtering function is realized in the mobile devices, status of the available resources, processing time and network bandwidth should be transmitted from edge MEC to mobile devices, introducing extra transmission latency. Therefore, the filtering function should be located in MEC nodes to achieve the global optimization for IoT devices.

7.4 Emerging Application Scenarios

7.4.1 Location Service

At present, the location service is mainly provided by the GPS satellite system, which provides accurate navigation information for outdoor navigation and has been widely used in daily life, unmanned aerial vehicle (UAV) use, Google maps, autonomous driving, and so on. However, as the population increases and human living habits change, the demand for indoor positioning is increasing. Its accuracy cannot be satisfied by the GPS system. The commonly used indoor navigation technologies are generally based on Wi-Fi and Bluetooth although these technologies have not been practically deployed due to the issue of accuracy and reliability. In this area, the MEC system has been proposed as a promising technology to estimate the location of end devices.

The advantage of the MEC system is that its geographical location is close to the BSs and end devices. The whole process is shown in Figure 7.7. The network measurement data of BSs and user data can be accessed and read with relatively small cost. Based on the data received, the algorithms in MEC stations could conduct comprehensive analysis of measurement data to accurately estimate the position of end devices, providing location services for IoT applications.

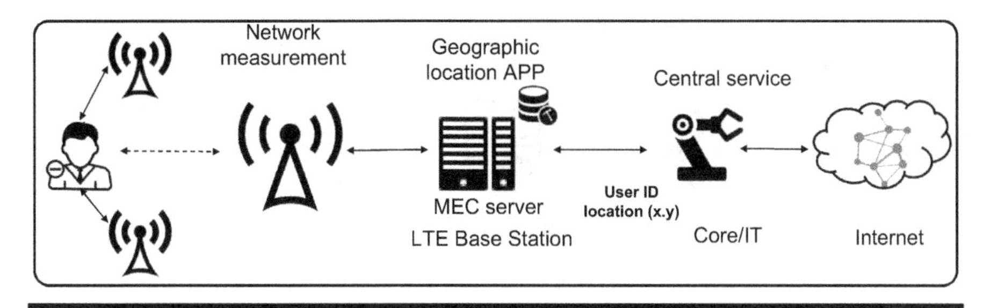

Figure 7.7 Location service.

7.4.2 Internet of Things Gateway

Constrained by the practical environment, IoT devices located at the end of the network are allocated with limited resources in computing and storage and can usually run simple communication protocols. To support resource-hungry IoT applications, IoT gateways are deployed to perform network protocol conversion. The IoT Gateway is a core equipment for the wireless sensor network to deploy IoT applications. It is responsible for building the links between a connection-aware network and a traditional communication network.

With the increasing popularity of IoT devices, various IoT applications—especially those that are sensitive to latency—are beginning to emerge. The MEC system provides conditions for supporting such IoT applications in a complex network environment as shown in Figure 7.8. By deploying the IoT gateway in the MEC-enabled 5G system, the processing power of the edge MEC servers can be used by the IoT gateway to support various IoT protocols with lower latency and higher agility, significantly enhancing the capability of service provisioning in IoT networks.

7.4.3 Internet of Vehicles

The increasing number of Internet-connected cars has promoted the development of the Internet of Vehicles (IoV), for example, various vehicle-related applications are appearing, automatic driving, smart road construction, and intelligent transportation. These emerging applications in the IoV pose challenging issues in network design by, for example, requiring low-latency transmission and large-scale data processing, and by the huge amount of IoT-connected devices. By migrating the remote cloud resources to nearby and distributed mobile BSs, MEC brings network operators a promising solution to deal with the strict performance requirements of the vehicle-related applications. The data could be sorted and processed in a nearby MEC server and key decisions made based on the analysis results, enabling higher processing capability and lower transmission latency, as shown in Figure 7.9.

Other vehicle networking applications that the MEC system can support include road fault notification, traffic congestion prompts, optimal path selection,

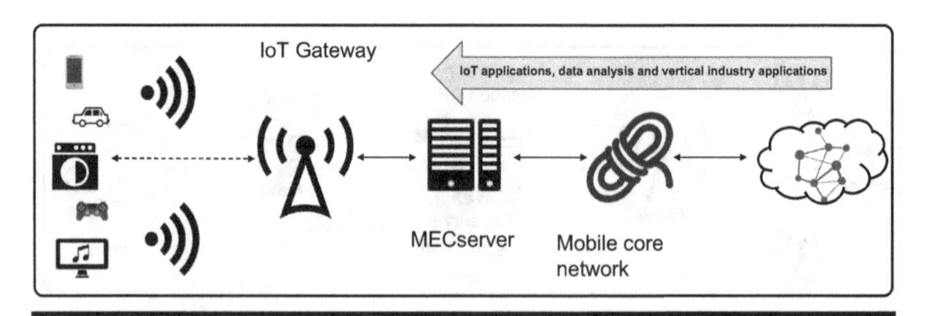

Figure 7.8 IoT gateway.

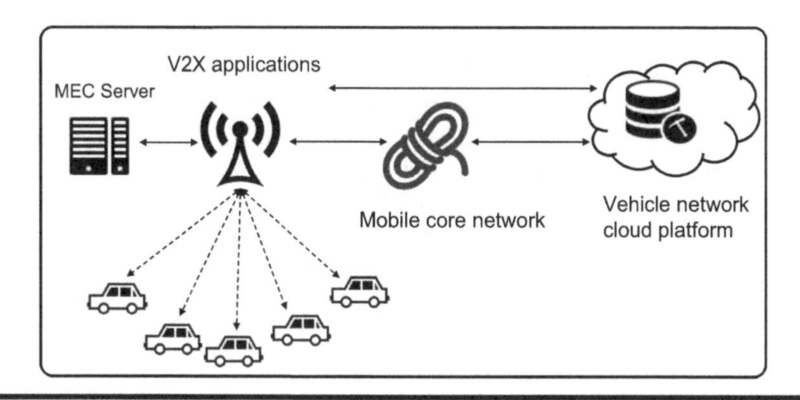

Figure 7.9 Internet of Vehicles based in MEC. Note: V2X, vehicle to everything.

and location tracking. These applications will greatly improve the safety, efficiency and convenience of the transportation system. Therefore, the combination of the IoV and MEC is a complementary solution and will have significant impact for smart vehicle applications.

7.5 Future Research Directions

The MEC-enabled 5G network is a multilevel computing and communication architecture. Compared with the traditional mobile communication networks, which focus purely on data transmission, MEC-enabled 5G networks cooperatively combine the data transmission, data storage and data processing into a single network architecture, greatly reducing the end-to-end transmission latency, which is a key challenge that the 5G communication network struggles to handle. Although the MEC-enabled 5G network brings various benefits for network design and has attracted tremendous research interest, there are still some challenges that need further research in order to realize the ambitions that MEC-enabled 5G networks promise. This subsection will briefly introduce the main challenges of MEC-enabled 5G networks, including architecture innovation, practical deployment, and security- and privacy-related issues.

7.5.1 Architecture Innovation

As discussed above, the MEC-enabled 5G network will be an intelligent information system integrating computing, storage and networking resources. However, the theory for the practical deployment of this network architecture has not been fully investigated. For example, how to globally organize, manage and optimize the network, storage and computing resources is a key issue to address. A unified management architecture is needed to allocate the system resources according to the upper-level application requirements. In addition, how to define and model network utilization capacity, how to convert network utility capacity and computing power, and how to use virtualization and network technology to achieve effective resource allocation, for example, also need more scientific work. These issues all place great demands on the effort to deepen and refine the design of the MEC-enabled 5G architecture. For instance, how to improve the infrastructure so that it can balance the communication and computation is an important research issue for MEC-enabled 5G architecture.

7.5.2 Computation Optimization

For MEC-enabled 5G networks, the collaborative optimization of communication, storage and computing resources is one of the key issues. The IoT applications running on the UE devices have specific QoS requirements such as end-to-end latency. On the other hand, fulfilling the function of IoT applications requires storage and computation resources that UE devices cannot meet due to limitations related to their size, cost, and energy-consumption. Remote cloud computing is an option for UE devices to support IoT applications. Because long-distance transmission results in large transmission latency, which cannot meet the QoS requirements, it is necessary to shorten the distance between the cloud resource and UE device or to leverage other affordable computing nodes that are nearby to conduct the calculation tasks. By integrating MEC to mobile communication networks, 5G networks gain multilevel computing, storage and networking capabilities. However, some research challenges that remain to be addressed regarding the limited computing resources in UE nodes. For instance, how to assign the computing tasks to multiple nodes with different computing powers. Computing task is a challenging issue and needs further research efforts. Compared with the traditional offloading technologies of computing task, task offloading in MEC-enabled 5G network should take not only the computation resources, but also the network transmission and storage capacity, into account. For instance, radio resources such as bandwidth and transmission power are strictly limited in wireless communication. Task offloading should consider the quality of channel transmission and storage capability in distributed servers and conduct the task division and allocation among multiple nodes for optimizing the overall system performance.

7.5.3 Management and Orchestration

In a reliable system, a systematic service management and orchestrator is responsible for managing the network topology, optimizing the network resources and providing end-to-end network services [21]. In the MEC-enabled 5G architecture, the diversity of the edge resources, the IoT devices and upper-level applications highlight the importance of service orchestration. Under the unified management of service orchestration, the components can communicate with each other and send the tasks and data to edge nodes. In addition, the network operators can also deploy fault detection and access control and introduce new services into the networks with the guidance of service orchestration.

The diversity of data also increases the difficulty for service management. For IoT applications, a variety of IoT devices will be deployed and connected to MEC-enabled 5G networks. These devices continuously generate heterogeneous data that pour into the network domain. For the service orchestrator, how to exploit the intelligent algorithms and network resources to handle the huge of heterogeneous data, analyze valuable information and optimize the network resources is an urgent problem to be solved.

7.5.4 Resource Allocation

In the MEC-enabled 5G IoT system, MEC is applied to the edge of the network to improve the performance of the system. However, MEC resources are distributively deployed and controlled by different management entities [22]. The examples of these management entities include the UE terminal equipment, BS control software, core network routers, and the edge servers. The resource management approaches in traditional cloud computing are mainly based on a logically centralized node, which makes it difficult to directly implement the resource management of cloud computing in MEC 5G networks. Therefore, new resource management solutions are required to develop a global resource management system for the MEC 5G system. Virtualization technologies have been considered a good technology in future management systems to virtualize the underlying network infrastructure and improve resource utilization. However, network virtualization brings new challenges to management systems—including complexity and performance guarantee issues—which need more research in regard to MEC-enabled 5G scenarios.

7.5.5 Security and Privacy

In MEC-enabled 5G networks, a large amount of IoT devices will be connected to the network infrastructure. For the security of the entire network and IoT applications running this infrastructure, the network management system should authenticate these devices before providing network service [23]. However, many resource-constrained terminals do not have enough storage and computation resources to perform the encryption operations required by the authentication protocol. Although a traditional public key infrastructure such as Key Performance Indicator (KPI)-based authentication can solve the problem of secure communication, it was not originally designed for IoT application and faces serious security issues challenges posed by the unique features of the IoT system, including the large scale of device connections. In addition, malicious computing nodes may pretend to be legitimate nodes to steal data generated by other IoT devices, and abuse user data, or provide malicious data to neighboring nodes to disrupt their behavior. Due to the complexity of trust management, dealing with this issue in the IoT environment can become more complicated.

In this context, novel permission authentications are needed to enhance the security and privacy of IoT system and application. For IoT applications, different devices have different permissions for data transmission and sharing. New methods are expected to detect and analyze the data received from IoT devices. The management system could be enhanced to conduct the task of authenticating and authorizing the requests from the normal nodes and rejecting the packets from malicious nodes.

7.6 Conclusion

This chapter mainly introduced a new network architecture to support IoT applications: the MEC-enabled 5G network. First, we discussed the basic architecture, the main components, and the network functions of 5G and MEC. Based on the analysis of complementary aspects of 5G and MEC architectures, a novel network architecture was presented in this chapter to integrate MEC into 5G architecture to improve the service provisioning ability for IoT applications. We presented the advantages and benefits of the proposed network architecture for IoT applications. In addition, the technology considerations of providing IoT service in the MEC-enabled 5G networks were introduced, including D2D communications, cross-layer communications, computation offloading and customized network services. Following the system design considerations, we investigated three use cases of how to use the MEC-enabled 5G network to support IoT applications. Finally, the challenges and future research directions were discussed with the aim of guiding and inspiring other researchers to make progressive contributions in this area.

References

1. M. Carugi, "Key features and requirements of 5G/IMT-2020 networks," *Report*, 2018.
2. L. Tian, C. Yan, W. Li, Z. Yuan, and W. Cao, "On uplink non-orthogonal multiple access for 5G: Opportunities and challenges," *China Communications*, 2017, 14(12): 142–152.
3. Y. Nakamura, "Japan's 5G Policy Perspectives," *Report*, 2016.
4. S. Ludwig, M. Karrenbauer, and A. Fellan, "A 5G architecture for the factory of the future," *Networking and Internet Architecture*, 2018: 1–7.
5. Statista, "Internet of Things (IoT) connected devices installed base worldwide from 2015 to 2025 (in billions)." https://www.statista.com/statistics/471 264/iot-number-of-connected-devices-worldwide/.

6. W. Shi, J. Cao, Q. Zhang, Y. Li, and L. Xu, "Edge computing: Vision and challenges," *IEEE Internet of Things Journal*, 2016, 3(5): 637–646.
7. E. Bastug, M. Bennis, and M. Debbah, "Living on the edge: The role of proactive caching in 5G wireless networks," *IEEE Communications Magazine*, 2014, 52(8): 82–89.
8. H. Pang and K. Tan, "Authenticating query results in edge computing," *International Conference on Data Engineering*, 2004, 20: 560–571.
9. M. Satyanarayanan, "The emergence of edge computing," *Computer*, 2017, 50(1): 30–39.
10. Y. Mao, C. You, J. Zhang, K. Huang, and K. Letaief, "A survey on mobile edge computing: The communication perspective," *IEEE Consumer Electronics Magazine*, 2016, 5(4): 732322–732358.
11. M. Palattella, M. Dohler, A. Grieco, G. Rizzo, and J. Torsner, "Internet of things in the 5G Era: Enablers, architecture, and business models," *IEEE Journal on Selected Areas in Communications*, 2016, 34(3): 510–527.
12. M. Du, K. Wang, Y. Chen, X. Wang, and Y. Sun, "Big data privacy preserving in multi-access edge computing for heterogeneous internet of things," *IEEE Communications Magazine*, 2018, 56(8): 62–67.
13. D. He, S. Chan, and M. Guizani "Security in the internet of things supported by mobile edge computing," *IEEE Communications Magazine*, 2018, 56(8), 56–61.
14. S. Li, L. Xu, and S. Zhao, "5G internet of things: A survey," *Journal of Industrial Information Integration*, 2018, 10: 1–9.
15. M. Song and H. Tian, "FreeBee: Cross-technology communication via free side-channel," *International Conference on Mobile Computing and Networking*, 2015: 317–330.
16. D. Duolikun, T. Enokidoy, and M. Takizawa, "A process migration approach to energy-efficient computation in a cluster of servers," *International Conference on Broadband and Wireless Computing*, 2016: 191–198.
17. R. Zhang, D. Zhang, and Z. Kan, "Using mobile agent techniques for distributed manufacturing network management," *International Conferences on Info-tech and Info-net*, 2001.
18. A. Milanes, N. Rodriguez, and B. Schulze, "A classification for the implementations of heterogeneous strong migration of computations," *IEEE International Symposium on Cluster Computing and the Grid*, 2007: 858–868.
19. C. Gao and V. C. S. Lee, "Energy efficient mobile computation offloading through workload migration," *IEEE International Conference on Smart City*, 2015: 1147–1150.
20. A. Milans, N. Rodriguez, and R. Lerusalimschy, "Reflection-based heterogeneous migration of computations," *Brazilian Symposium on Computer Networks and Distributed Systems*, 2014: 223–230.
21. P. Shantharama, A. Thyagaturu, N. Karakoc, L. Ferrari, M. Reisslein, and A. Scaglione, "LayBack: SDN management of Multi-Access Edge Computing (MEC) for network access services and radio resource sharing," *IEEE Access*, 2018, 6: 57545–57561.
22. A. Kiani and N. AnsariA, "Edge computing aware NOMA for 5G networks," *IEEE Internet of Things Journal*, 2018, 5(2): 1299–1306.
23. W. Zhou, Y. Jia, A. Peng, Y. Zhang, and P. Liu, "The effect of IOT new features on security and privacy: New threats, existing solutions, and challenges yet to be solved," *IEEE Internet of Things Journal*, 2018.

Chapter 8

Millimeter-Wave 5G-Enabled Internet of Things

Turker Yilmaz, Naveed A. Abbasi, and Ozgur B. Akan

Contents

8.1 Introduction

The regulatory activities for the fifth generation (5G) mobile systems within the International Telecommunication Union (ITU) Radiocommunication Sector (ITU-R) began in 2012, principally under the Working Party 5D. The three key usage scenarios of International Mobile Telecommunications (IMT) for 2020 and beyond were determined as enhanced mobile broadband, ultra-reliable and low latency communications and massive machine-type communications (mMTCs). mMTCs use cases mainly cover the very high number of connected devices that are required to be inexpensive with a very long battery life, resulting in low volume of non-delay sensitive data transmission [1]. In addition, the minimum requirement for connection density was set to be 1M devices/km^2 [2], or 1 device/m^2 on average.

Several 5G-related articles in the literature mention an expectation of a thousandfold growth in global mobile data traffic between 2010 and 2020, some of which refer to a 2011 white paper by Nokia Networks [3]. However, that document does not contain any scientific analysis. UMTS Forum's market study, which was also published in 2011, forecast 33 times growth for the same decade [4]. ITU also estimated 2020 traffic volumes, using different forecasts of external sources; their estimations are available in ITU-R M.2243. Taking the lowest and highest approximations into account and extrapolating them to 2020 yielded approximate growths of 25 and 100 times, respectively, and this span was accepted as the estimated range to assist future spectrum allocations. Expectations of 44- and 80-fold growth, corresponding to 25%–75% of the range, were then used to specify the exact 2020 market attributes for the lower and higher user density settings, respectively [5]. The accuracies of these forecasts can be checked using Cisco Systems' annual "Visual Networking Index" white papers. Cisco reported the total mobile data traffic as 0.237 EB/month in 2010 [6] and 7.241 EB/month in 2016, with a projection of 34.59 EB/month for 2020 [7]. These values result in growth estimates of 30.59 and 146.1 times in periods of 6 and 10 years, respectively.

The constant increase in the amount of Internet-connected user equipment (UE) already has a substantial congesting effect on the sub–6-GHz legacy bands. Achieving either of the IMT-2020s 20 Gb/s downlink (DL) or 10 Gb/s uplink peak data rate minimum requirements is challenging enough under ideal conditions, let alone under realistic test environments. The most researched and utilized technique to increase the data rate and network capacity has been spectral efficiency enhancement. However, because modulation and coding schemes (MCSs) are already vastly advanced, the returns of additional development efforts are diminishing. To illustrate, keeping the coding parameters the same, an ascent from binary phase-shift keying (BPSK) to quadrature PSK (QPSK) results in a doubling of the bit rate. However, the jump from 64-ary quadrature amplitude modulation (64-QAM) to 256-QAM leads to an increase of only 33%. A further hypothetical leap to 1024-QAM would result in an even lesser increase of 25% [8].

Another possible method to attain the target rates is increasing the operation bandwidth via transmitting in higher bands. This approach is yet to be commonly adopted, mainly due to the difficulties in generating signals with high enough output power, and higher free space and atmospheric attenuations in the millimeter-wave (mm-wave) band (30–300 GHz). However, the forthcoming device densification and subsequent decrease in mean link distances offer novel opportunities to wireless networks. Consequently, investigation of new frequency spectra for wireless communications have already started for 5G mobile communications concurrently with the respective standardization activities [9–12], and Chapter 8 describes the first steps toward a resolution of this predicament. The section continues by describing some of the foremost mm-wave 5G Internet of Things (IoT) usage scenarios. Section 8.3 describes the steps taken in wireless local area network (WLAN) standardization activities towards the initiation of mm-wave for 5G IoT purposes, and Section 8.4 explains the electromagnetic (EM) wave channel characteristics in the 30–300 GHz frequency range. Future research directions are outlined prior to the concluding remarks.

8.2 Applications of Millimeter-Wave 5G-Enabled Internet of Things

Mm-wave offers an array of opportunities for prominent IoT applications. Some important directions are listed as follows:

- The high data rates attainable in the mm-wave band are especially suitable for the applications that involve users in indoor scenarios. Additionally, the short range of mm-wave allows a higher frequency reuse and the device densification. In [13–15], the case for mm-wave and low-THz bands in such scenarios are presented, showing that very high data rates can reliably be supported. The authors of [16] further suggest that IoT can utilize the small size and cost efficiency offered in the mm-wave domain.

  The characterization of indoor mm-wave systems has been established by a number of publications. However, some of the recent studies show that these characteristics may vary significantly between various indoor areas. For instance, [17] shows that the properties of mm-wave in indoor environments, like factories which usually have a high number of metallic reflective surfaces, are quite different from other indoor scenarios. This issue is quite important because apart from houses, the IoT particularly targets industrial deployment of sensor networks that make the entire system efficient and smart.
- While many IoT applications target indoor deployment, mm-wave has interesting outdoor usages too. Primary among these is the Internet of Vehicles (IoV). This direction is very noteworthy because traffic data requirements are

increasing exponentially, and traditional networks may not be able to support the IoV once the era of driverless vehicles fully arrives. Moreover, these systems may be further modified to include applications of high-speed trains and autonomous vehicles in the future.

- Mm-wave based systems are envisioned to be the fronthauls of traditional networks because current technologies will not be able to satisfy the technical performance requirements of IMT-2020, apart from the latency. Preliminary work by [18] showed that the use of mm-wave in the discussed circumstances causes an 11% reduction of resources in comparison to common networks. Using these results, mm-wave can be envisioned as the interface between traditional and IoT networks to mitigate the issues related to interfacing and maintain small antenna size and high data rates for IoT devices.
- Many efforts for future IoT are directed toward the creation of smart networks. Smart IoT networks cover a vast range of applications that include smart business, education, agriculture, monitoring, security, living environment, health, traffic, grids and wearables. Having intelligence in IoT provides those networks with unique qualities such as heterogeneity and resilience against errors and attacks and the ability to conduct data rate management, advance analytics, spectrum management and latency reduction. To achieve these qualities in 5G, mm-wave networks with an artificial neural network component have been proposed [19].
- Easily producible and very low-cost mm-wave IoT applications have recently started appearing more commonly in the literature. Among these is the one described in [20], where the authors present inkjet-printable flexible mm-wave arrays that may be used in chipless radio frequency identification (RFID) implementations targeting smart skins. These skins can operate within relatively long distances, up to 30 m. Moreover, they also have the ability to support multitag and multisensing.

8.3 Millimeter-Wave IEEE 802.11 Standardization Activities

In an effort to develop WLAN physical layer and Media Access Control (MAC) specifications, since 1991, IEEE 802.11 has published 5 standards, 29 amendments, 2 recommended practices and 1 corrigendum in 27 years' time, with 5 amendments and 1 standard currently under development. After the second standard, P802.11-1999, the standard issuing practice turned into a periodic process of accumulated maintenance, where amendments published following the last standard are included in the current version. The penultimate standard, P802.11-2012, features the 10 amendments published since the previous release in 2007, whose details are available in Table 8.1.

Table 8.1 Incorporated Baselines of IEEE P802.11-2012

Project	*Title*	*Final Approval*
802.11s-2011	Mesh Networking	August 1, 2011
802.11u-2011	InterWorking with External Networks	February 25, 2011
802.11v-2011	Wireless Network Management	February 9, 2011
802.11z-2010	Extensions to Direct Link Setup	October 1, 2010
802.11p-2010	Wireless Access for the Vehicular Environment	July 15, 2010
802.11w-2009	Protected Management Frames	September 11, 2009
802.11n-2009	High Throughput	September 11, 2009
802.11y-2008	3650–3700 MHz Operation in United States	November 6, 2008
802.11r-2008	Fast Roaming	May 9, 2008
802.11k-2008	Radio Resource Measurement	May 9, 2008

Five amendments were merged to develop the current standard, P802.11-2016. The oldest one, 802.11ae-2012, is titled "Prioritization of Management Frames." Quality of service (QoS) mechanisms have been defined for IEEE 802.11 since the 2005 amendment, 802.11e-2005. 802.11ae extends this capability toward management frames. As the 802.11 continued to expand through the years, the number of management frame categories also expanded. Currently, 14 of the 16 possible management subtypes are allocated for various functions, whereas additional actions are set via the 1-byte-long action field [21]. This rise in the variation of management frames made those a rational candidate for a QoS-type mechanism implementation. In fact, the introduction of QoS management frame (QMF) service with 802.11ae is an attempt to meet this target. The QMF determines the enhanced distributed channel access (EDCA) access category (AC). Default QMF policy sets ACs for 32 of the management tasks to best effort (BE), video and voice, with BE also being the default category for the rest. QMF policy is also dynamic and can be changed between a station (STA) and a non-access point (non-AP) STA or an STA and an AP but only within infrastructure basic service sets (BSSs) or mesh BSSs.

The second oldest incorporated amendment is 802.11aa-2012. It obtained the final approval on June 1, 2012, with the full title "MAC enhancements for robust audio video streaming." Global data traffic is growing unendingly, and video streaming is no exception. Even though it is the largest global mobile traffic generator, holding a share of 60% as of 2016, video is still expected to outgrow the audio, data and file sharing categories for the period 2016–2021. The prediction is that video will achieve a Compound Annual Growth Rate (CAGR) of 54% until 2021, higher than the CAGR of the overall average mobile traffic, which is 47%; thus, increasing its share to 78% of the total mobile traffic [7].

Apart from a mere categorization [22], which appears rather coarse a decade after its introduction, no specific attention has been given to video streaming, and 802.11aa rectifies this error for WLANs. With this standard, many new policies are defined, one of them being for stream classification service (SCS). SCS allows finer classification of unicast MAC service data units such that voice and video streams can be buffered in primary or alternate transmit queues before being transmitted to the standard EDCA function. Another new mechanism is group-addressed transmission service (GATS). The aim is the improved delivery of multicast transmissions, and the retransmission policy of GATS can be set to groupcast with retries (GCR) unsolicited retry or GCR Block Ack, in addition to the directed multicast service, which was first identified in 802.11v.

802.11af-2013, "TV White Spaces Operation," was approved on December 11, 2013. Low frequency bands are valuable for many reasons, the foremost being the preferable transmission characteristics because material absorption is lower [23], atmospheric attenuation is nearly nonexistent, and power disperses less in free space [24–26]. Legacy bands are also more coveted because necessary communications infrastructure are already deployed and operational. The economic result of these facts is a spectrum that is nearly five times more valuable in 800 MHz compared to 2.6 GHz [27].

The main aim of 802.11af is better utilization of lower frequency bands, which are primarily licensed for broadcast TV. Historically, analog TV broadcasting has been performed on sub–1-GHz frequency bands. In Europe, the 470–790 MHz bandwidth was reserved for this purpose, whereas in the United States, sixty 6 MHz channels are placed intermittently in the 54- to 758-MHz range. Digital TV switchover frees part of this spectrum, and the remainder is generally underused. 802.11af has been devised to better utilize the spectrum allocated for broadcast TV but not assigned at a specific location, formally termed TV white space (TVWS).

The core component of the standard's new architecture is the geolocation database (GDB). To ensure the protection of licensed usage of TVWS, it is directed by regulatory domains and stores and distributes the white space maps (WSMs) of geographic locations. A WSM is a list that contains available channels and corresponding operation parameters, such as validity duration, maximum power level or channel bandwidth. Depending on the particular system design, which is not within the scope of the standard, GDB can either directly communicate with a geolocation database-dependent (GDD)–enabling STA that is the complement of an AP within 802.11af or through a registered location secure server (RLSS). Again, depending on the specific architecture, RLSS can convey the current channel usage states of all the BSSs it communicates with to GDBs, communicate with the controllers of other TVWS networks to coordinate their operations, or be queried by GDD STAs (which are the complements of regular STAs within 802.11af) about channel usage and WSMs.

The amendment with the closest ties to the 802.11ad is the 802.11ac-2013: "Enhancements for Very High Throughput for Operation in Bands below 6 GHz."

It was approved nearly a year later than 802.11ad, on December 11, 2013; however, both were originated from the same 802.11 Very High Throughput (VHT) Study Group (SG). VHT SG was originated in May 2007 to further develop the throughput advancements achieved by 802.11n-2009: "Enhancements for Higher Throughput." Operation in 40 MHz channels was introduced in 802.11n. The next step would be 80-MHz channels, but because there is not enough bandwidth within the whole 2.4 GHz band as it includes thirteen 5 MHz channels whose center frequencies uniformly increase from 2412 to 2472 MHz, 802.11ac focused solely on the 5-GHz band. This spectrum search of the SG also led to another standard: 802.11ad.

Setting the aim at even higher throughput, 802.11ac instituted many new features. Highest modulation is increased from 64-ary quadrature amplitude modulation (64-QAM) to 256-QAM. According to the new channelization, 80 and 160 MHz contiguous and 80 + 80 MHz non-contiguous channels are defined. DL multi-user multiple input–multiple output (DL-MU-MIMO) technique is introduced, allowing an AP with multiple antennas to transmit to multiple STAs over the same spectrum. In the standard, the maximum number of users and spatial streams per user are limited to four, whereas maximum total number of spectral streams is limited to eight for the DL-MU-MIMO. Putting together all the mechanisms, the VHT MCS of 256-QAM, 5/6 forward error correction coding rate, eight spatial streams and 400 ns guard interval over a 160-MHz contiguous or 80 + 80 MHz non-contiguous channel results in a maximum theoretical data rate of 6933.3 Mb/s.

To conclude, the currently in-process and published but unincorporated amendments are listed in Table 8.2.

Table 8.2 In-process and Published but Unincorporated IEEE 802.11 Amendments

Project	*Title*	*Final Approval*
802.11bb	Light Communications	***Predicted*** July 2021
802.11ba	Wake Up Radio	***Predicted*** September 2020
802.11az	Next Generation Positioning	***Predicted*** March 2021
802.11ay	Next Generation 60 GHz	***Predicted*** December 2019
802.11ax	High Efficiency WLAN	***Predicted*** December 2019
802.11aq	Pre-Association Discovery	June 14, 2018
802.11ak	General Link	March 8, 2018
802.11aj	China Millimeter Wave	January 24, 2018
802.11ah	Sub 1 GHz	December 7, 2016
802.11ai	Fast Initial Link Setup	December 7, 2016

8.4 Millimeter-Wave Channel Properties and Models

Modeling approaches for the mm-wave band can be broadly divided into two categories: Deterministic and stochastic channel models [28]. In deterministic channel models, propagation of individual EM waves is predicted in a detailed manner using propagation mathematics such as Maxwell's equations. Thus, these approaches have high computational complexities and require very detailed information from the propagation network. Because these models are directly based on the environment, they cannot be easily adapted to other scenarios. Typical examples of these models include ray tracing and map-based modeling schemes.

Contrariwise, in stochastic channel modeling, channel parameters are described using various probability distributions. Although these methods are generalizable and they reduce the computational complexity, their accuracy is lower in comparison with deterministic models. Stochastic models may be further split into two categories, according to whether or not the model is based on a network geometry. The selection of one or the other set of methods depends on the usage and intended applications. Hence, use case is significant for mm-wave 5G IoT channel modeling.

8.4.1 Millimeter-Wave Channel Measurement Scenarios

Measurement scenarios for mm-wave channels can very broadly be separated in two major categories: indoor and outdoor scenarios. Each of these sets has a large variety of applications targeting IoT, and they have their own special considerations. Detailed discussions of these are provided below.

8.4.1.1 Indoor Scenarios

Many IoT applications target indoor deployment scenarios in residential and work areas. Because the requirements of users to get the highest data rates typically occur in these environments, mm-wave communication is naturally considered for most 5G indoor IoT use cases.

A typical mm-wave indoor network is illustrated in Figure 8.1, where a central AP covers a room having an office floor outlook.

Key considerations and properties for indoor networks are noted below:

- The presence of opaque objects, such as walls, cause reflections to occur within the networks, giving rise to a large number of multipath components (MPCs).
- The geometry of the network is relatively stable in comparison to outdoor networks. Conversely, the stable geometry forces considering the architecture of the coverage area for accurate network modeling.

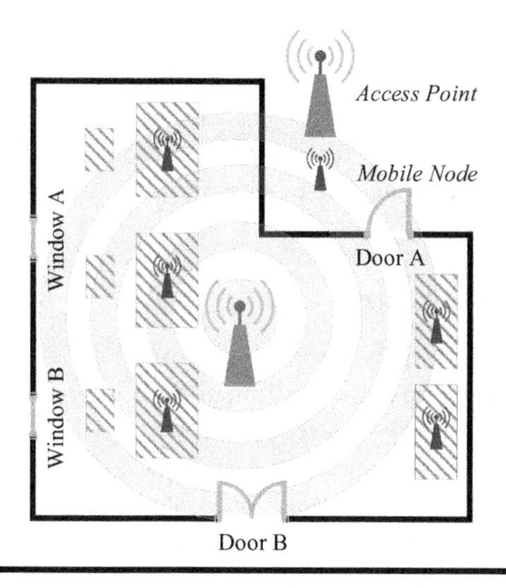

Figure 8.1 A typical indoor scenario for mm-wave deployment.

- UEs in indoor networks either remain stationary or move slowly, because their movements are constrained by the room boundaries. The same is also true for blocking objects.
- Power resources, such as electric power mains, are usually accessible or available to indoor IoT. Therefore, the power requirements are not very demanding.

The short range of mm-wave is especially suited for indoor networks. The presence of walls and other impervious objects ensure reflections within the confines of the network coupled with small absorption losses. Hence, a higher number of MPCs usually exists in indoor mm-wave scenarios. This effect can be both productive and destructive. It can be useful because the network area can be covered with less power and areas that are blocked by objects may be accessible as a result of the reflections. However, this allows for higher multipath fading within the channel, which may limit the overall channel utility. Thus, a balance is required between both of the considerations in order to achieve optimum network performance.

Similar trade-offs occur as a result of slow moving objects. While the slow speed ensures a smaller Doppler spread, the rapid decrease in signal-to-noise ratio or outages as a result of blocking objects may also last longer. Such scenarios can be managed by intelligent burst formatting; however, additional network resources will be required. Deterministic models are very accurate for indoor channels, especially when they are based on geometry. Yet, those cannot be generalized for all permutation of events.

Several key studies on indoor deployment of mm-wave communications have targeted the use of massive MIMO technology. The use of a very high number of antennas in massive MIMO make mm-wave systems resistant to interference and provide the network with a higher number of possible signal paths, and thus the performance in terms of data rate and link reliability improves. As a result, massive MIMO networks can multiply the capacity of a wireless connection without requiring more spectrum. Early reports point to considerable capacity improvements that could potentially increase by severalfold in the future [29,30].

8.4.1.2 Outdoor Scenarios

A large body of work targets mm-wave channels in outdoor environments. Although the ranges for mm-wave are limited for outdoor deployment over large distances, small cells with very high repeatability can be used for both UE and 5G IoT. Crowded urban areas are especially very important application scenarios that are often kept in consideration. In addition, key applications exist in locomotive communications such as vehicle-to-vehicle (V2V) communications [31] and high-speed train (HST) networks [32], which have essential uses in IoV. Typical V2V and HST networks are shown in Figure 8.2.

Some key characteristics and considerations for outdoor networks are reviewed below:

- The overall geometry of a typical coverage area is very dynamic, and the presence of obstacles may change more frequently than in the indoor case. Cell boundaries cannot be sharply defined because there are no walls around the coverage area, unlike the indoor network scenarios. However, the presence of buildings in urban scenarios can create large areas of blockage.
- The speed of the users is not limited by the area constraints, and thus higher average UE speeds can be expected.

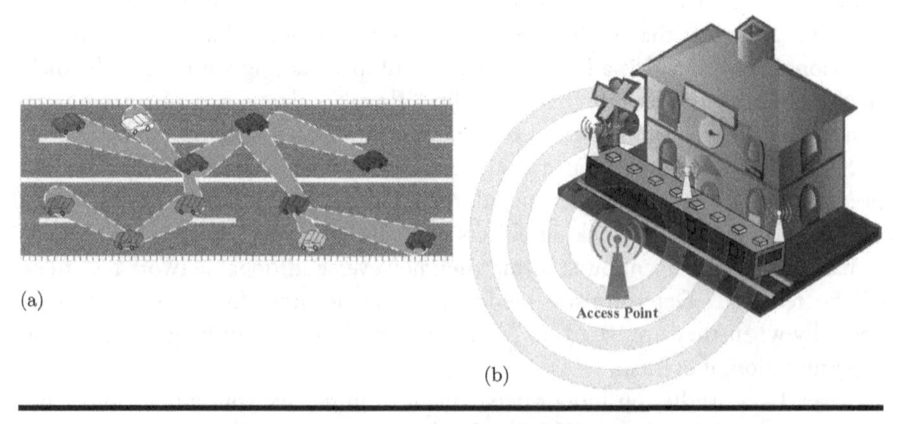

Figure 8.2 Outdoor measurement scenarios for mm-wave applications. (a) V2V communication network and (b) An exemplary HST network.

- Because outdoor 5G IoT hardware will have to be deployed independently away from a power source in many of the cases, power is a key restraint on these devices. They should be able to harvest energy or their power consumption should be kept minimal [33,34].

Probabilistic models can be used to evaluate an expected behavior for outdoor networks more easily because the network geometry is active. This ensures a wide range of applications as well as simpler computation. However, the challenges associated with typical outdoor communications may not be easy to deal with for mm-wave frequencies. This is especially true if large blocking objects, such as buildings, are present in a coverage area such as a conventional urban setting. Considering IoV applications, outdoor communications also suffer from the effects of other large-scale fading such as Doppler spread, because the users are moving at high speeds with reference to both each other and their environment. These effects are discussed further in Section 8.4.2. Finally, when the short range of mm-wave communications is coupled with the low power usage requirement of outdoor 5G IoT applications, an opportunity for high reusability arises. This ensures a more efficient use of frequency resources. Consequently, this requirement also warrants devices with smaller sizes.

8.4.2 Millimeter-Wave Channel Parameters

Based on the discussed scenarios, typical channel parameters for mm-wave spectra have been reported extensively in the literature. Because these channel parameters form the basis of design for any mm-wave 5G IoT application, their study is very important [35]. Therefore, some of the key parameters and their significance for the mm-wave 5G IoT is examined.

8.4.2.1 Typical Measurement Setups

The diagram of a typical mm-wave band measurement setup is provided in Figure 8.3. On the transmitter side (TX), a baseband signal, that is generated by an arbitrary waveform generator (AWG), is mixed with a high frequency signal via an extender and is then transmitted over the channel. On the receiver end (RX), the received signal is demodulated using a similar frequency extension setup, and the overall results are analyzed on either a type of scope or a computer. Normally, after a large set of measurements, considerations of geometry and deployment are used to evaluate the required channel parameters.

The functions for the AWG of the TX and the scope of the RX can be combined by using a vector network analyzer (VNA). The use of a single VNA offers both advantages and disadvantages. More accurate models and excellent synchronization between various components can be listed as some of the pros. However, a single

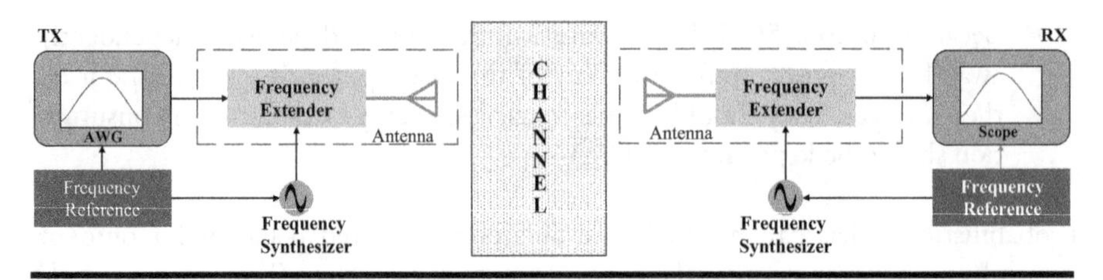

Figure 8.3 Key components of a typical mm-wave band measurement setup.

device cannot be used over large networks because the manipulation of the cables themselves can become a major issue. Thus, it is recommended to use a VNA for small indoor networks, whereas for larger outdoor networks, a deconstructed VNA or separate set of TX and RX are required.

Highly directional antennas are commonly used on both the TX and RX to obtain additional gain for the communication link. MIMO and massive MIMO systems make use of multiple directional antenna arrays to achieve better performance [29]. Several massive MIMO experiments make use of virtual MIMO arrays that also give a good estimate for the behavior of MIMO systems with smaller setups [36,37].

For the channel itself, a number of options can be utilized. For instance, measurements can be performed in any one of the scenarios discussed earlier for various cases under controlled or uncontrolled environments. A number of studies use anechoic chambers to conduct measurements so that the intricacies of a particular scenario can be deciphered without external interference. The movement of the TX and RX can be emulated using linear movement and rotational stages. Because the mm-wave band covers a large number of frequencies from 30 to 300 GHz, various frequency extenders can be used to reach a large number of distinct operational ranges. Currently 28, 38, 60, 71–76 and 81–86 GHz frequency bands are the conventional targets for mm-wave measurement studies [28].

8.4.2.2 Channel Impulse Response

The process to describe the channel impulse response (CIR) for a mm-wave band channel is by providing the channel with an impulse and recording its response. The impulse response is a useful characterization of the channel, because it may be used to predict and compare the performance of many different mobile communication systems and transmission bandwidths for a particular mobile channel condition. Most high-level channel estimations are based on the calculation of CIR. Because the frequencies will be high for mm-wave 5G IoT systems, impulse responses will usually have high attenuation. In addition, atmospheric absorption causes some additional losses in the mm-wave band channels too.

8.4.2.3 Reflection and Diffractions

Reflection and diffractions are very important parameters from a modeling perspective because they describe how the network area may be covered efficiently and point out key limitations of a particular scenario. In [38], the authors conducted measurement campaigns for indoor and outdoor scenarios of a 28-GHz channel in the New York City. The key focus of the measurements was on reflection coefficients and penetration losses for different materials in these scenarios. Their results showed that outdoor materials, such as concrete, usually have higher reflection coefficients compared to indoor materials such as clear glass. Furthermore, penetration loss for outdoor materials such as tinted glass is generally higher than its indoor counterpart clear glass. The high reflection coefficients and penetration losses of outdoor materials make the boundaries between indoor and outdoor networks very critical. Therefore, hand-offs between any two such networks should be specifically considered in deployments.

Diffractions assist signal EM waves to travel behind objects and thus reduce losses due to shadowing. The high frequency and short wavelength of mm-wave band results in weak diffractions that makes mm-wave spectra exhibit properties similar to optical waves. Thus, most of the propagation power is contributed by the line-of-sight components along with some low order reflected components [39]. These show that outdoor mm-wave scenarios face a big challenge as diffraction-assisted communications may not be efficiently operable.

8.4.2.4 Atmospheric Attenuation

Recommendation ITU-R P.676-9 offers two calculation methods for the atmospheric attenuation. The first and the exact one is a summation of the main factors, which are the resonance lines of oxygen and water vapor, and secondary factors. The resulting attenuations are shown in Figure 8.4 up to 500 GHz. The other technique possesses less calculation complexity and provides an approximation of the correct results calculable by the first method within the 1–350 GHz range.

Accepting the outcomes from the first method as true values and the second method as estimations, Figure 8.5 provides the absolute and squared errors at each point spaced at 1-MHz intervals up to 350 GHz, together with the absolute error percentage and mean squared error, which is computed to be 0.007035 dB/km, again, throughout the range.

The maximum difference between the two methods occur at 59.16 GHz as 0.7685 dB/km. As expected, local maxima of the errors and gaseous attenuation overlap. Absolute error percentage, which is calculated by dividing the absolute error to the true attenuation value at each point, is at most 10.7454% at 53.007 GHz and generally around or below 1%. Therefore, the second and computationally less intensive method can be preferred to the first one everywhere within the 350 GHz range, except for the 60 GHz band, where the absolute error peaks and the error percentage average over the 9 GHz spectrum escalates to 2.7712%.

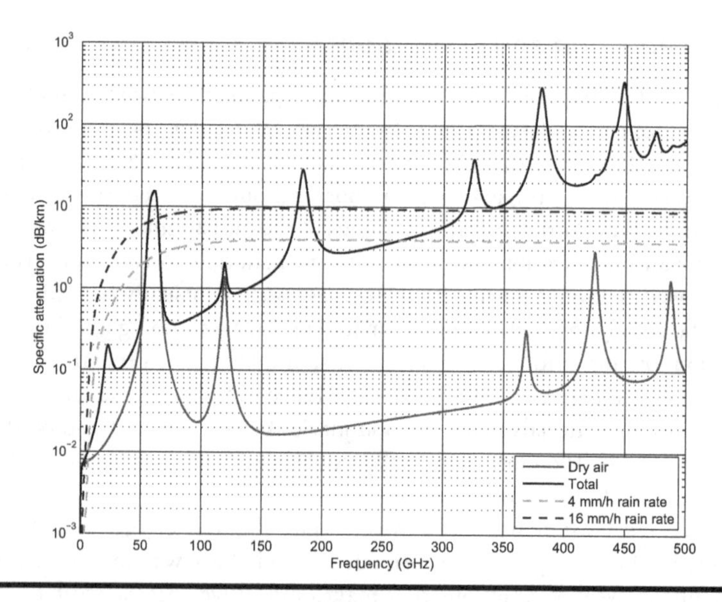

Figure 8.4 Attenuation due to atmospheric gases and rain, calculated up to 500 GHz and for rain rates of 4 and 16 mm/h, as per ITU-R P.676-9 and P.838-3.

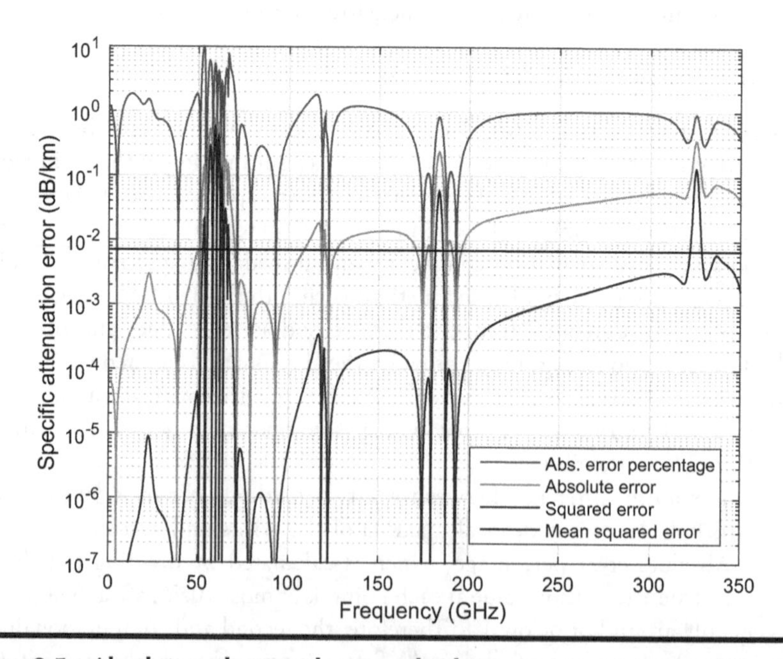

Figure 8.5 Absolute and squared errors, absolute error percentage and the mean squared error between the two atmospheric attenuation calculation methods provided in the ITU-R P.676.

8.5 Future Research Directions

5G mm-wave IoT systems provide a large number of opportunities for current and future applications. Some of the related significant research areas are noted below:

- Various modeling techniques developed for mm-wave systems can be merged together to bring forth newer schemes. For instance, a trade-off between accuracy and computational complexity can be based on the combination of deterministic and stochastic channel modeling techniques such as in [40]. These hybrid models are both flexible and provide scalable alternatives to both of the methods.
- So far, most of the studies on mm-wave IoT have focused on channel modeling and very few target the network or data link layers. In this regard, networking schemes, power allocation in networks, network mapping, burst formatting, scheduling, latency management and so forth are key issues from these layers that require additional studies.
- THz band (0.1–10 THz) offers very high data rates. However, its range of operation is very limited [41,42]. One possible use of it is high speed interfaces for 5G systems operating on mm-wave technology, provided appropriate interface technologies can be designed. Apart from this, the modeling techniques developed for mm-wave technology can be easily adapted to THz communications because the frequencies for both of the spectra are adjacent.
- Wireless networks can significantly improve on board communications and help create room in tight indoor areas of a spaceship. Moreover, many applications of space sensors require the use of data rates on the orders offered by 5G mm-wave networks. The technology is ready for deployment of 5G mm-wave IoT-connected sensors on board in space shuttles and stations. Furthermore, these can be used in future space explorations in the solar system and in deep space.
- Because nanomachines are very simplistic in nature, to perform complex tasks, they need to operate cooperatively by forming networks with each other. Molecular communication offers the most feasible way to realize this type of nanonetworks [43,44]. Many future Internet of Bio-Nano Things (IoBNT) applications are based on the concept of molecular communications [45]. However, most of these applications currently do not interface to macro-scale systems, and the few that do are very limited in performance [46]. Therefore, a number of applications that improve such interfaces between 5G IoT and IoBNT systems can be envisioned.

8.6 Conclusions

The continuously increasing user demand for higher data rate and network capacity necessitates the utilization of the mm-wave frequencies in addition to the sub–6-GHz bands, for 5G-enabled IoT applications. In line with this, IEEE has approved the project authorization request for the "Very High Throughput in the 60 GHz Band" amendment on December 10, 2008, and has officially begun the WLAN standardization activities in the mm-wave band. EM wave propagation characteristics of the mm-waves are notably different from the legacy wireless communications spectra. Finally, many open research issues remain to be solved to enable widespread mm-wave 5G IoT adoption by the general public.

Acknowledgment

This work was supported in part by the Scientific and Technological Research Council of Turkey (TUBITAK) under grant #113E962.

References

1. Recommendation ITU-R M.2083-0: IMT Vision—Framework and overall objectives of the future development of IMT for 2020 and beyond. *ITU, Geneva, CHE*, 2015.
2. Recommendation ITU-R M.2410-0: Minimum requirements related to technical performance for IMT-2020 radio interface(s). *ITU, Geneva, CHE*, 2017.
3. 2020: Beyond 4G—Radio evolution for the gigabit experience. Report, Nokia Networks, 2011.
4. Mobile traffic forecasts 2010–2020: Report, UMTS Forum, 2011.
5. Report ITU-R M.2290: Future spectrum requirements estimate for terrestrial IMT. *ITU, Geneva, CHE*, 2014.
6. Cisco Visual Networking Index: Global mobile data traffic forecast update, 2010–2015. Report, Cisco Systems, Inc., 2011.
7. Cisco Visual Networking Index: Global mobile data traffic forecast update, 2016–2021. Report, Cisco Systems, Inc., 2017.
8. T. Yilmaz, G. Gokkoca, and O. B. Akan. *Millimetre Wave Communication for 5G IoT Applications*, pp. 37–53. Springer International Publishing, Cham, Switzerland, 2016.
9. T. Yilmaz and O. B. Akan. *Millimeter-Wave Communications for 5G Wireless Networks*, pp. 425–440. CRC Press, 2016. doi:10.1201/b19698-20.
10. T. Yilmaz. *Advanced Image Coding Algorithms: Beyond JPEG2000*. MSc thesis, Dept. Electron. Elect. Eng., Univ. College London, London, UK, 2009.
11. T. Yilmaz. *On the Use of Low Terahertz Band for Wireless Communications*. Ph. D. dissertation, Dept. Elect. Electron. Eng., Koç Univ., Istanbul, Turkey, 2018.
12. O. Erturk and T. Yilmaz. A hexagonal grid based human blockage model for the 5G low terahertz band communications. In *2018 IEEE 5G World Forum (5GWF)*, pp. 395–398, 2018. https://ieeexplore.ieee.org/abstract/document/8516978.
13. T. Yilmaz, E. Fadel, and O. B. Akan. Employing 60 GHz ISM band for 5G wireless communications. In *2014 IEEE International Black Sea Conference on Communications and Networking (BlackSeaCom)*, pp. 77–82, 2014. https://ieeexplore.ieee.org/abstract/document/6849009.

14. T. Yilmaz and O. B. Akan. Utilizing terahertz band for local and personal area wireless communication systems. In *2014 IEEE 19th International Workshop on Computer Aided Modeling and Design of Communication Links and Networks (CAMAD)*, pp. 330–334, 2014. https://ieeexplore.ieee.org/abstract/document/7033260.
15. T. Yilmaz and O. B. Akan. On the use of the millimeter wave and low terahertz bands for Internet of Things. In *2015 IEEE 2nd World Forum on Internet of Things (WF-IoT)*, pp. 177–180, 2015. https://ieeexplore.ieee.org/abstract/document/7389048.
16. S. Saponara and B. Neri. mm-wave integrated wireless transceivers: Enabling technology for high bandwidth connections in IoT. In *2015 IEEE 2nd World Forum on Internet of Things (WF-IoT)*, pp. 149–153, 2015. https://ieeexplore.ieee.org/abstract/document/7389043.
17. D. Solomitckii, A. Orsino, S. Andreev, Y. Koucheryavy, and M. Valkama. Characterization of mm-wave channel properties at 28 and 60 GHz in factory automation deployments. In *2018 IEEE Wireless Communications and Networking Conference (WCNC)*, pp. 1–6, 2018. https://ieeexplore.ieee.org/abstract/document/8377337.
18. M. Artuso, A. Marcano, and H. Christiansen. Cloudification of mm-wave-based and packet-based fronthaul for future heterogeneous mobile networks. *IEEE Wireless Communications*, 22(5): 76–82, 2015.
19. R. Li, Z. Zhao, X. Zhou, G. Ding, Y. Chen, Z. Wang, and H. Zhang. Intelligent 5G: When cellular networks meet artificial intelligence. *IEEE Wireless Communications*, 24(5): 175–183, 2017.
20. J. G. D. Hester and M. M. Tentzeris. Inkjet-printed flexible mm-wave van-atta reflectarrays: A solution for ultralong-range dense multitag and multisensing chipless RFID implementations for IoT smart skins. *IEEE Transactions on Microwave Theory and Techniques*, 64(12): 4763–4773, 2016.
21. IEEE Standard for Information technology–Telecommunications and information exchange between systems Local and metropolitan area networks–Specific requirements Part 11: Wireless LAN Medium Access Control (MAC) and Physical Layer (PHY) Specifications. In *IEEE Std 802.11-2012 (Revision of IEEE Std 802.11-2007)*, pp. 1–2793, 2012. https://ieeexplore.ieee.org/document/6178212.
22. G. Daqing and Z. Jinyun. QoS enhancement in IEEE 802.11 wireless local area networks. *IEEE Communications Magazine*, 41(6): 120–124, 2003.
23. W. C. Stone. NIST construction automation program Report No. 3: Electromagnetic signal attenuation in construction materials. In *Building Fire Res. Lab., Nat. Inst. Standards Technol., Gaithersburg, MD, Tech. Rep. NISTIR*, 6055, 1997. https://www.nist.gov/publications/electromagnetic-signal-attenuation-construction-materials.
24. T. Yilmaz and O. B. Akan. On the use of low terahertz band for 5G indoor mobile networks. *Computers & Electrical Engineering*, 48: 164–173, 2015.
25. T. Yilmaz and O. B. Akan. State-of-the-art and research challenges for consumer wireless communications at 60 GHz. *IEEE Transactions on Consumer Electronics*, 62(3): 216–225, 2016.
26. T. Yilmaz and O. B. Akan. On the 5G wireless communications at the low terahertz band. *arXiv preprint arXiv:1605.02606*, 2016.
27. Spectrum value of 800 MHz, 1800 MHz and 2.6 GHz: Report, DotEcon Ltd., Aetha Consulting Ltd., 2012.
28. C. Wang, J. Bian, J. Sun, W. Zhang, and M. Zhang. A survey of 5G channel measurements and models. *IEEE Communications Surveys & Tutorials*, pp. 1–1, 2018.

29. J. Huang, C. Wang, R. Feng, J. Sun, W. Zhang, and Y. Yang. Multi-frequency mm-wave massive MIMO channel measurements and characterization for 5G wireless communication systems. *IEEE Journal on Selected Areas in Communications*, 35(7): 1591–1605, 2017.
30. O. Martnez, E. De Carvalho, and J. Nielsen. Towards very large aperture massive MIMO: A measurement based study. In *2014 IEEE Globecom Workshops (GC Wkshps)*, pp. 281–286, 2014. https://ieeexplore.ieee.org/document/7063445.
31. R. He, A. F. Molisch, F. Tufvesson, Z. Zhong, B. Ai, and T. Zhang. Vehicle-to-vehicle propagation models with large vehicle obstructions. *IEEE Transactions on Intelligent Transportation Systems*, 15(5): 2237–2248, 2014.
32. C. Wang, A. Ghazal, B. Ai, Y. Liu, and P. Fan. Channel measurements and models for high-speed train communication systems: A survey. *IEEE Communications Surveys & Tutorials*, 18(2): 974–987, 2016.
33. N. Khalid, T. Yilmaz, and O. B. Akan. Energy-efficient modulation scheme for THz-band 5G femtocell internet of things. In *2017 International Balkan Conference on Communications and Networking (BalkanCom)*, 2017. http://www.balkancom.info/2017/index.html.
34. N. Khalid, T. Yilmaz, and O. B. Akan. Energy-efficient modulation and physical layer design for low terahertz band communication channel in 5G femtocell Internet of Things. *Ad Hoc Networks*, 79: 63–71, 2018.
35. T. S. Rappaport, S. Sun, R. Mayzus, H. Zhao, Y. Azar, K. Wang, G. N. Wong, J. K. Schulz, M. Samimi, and F. Gutierrez Jr. Millimeter wave mobile communications for 5G cellular: It will work! *IEEE Access*, 1: 335–349, 2013.
36. S. Payami and F. Tufvesson. Channel measurements and analysis for very large array systems at 2.6 GHz. In *2012 6th European Conference on Antennas and Propagation (EUCAP)*, pp. 433–437, 2016. https://ieeexplore.ieee.org/abstract/document/6206345.
37. J. Chen, X. Yin, X. Cai, and S. Wang. Measurement-based massive MIMO channel modeling for outdoor LoS and NLoS environments. *IEEE Access*, 5: 2126–2140, 2017.
38. H. Zhao, R. Mayzus, S. Sun, M. Samimi, J. K. Schulz, Y. Azar, K. Wang, G. N. Wong, F. Gutierrez, and T. S. Rappaport. 28 GHz millimeter wave cellular communication measurements for reflection and penetration loss in and around buildings in New York city. In *2013 IEEE International Conference on Communications (ICC)*, pp. 5163–5167, 2013. https://ieeexplore.ieee.org/document/6655403.
39. A. Maltsev, R. Maslennikov, A. Sevastyanov, A. Lomayev, and A. Khoryaev. Statistical channel model for 60 GHz WLAN systems in conference room environment. In *2010 Proceedings of the Fourth European Conference on Antennas and Propagation (EuCAP)*, pp. 1–5, 2010. https://ieeexplore.ieee.org/document/5504964.
40. V. Nurmela et al. METIS Channel Models. Report, 2015.
41. N. Khalid, N. A. Abbasi, and O. B. Akan. 300 GHz broadband transceiver design for low-THz band wireless communications in indoor internet of things. In *2017 IEEE International Conference on Internet of Things (iThings) and IEEE Green Computing and Communications (GreenCom) and IEEE Cyber, Physical and Social Computing (CPSCom) and IEEE Smart Data (SmartData)*, pp. 770–775, 2017. https://ieeexplore.ieee.org/document/8276837.

42. N. Khalid, N. A. Abbasi, and O. B. Akan. Capacity and coverage analysis for FD-MIMO based THz band 5G indoor internet of things. In *2017 IEEE 28th Annual International Symposium on Personal, Indoor, and Mobile Radio Communications (PIMRC)*, pp. 1–7, 2017. https://ieeexplore.ieee.org/document/8292725.
43. O. B. Akan, H. Ramezani, T. Khan, N. A. Abbasi, and M. Kuscu. Fundamentals of molecular information and communication science. *Proceedings of the IEEE*, 105(2): 306–318, 2017.
44. N. A. Abbasi and O. B. Akan. A queueing-theoretical delay analysis for intra-body nervous nanonetwork. *Nano Communication Networks*, 6(4): 166–177, 2015.
45. N. A. Abbasi and O. B. Akan. An information theoretical analysis of human insulin-glucose system toward the internet of bio-nano things. *IEEE Transactions on NanoBioscience*, 16(8): 783–791, 2017.
46. N. A. Abbasi, D. Lafci, and O. B. Akan. Controlled information transfer through an in vivo nervous system. *Scientific Reports*, 8(1): 2298, 2018.

Chapter 9

Algorithms and Performance Analysis for Narrowband Internet of Things and Broadband Long-Term Evolution Coexisting System

Bowen Yang, Lei Zhang, Yansha Deng,
Deli Qiao, and Muhammad Imran

Contents

9.1 Introduction

It is predicted that by 2021 there will be more than 1 billion cellular-based IoT devices in the world [1]. To cater to that huge market, NB-IoT was developed by the Third Generation Partnership Project (3GPP) as a sustainable technology to support the connections among these billions of devices. With only 180 kHz operation bandwidth, the NB-IoT device applies energy-efficient modules and bears only 15% complexity compared with the normal LTE UE, which brings ultra low-cost (5 USD per device) and up to 10 years of battery life [2,3].

In order to achieve the possibility of quick deployment, the NB-IoT is designed to operate on existing cellular networks, for example, the Evolved Universal Mobile Telecommunications System Terrestrial Radio Access (E-UTRA) and Global System for Mobile Communication (GSM). In addition, three different modes could be assigned for NB-IoT operation—stand-alone mode, in-band mode, and guard-band mode—so that the spectrum resource could be utilized efficiently and flexibly. The stand-alone mode is shown in Figure 9.1a, where the frequency band for GSM as well as the scattered spectrum for potential IoT deployment will be allocated for NB-IoT operation. The in-band mode, shown in Figure 9.1b, allows NB-IoT devices to utilize the resource blocks (RBs) within LTE carriers, whereas in the guard-band mode, shown in Figure 9.1c, NB-IoT devices will be assigned to operate on the LTE carriers' guard band [3]. The in-band and guard-band modes enable the reuse of the LTE BS only by updating their software. Nevertheless, for the sake of low cost (NB-IoT devices have much lower sampling rate than that of LTE BS), the reuse of the LTE BS radio frequency (RF) and baseband processing chain may destroy the orthogonality of the orthogonal frequency division multiplexing (OFDM) system. In consequence, the extensively used algorithms (e.g., channel equalization and synchronization) and performance analysis method might no longer be valid.

Most of the NB-IoT–related research reported in the literature has focused on frame structure design [4], physical layer analysis [5,6], random access network

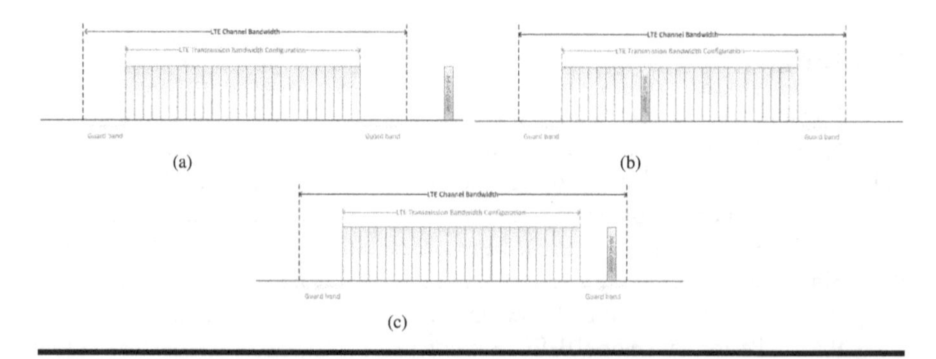

Figure 9.1 Operation modes of NB-IoT: (a) standalone operation, (b) in-band operation, and (c) guard-band operation.

Table 9.1 List of Notations

Notation	*Meaning*
$\{\cdot\}^H$	Hermitian conjugate
$\{\cdot\}^T$	Transpose operation
$\mathbb{E}\{\mathbf{A}\}$	Expectation of matrix **A**
diag{**A**}	Form a diagonal matrix by taking the diagonal elements of matrix **A**
diag{**a**}	Form a diagonal matrix taking each elements of vector **a**
$\mathbf{I}_M$	M length identity matrix
$\mathbf{0}_{M\times N}$	$M \times N$ zero matrix

[7,8] and scheduling [9], for example. In addition, studies evaulating the co-existence between different slices (i.e., services with different network configurations) have also been performed (e.g., [10] and [11] proposed the concept of multi-service, which can accommodate services with different RF and baseband configurations). Furthermore, performance analysis on inter-slice interference and cancellation have also been proposed in [12] and [13].

For the purpose of investigating how the mismatched sampling rate between NB-IoT and LTE could affect the coexisting system, a comprehensive uplink system model is described in this chapter, based on which the mathematical expressions of the equivalent channel frequency response (CFR) for NB-IoT and LTE are derived, and the interference between the two services is analyzed. In addition, an arbitrary sample duration (or sample length) for the NB-IoT device is considered to construct the foundation of sample duration optimization. The meanings of notations used in this chapter are listed in Table 9.1.

9.2 System Model

The uplink system model for LTE and NB-IoT coexistence is shown in Figure 9.2. In order to differentiate the parameters of two services, different subscripts are applied for NB-IoT ($\{\cdot\}_I$) and LTE ($\{\cdot\}_L$).

In Figure 9.2, the sampling rate S_L for LTE UE and BS follows the normal LTE RF and baseband configuration, whereas the NB-IoT UE bears a much smaller sampling rate S_I to reduce complexity and energy consumption. Without losing generality, it is assumed one RB is allocated for each service, and each RB contains M subcarriers. The non-overlapping subcarrier indices of LTE and NB-IoT

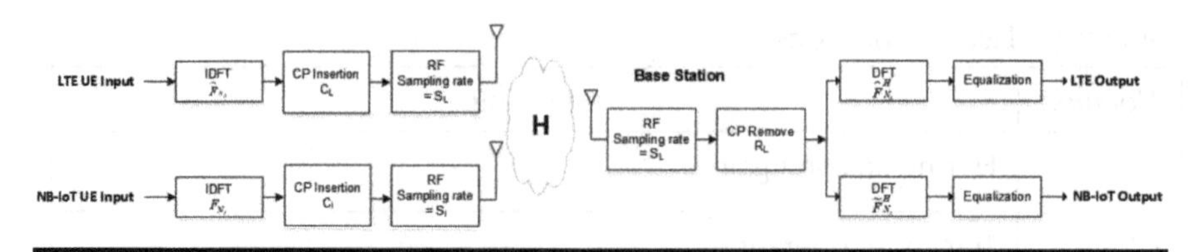

Figure 9.2 Uplink LTE/NB-IoT coexistence system.

are $\{M_L, M_L+1, \cdots, M_L+M-1\}$ and $\{M_I, M_I+1, \cdots, M_I+M-1\}$, respectively. Hence, the guard band between them can be calculated as

$$B_G = |M_I - M_L| - M. \tag{9.1}$$

Another important parameter is the up-sampling rate, G, for the NB-IoT device (i.e., when the low sampling rate NB-IoT signals are received at the high sampling rate LTE BS). By defining N_L and N_I as the Discrete Fourier Transform (DFT) size of LTE and NB-IoT, respectively, the up-sampling rate can be calculated as $G = N_L / N_I$ with $G \in \mathbb{N}^+$ for derivation simplicity. Furthermore, to avoid introducing trivial subcarrier mapping/selection and focusing on mathematical analysis, it is assumed that $N_I = M$. However, it should be noticed the results are still valid when $N_I > M$.[1]

If we define $\mathbf{x}_L$ and $\mathbf{x}_I$ as the modulated LTE and NB-IoT input signals, after IDFT processing, they can be written as $\hat{\mathbf{F}}_{N_L}\mathbf{x}_L$ and $\mathbf{F}_{N_I}\mathbf{x}_I$, respectively. $\mathbf{F}_{N_L}$ and $\mathbf{F}_{N_I}$ are the corresponding normalized Inverse Discrete Fourier Transform (DFT) matrices of LTE and NB-IoT, respectively, and $\hat{\mathbf{F}}_{N_L}$ takes the $M \in \{M_L, M_L+1, \cdots, M_L+M-1\}$ columns of $\mathbf{F}_{N_L}$. For the addition of the cyclic prefix (CP), we can form the CP insertion matrices as

$$\mathbf{C}_I = [\mathbf{0}_{L_{CP,I} \times (N_I - L_{CP,I})}, \mathbf{I}_{L_{CP,I}}; \mathbf{I}_{N_I}] \tag{9.2}$$

$$\mathbf{C}_L = [\mathbf{0}_{L_{CP,L} \times (N_L - L_{CP,L})}, \mathbf{I}_{L_{CP,L}}; \mathbf{I}_{N_L}] \tag{9.3}$$

where $L_{CP,L}$ and $L_{CP,I}$ are the *CP* lengths of LTE and NB-IoT, respectively, in a unit of their corresponding sample duration. In order to eliminate the inter-symbol interference, it is assumed that both $L_{CP,L}$ and $L_{CP,I}$ are larger than the delay spread of their channels. Hence, we can represent the transmitted signals for both services as

$$\mathbf{x}_{I,tx} = \mathbf{C}_I \mathbf{F}_{N_I} \mathbf{x}_I + \tilde{\mathbf{n}}_I, \tag{9.4}$$

[1] According to the Nyquist sampling theorem, N_I should be no less than M.

$$\mathbf{x}_{L,tx} = \mathbf{C}_L \hat{\mathbf{F}}_{N_L} \mathbf{x}_L + \tilde{\mathbf{n}}_L, \tag{9.5}$$

where the terms $\tilde{\mathbf{n}}_L$ and $\tilde{\mathbf{n}}_I$ are white noises.

The channel impulse response (CIR) of LTE and NB-IoT can be denoted as $\mathbf{h}_L \in \mathbb{C}^{1 \times L_{CH,L}}$ and $\mathbf{h}_I \in \mathbb{C}^{1 \times L_{CH,I}}$, respectively, in which $L_{CH,L}$ and $L_{CH,I}$ are their corresponding channel lengths. Without losing generality, it is assumed that the two channels are power normalized, that is,

$$\mathbb{E}\left\{\sum_{i=1}^{L_{CH,L}} |h_L(i)|^2\right\} = \mathbb{E}\left\{\sum_{i=1}^{L_{CH,I}} |h_I(i)|^2\right\} = 1. \tag{9.6}$$

Because the convolution of the transmitted signal and its CIR is equivalent to multiplex the signal with the corresponding Toeplitz matrix, we define $\mathbf{A}_I$ and $\mathbf{A}_L$ to be the two Toeplitz matrices comprising of $\mathbf{h}_I$ and $\mathbf{h}_L$, respectively.

At the receiver side, as the BS processes the two signals with the same RF processing chain, the low sampling rate NB-IoT signal would be up-sampled immediately when it arrives at the BS. Then, after CP removal and FFT processing, the output NB-IoT and LTE signals can be written as

$$\mathbf{y}_I = \tilde{\mathbf{F}}_{\mathbf{N_L}}^{\mathbf{H}} \mathbf{R}_L \Phi_I \mathbf{U} \mathbf{A}_I \mathbf{C}_I \mathbf{F}_{\mathbf{N_I}} \mathbf{x}_I + \mathbf{v}_I + \mathbf{n}_I, \tag{9.7}$$

$$\mathbf{y}_L = \hat{\mathbf{F}}_{\mathbf{N_L}}^{\mathbf{H}} \mathbf{R}_L \mathbf{A}_L \mathbf{C}_L \hat{\mathbf{F}}_{\mathbf{N_L}} \mathbf{x}_L + \mathbf{v}_L + \mathbf{n}_L, \tag{9.8}$$

where the first terms in Eqs. 9.7 and 9.8 are the desired NB-IoT and LTE signals, and the second and third terms are interference and random Gaussian noises, respectively. Specifically, for the desired NB-IoT signal, $\mathbf{U}$ is the up-sampling matrix with the size of $(N_L + L_{CP,L}) \times (N_I + L_{CP,I})$. By setting $\mathbf{U}$ in an appropriate way, we can easily achieve the arbitrary NB-IoT sample duration. Here we denote b as the NB-IoT sample length in a unit of LTE sample length,[2] hence we can formulate the matrix $\mathbf{U}$ as

$$\mathbf{U} = \begin{pmatrix} B & 0 & \cdots & 0 \\ 0 & B & \cdots & 0 \\ \vdots & \vdots & \ddots & \vdots \\ 0 & 0 & \cdots & B \end{pmatrix}, \tag{9.9}$$

[2] Hence, the square wave duty cycle of NB-IoT signal is b/G. Here we assume b is an integer to simplify the derivation, but the results are still valid if b is a non-integer value.

where $B =_{1\times b}^{T} \underbrace{[1,1,\cdots,1}{},0,0,\cdots]_{1}^{T} \times$ G. $\mathbf{\Phi}_I \in \mathbb{C}^{(N_L+L_{CP,L})\times(N_L+L_{CP,L})}$ is a phase shift matrix to move the NB-IoT signal to its assigned subcarriers. The ith diagonal element of $\mathbf{\Phi}_I$ is

$$\Phi_I(i) = e^{j2\pi(i-L_{CP,L}-1)(M_I-1)/N_L}. \tag{9.10}$$

Apart from the above two NB-IoT dedicated parameters, $\mathbf{R}_I$ and $\mathbf{R}_L$ are the CP removal matrices that can be formed as follows:

$$\mathbf{R}_I = [\mathbf{0}_{N_I \times L_{CP,I}}, \mathbf{I}_{N_I}], \tag{9.11}$$

$$\mathbf{R}_L = [\mathbf{0}_{N_L \times L_{CP,L}}, \mathbf{I}_{N_L}]. \tag{9.12}$$

$\hat{\mathbf{F}}_{N_L}^H$ and $\tilde{\mathbf{F}}_{N_L}^H$ are two sub-matrices of $\mathbf{F}_{N_L}^H$, where the former one takes the $M \in \{M_L, M_L+1,\cdots,M_L+M-1\}$ rows of $\mathbf{F}_{N_L}^H$ and the latter one takes the $M \in \{M_I, M_I+1,\cdots,M_I+M-1\}$ rows of $\mathbf{F}_{N_L}^H$.

Similarly, the interference terms can be represented as

$$\mathbf{v}_I = \tilde{\mathbf{F}}_{N_L}^H \mathbf{R}_L \mathbf{A}_L \mathbf{C}_L \hat{\mathbf{F}}_{N_L} \mathbf{x}_L, \tag{9.13}$$

$$\mathbf{v}_L = \hat{\mathbf{F}}_{N_L}^H \mathbf{R}_L \mathbf{\Phi}_I \mathbf{U} \mathbf{A}_I \mathbf{C}_I \mathbf{F}_{N_I} \mathbf{x}_I. \tag{9.14}$$

9.3 Desired Signal and Interference Analysis

In this section, we first derive the general expressions of the received NB-IoT and LTE signals by setting an arbitrary $b \in [1, G]$, then a special case with $b = 1$ is investigated as an example, according to which we can construct a better understanding of how the sample length would affect the system performance.

Before we start the derivation, let us define T_c as the LTE symbol duration, while $T_{s,L}$ and $T_{s,I}$ are the sample length of LTE and NB-IoT, respectively. It is easy to get $T_{s,L} = T_c / N_L$. Recalling that the NB-IoT sample length is b times the length of LTE sample, we can have $T_{s,I} = bT_{s,L} = bT_c / N_L$. Figure 9.3 demonstrates the LTE and NB-IoT symbol structure with $b = 2$ and $G = 4$.

9.3.1 General Expression

Considering the previous assumption about the relationship between LTE and NB-IoT sample length, we can easily get $L_{CP,L} = GL_{CP,I}$. Hence, the desired NB-IoT signal $\mathbf{y}_{I,des}$ can be written as

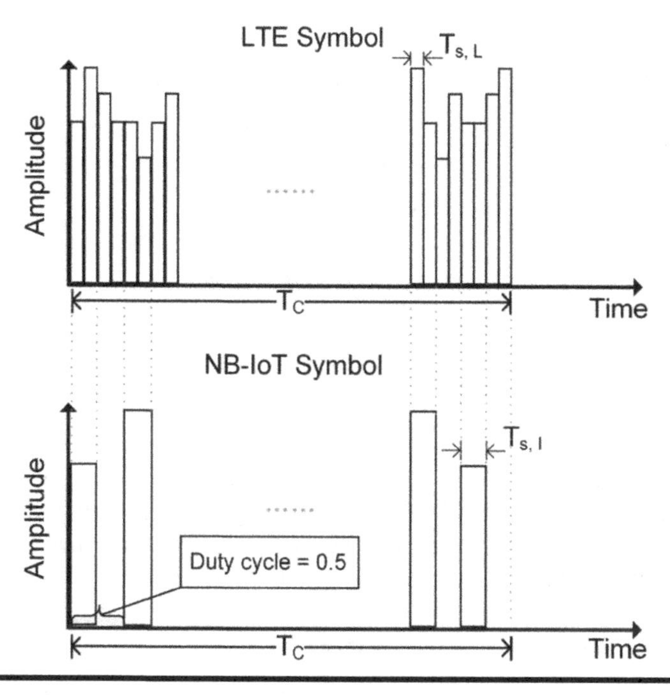

Figure 9.3 LTE versus NB-IoT symbol structure with $b = 2$ and $G = 4$.

$$\begin{aligned} \mathbf{y}_{I,des} &= \tilde{\mathbf{F}}_{N_L}^H \bar{\mathbf{\Phi}}_I \bar{\mathbf{U}} \mathbf{R}_I \mathbf{A}_I \mathbf{C}_I \mathbf{F}_{N_I} \mathbf{x}_I \\ &= \bar{\mathbf{F}}_{N_L}^H \bar{\mathbf{U}} \mathbf{R}_I \mathbf{A}_I \mathbf{C}_I \mathbf{F}_{N_I} \mathbf{x}_I, \end{aligned} \tag{9.15}$$

where $\bar{\mathbf{U}}$ takes the last N_L rows of $\mathbf{U}$, $\bar{\mathbf{\Phi}}_\mathbf{I}$ comprises the last N_L rows of $\mathbf{\Phi}_\mathbf{I}$, and $\bar{\mathbf{F}}_{N_L}^H$ takes the first M rows of $\mathbf{F}_{N_L}^H$. As is known that when CP length is larger than the length of the channel, the CP insertion and removal convert the multi-path channel to a circular one, that is, $\mathbf{R}_I \mathbf{A}_I \mathbf{C}_I = \mathbf{A}_{cir,I}$. Thus, $\mathbf{y}_{I,des}$ can be further modified as

$$\mathbf{y}_{I,des} = \frac{1}{\sqrt{G}} \sum_{i=1}^{b} \bar{\mathbf{F}}_{N_{L,i}}^H \mathbf{A}_{cir,I} \mathbf{F}_{\mathbf{N_I}} \mathbf{x}_I, \tag{9.16}$$

where $\bar{\mathbf{F}}_{N_{L,i}}^H$ takes every Gth column of $\bar{\mathbf{F}}_{N_L}^H$, starting from the ith column, that is, the ith, $(i+G)$th, $\cdots$, $[i+(N_I-1)G]$th column. Taking the periodicity of the DFT matrix into account, it can be proved that $\bar{\mathbf{F}}_{N_{L,i}}^H = \Psi_i \mathbf{F}_{N_I}^H$, where the kth ($k \in [1,2,\cdots,M]$) diagonal element of the phase shift matrix Ψ_i can be represented as

$$\Psi_i(k) = e^{-j2\pi(i-1)(k-1)/N_L}. \tag{9.17}$$

If we define the CFR from the NB-IoT UE to the BS as $\mathbf{H}_I = \sqrt{N_I}\text{diag}(\mathbf{F}_{N_I}^H \tilde{\mathbf{h}}_I)$, where $\tilde{\mathbf{h}}_I = [\mathbf{h}_I, \mathbf{0}_{1\times(N_I - L_{CH,I})}]^T$, the desired NB-IoT signal can be simplified as

$$
\begin{aligned}
\mathbf{y}_{I,des} &= \frac{1}{\sqrt{G}}\sum_{i=1}^{b}\Psi_i \mathbf{F}_{N_I}^H \mathbf{A}_{cir,I}\mathbf{F}_{N_I}\mathbf{x}_I \\
&= \frac{1}{\sqrt{G}}\sum_{i=1}^{b}\Psi_i \mathbf{H}_I \mathbf{F}_{N_I}^H \mathbf{F}_{N_I}\mathbf{x}_I \\
&= \frac{1}{\sqrt{G}}\sum_{i=1}^{b}\Psi_i \mathbf{H}_I \mathbf{x}_I.
\end{aligned}
\tag{9.18}
$$

For the interference signal from LTE to NB-IoT (i.e., Eq. 9.13), it is easy to find $\mathbf{v}_I = 0$ because the transmitted LTE signals are assigned to the standard orthogonal subcarriers in the OFDM system, that is, $\tilde{\mathbf{F}}_{N_L}^H \hat{\mathbf{F}}_{N_L}$ is equal to a $\mathbf{0}$ matrix. By defining $\Upsilon = \frac{1}{\sqrt{G}}\sum_{i=1}^{b}\Psi_i$, the final NB-IoT output signal can be written as

$$
\mathbf{y}_I = \Upsilon \mathbf{H}_I \mathbf{x}_I + \mathbf{n}_I. \tag{9.19}
$$

When it comes to the received LTE signal $\mathbf{y}_L$, by defining $\mathbf{H}_L = \sqrt{N_L}\text{diag}(\hat{\mathbf{F}}_{N_L}^H \tilde{\mathbf{h}}_L)$, where $\tilde{\mathbf{h}}_L = [\mathbf{h}_L, \mathbf{0}_{1\times(N_L - L_{CH,L})}]^T$, we can write its desired part as

$$
\mathbf{y}_{L,des} = \mathbf{H}_L \mathbf{x}_L. \tag{9.20}
$$

Moreover, the interference signal from NB-IoT to LTE (i.e., Eq. 9.14) can be derived following similar procedures as for $\mathbf{y}_{I,des}$, that is,

$$
\begin{aligned}
\mathbf{v}_L &= \breve{\mathbf{F}}_{\mathbf{N_L}}^{\mathbf{H}} \bar{\mathbf{U}}\mathbf{A}_{cir,I}\mathbf{F}_{\mathbf{N_I}}\mathbf{x}_L \\
&= \frac{1}{\sqrt{G}}\sum_{i=1}^{b}\Theta_i \mathbf{H}_I \mathbf{x}_I.
\end{aligned}
\tag{9.21}
$$

In Eq. 9.21, $\breve{\mathbf{F}}_N^H$ takes the $M \in \{N - [mod(L_S - 1, N), mod(L_S - 1, N) mod(L_S - 1, N)]\}$ rows of F^H is a matrix that takes the $M \in [mod(L_s + 1, N_L), mod(L_s + 2, N_L), \cdots, mod(L_s + M, N_L)]$ rows of $\mathbf{F}_{N_L}^H$, and Θ_i is the circularly shifted version of a diagonal matrix whose kth ($k \in [1, 2, \cdots, M]$) diagonal element is $e^{j2\pi(i-1)(M-k+B_G+1)/N_L}$. Here L_s is the shift length of Θ_i and can be calculated as $L_s = M_L - M_I$.[3] By defining $\Gamma = \frac{1}{\sqrt{G}}\sum_{i=1}^{b}\Theta_i$, the received LTE signal can be written as

[3] It should be noticed that the shift direction of Θ_i depends on the sign of L_s, i.e., the shift is to the right if $L_s > 0$, whereas the shift is to the left if $L_s < 0$.

$$\mathbf{y}_L = \mathbf{H}_L\mathbf{x}_L + \Gamma\mathbf{H}_I\mathbf{x}_I + \mathbf{n}_L. \tag{9.22}$$

Based on the assumption that the channel is power normalized (i.e., Eq. 9.6), we can calculate the interference signal power from NB-IoT to LTE as

$$\begin{aligned}\mathbf{p} &= \mathbb{E}[\text{diag}(\mathbf{v}_L\mathbf{v}_L^H)] \\ &= \mathbb{E}[\text{diag}(\Gamma\mathbf{H}_I\mathbf{H}_I^H\Gamma^H)] \\ &= \text{diag}(\Gamma\Gamma^H).\end{aligned} \tag{9.23}$$

From the above derivations, we have the following remarks:

Remark 1: Eq. 9.19 reveals that the equivalent CFR for NB-IoT transmission is the multiplication of a phase shift matrix (i.e., Υ) and the Fourier transform of its channel impulse response. This enables the interference-free one-tap equalization. Hence, the receivers, which can only perform the traditional one-tap equalization method, could still be applied as long as the equalization coefficients are updated accordingly.

Remark 2: As the phase shift matrix Υ and the CFR of the NB-IoT signal (i.e., $\mathbf{H}_I$) are both diagonal matrices, we can have $\Upsilon\mathbf{H}_I\mathbf{x}_I = \mathbf{H}_I\Upsilon\mathbf{x}_I$. To balance the computational cost between the transmitter and the receiver, a precoding matrix $\mathbf{P}_{com} = \Upsilon^{-1}$ could be used to compensate the unevenly distributed power. Thus, signal $\tilde{\mathbf{x}}_\mathbf{I} = \mathbf{P}_{com}\mathbf{x}_I$ will be transmitted instead of $\mathbf{x}_I$.

Remark 3: Eq. 9.22 implies that the NB-IoT device will generate interference in the LTE signal, which may degrade the LTE performance. From Eq. 9.23, it can be found that the interference power level depends on the guard band B_G and the NB-IoT sample length.

Remark 4: Fundamentally, the interference level on LTE signal depends on the OFDM out-of-band emission (OoBE). Whereas one of the disadvantages of the OFDM system is the high OoBE level, a new low OoBE waveform can be implemented on top of the desired subcarrier (or RB) to attenuate the interference leakage from NB-IoT device to improve LTE UE performance. The optional new waveforms include the subcarrier filtered ones such as filter bank multi-carrier (FBMC) [14] and generalized frequency division multiplexing (GFDM) [15], subband filtered waveforms such as filtered OFDM (f-OFDM) [16] and universal filtered multi-carrier (UFMC) [12,17] and windowed OFDM (W-OFDM) [13].

9.3.2 Special Case

Let us assume the NB-IoT and LTE have the same sample duration, that is, $b = 1$. Thus, the received NB-IoT and LTE signals can be written as

$$\mathbf{y}_I = \frac{1}{\sqrt{G}}\mathbf{H}_I\mathbf{x}_I + \mathbf{n}_I, \tag{9.24}$$

$$\mathbf{y}_L = \mathbf{H}_L\mathbf{x}_L + \frac{1}{\sqrt{G}}\Theta_1\mathbf{H}_I\mathbf{x}_I + \mathbf{n}_L, \tag{9.25}$$

where Θ_1 is a circularly shifted version of $\mathbf{I}_{N_I}$. Accordingly, the interference signal power from NB-IoT to LTE can be calculated as

$$\begin{aligned}\mathbf{p} &= \frac{1}{G}\mathbb{E}[\mathrm{diag}(\Theta_1\mathbf{H}_I\mathbf{H}_I^H\Theta_1^H)] \\ &= \frac{1}{G}\mathbf{I}_{N_I}.\end{aligned} \tag{9.26}$$

From Eq. 9.24, we can find that by setting the NB-IoT sample length as short as that of LTE, the only influence of the sampling rate mismatching to NB-IoT signal detection is presented as a constant attenuation factor $1/\sqrt{G}$. In addition, according to Eq. 9.26, the power of interference from NB-IoT to LTE is only related to the up-sampling rate G and is evenly distributed in each assigned LTE subcarrier.

9.4 Simulation Results

To investigate how the parameters b and B_G could affect the uplink NB-IoT and LTE system performance, numerical results are illustrated in this section. In the first part, the LTE's performance (i.e., interference power and bit-error rate [BER]) is studied in terms of different sampling methods (i.e., b) and guard bands (i.e., B_G), while in the second part, the impacts of b and B_G on NB-IoT's performance (i.e., signal power selectivity and BER) are explored.

The extended typical urban (ETU) channel model [18] defined by 3GPP is considered in all simulations. Some of the general system configurations are illustrated in the Table 9.2:

Table 9.2 General System Configurations

System Configurations	*LTE*	*NB-IoT*
DFT size	360	12
CP length	30	1
Modulation scheme	16-QAM	QPSK

9.4.1 Performance of LTE System

In Figure 9.4, the average interference power over all LTE subcarriers are illustrated with $b = [1,2,20,30]$, respectively. It can be found that the simulated results perfectly match the analytical ones, verifing the effectiveness of the derivations in the previous section. In general, we can observe the reduction of the average interference level as the guard band increases, and the interference reaches its minimum point when NB_G (i.e., the normalized guard band) is around 0.5, after which the interference level starts to go up as the guard band keeps increasing. The rationale behind this variation is the circular property of the baseband processing. However, it is interesting to notice that the interference power remains unchanged over different NB_G when $b = 1$, which is aligned with the derivation in Eq. 9.26. Furthermore, with different NB-IoT sample methods, the interference power falling into LTE subcarriers varies differently. For example, among all the cases shown in Figure 9.4, when the normalized guard band NB_G is less than 0.2, the shortest NB-IoT sample length contributes to the lowest interference, whereas it introduces the highest interference power when the two services are placed in the

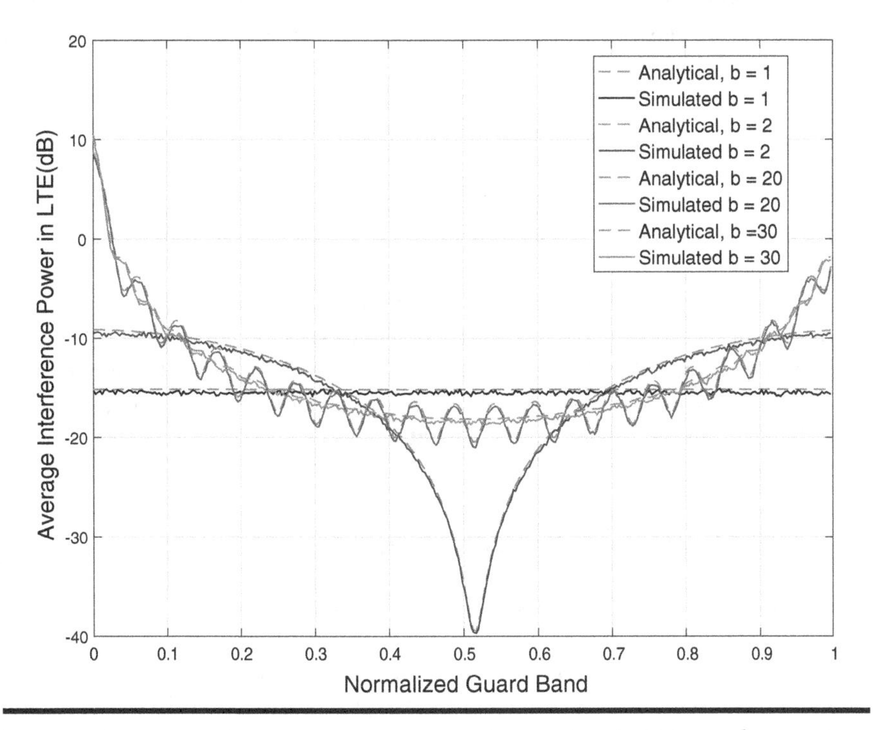

Figure 9.4 Average interference power in LTE signal with different b and NB_G.

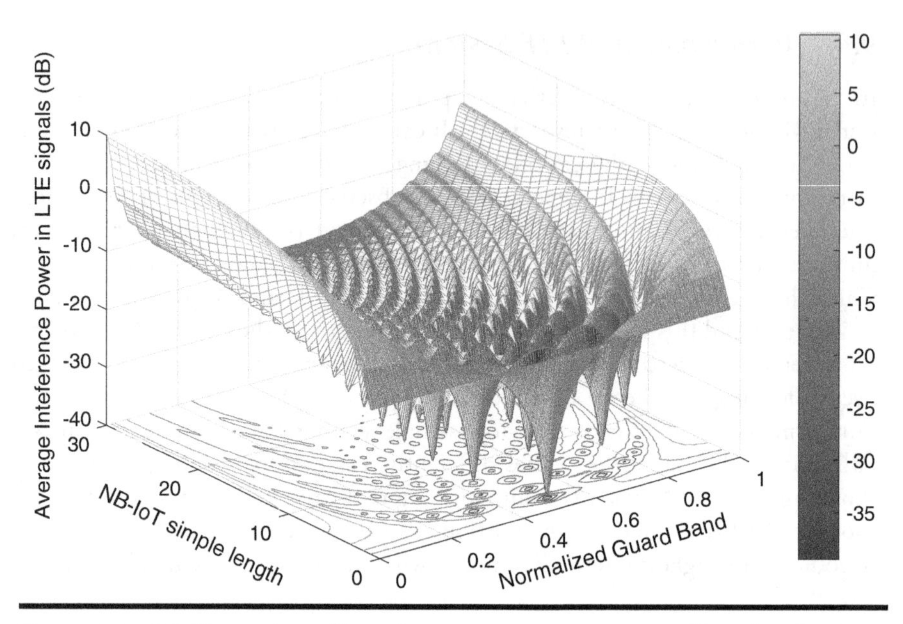

Figure 9.5 3D illustration of interference power versus guard band and NB-IoT sample length.

largest distance in terms of guard band (i.e., NB_G ⊕0.5). The three-dimensional illustration of the relationship among NB_G, b and the average interference power is shown in Figure 9.5.

The BER versus E_b / N_0 performance of LTE is shown in Figure 9.6. We can find that generally, the LTE system has better BER performance when the guard band is larger, that is, $NB_G = 0.5$, except in the case $b = 1$, where the BER remains unchanged with the guard band increasing. This is also aligned with Figure 9.4. By flexibly selecting the NB-IoT sample length, significant performance gain can be expected.

9.4.2 Performance of NB-IoT System

As for the NB-IoT's performance, we can see from Eq. 9.19 that the NB-IoT signal is free of interference from LTE UE; however, its power level and distribution will be affected by the NB-IoT sample length. Figure 9.7 demonstrates how the NB-IoT signal (i.e., power level and distribution) changes in each NB-IoT subcarrier with different b.

It can be seen that the elements' values in Υ are almost evenly distributed among subcarriers while b is small, nonetheless, increasing distortion is generated by Υ as b becomes larger. We call this the signal power selectivity property of Υ.

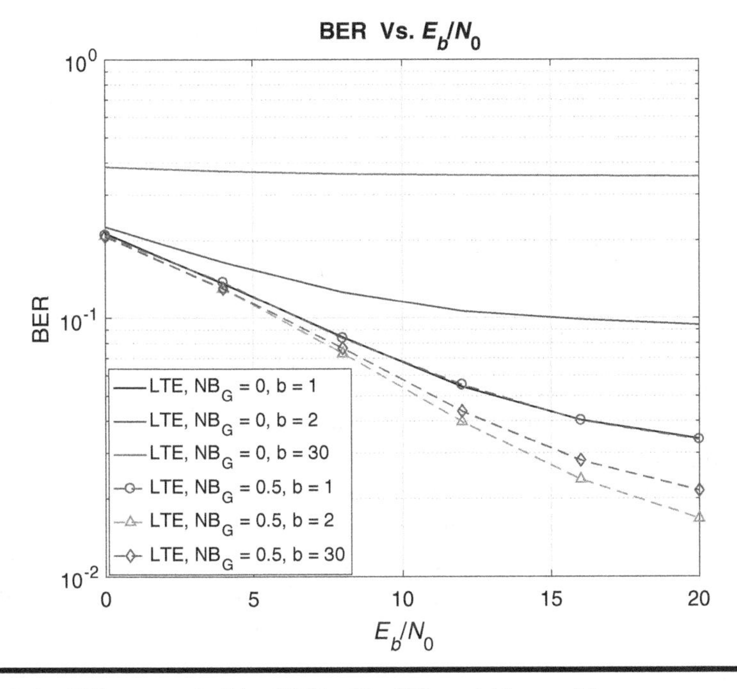

Figure 9.6 BER versus E_b/N_0 of LTE with different $N_{B,G}$ and b.

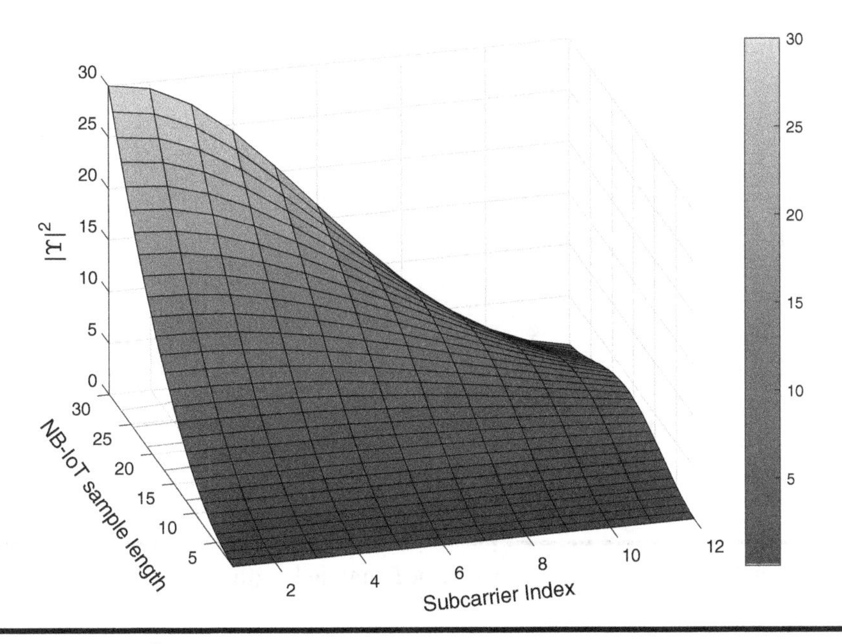

Figure 9.7 $|\Upsilon|^2$ versus b in different NB-IoT sub-carriers.

Theoretically, the distortion factor Υ can be perfectly canceled by implementing the one-tap equalization method mentioned in *Observation 1*, Section 9.3.1. However, the updated equalization coefficients will, in turn, enlarge the noise power unexpectedly. With zero forcing (ZF) equalization, the noise power after equalization can be calculated as

$$\mathbf{P}_n = \mathbb{E}\left(|\mathbf{E}_q \mathbf{n}_I|^2\right) = \mathbb{E}\left(|\frac{1}{\Upsilon \mathbf{H}_I}|^2\right) N_0 = |\frac{1}{\Upsilon}|^2 N_0, \tag{8.27}$$

where $\mathbf{E}_q$ is the equalization factor and N_0 is the power density of noise $\mathbf{n}_I$. To have a clear view on how the noise power changes with the equalization process, Figure 9.8 is depicted by averaging its coefficients (i.e., $1/|\Gamma|^2$) over all the NB-IoT sub-carriers. It can be observed that the average noise power reaches its smallest point when $b = 20 \sim 22$.

The BER versus E_b / N_0 performance of NB-IoT is shown in Figure 9.9. As can be expected, it is insensitive to the guard band. In addition, the worst and the best BER curves occur when b is equal to 1 and 20, respectively. This can be explained by Figure 9.8, that is, larger noise power results in a worse signal-to-noise ratio with fixed signal power.

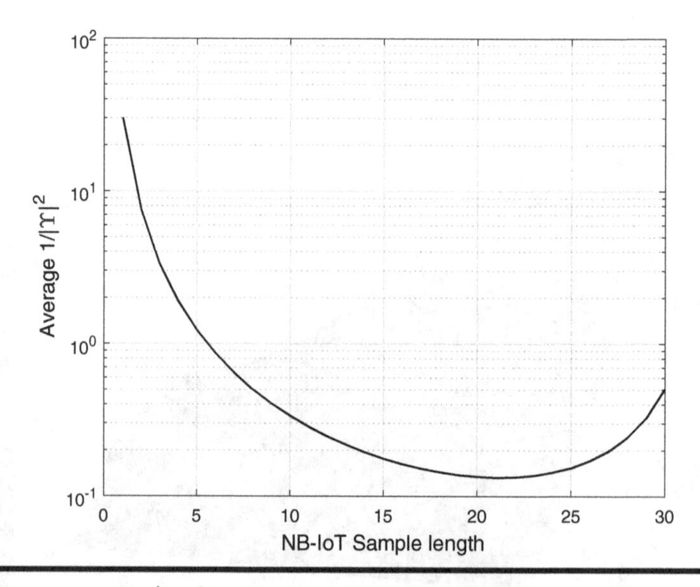

Figure 9.8 Average $1/|\Upsilon|^2$ versus NB-IoT sample length.

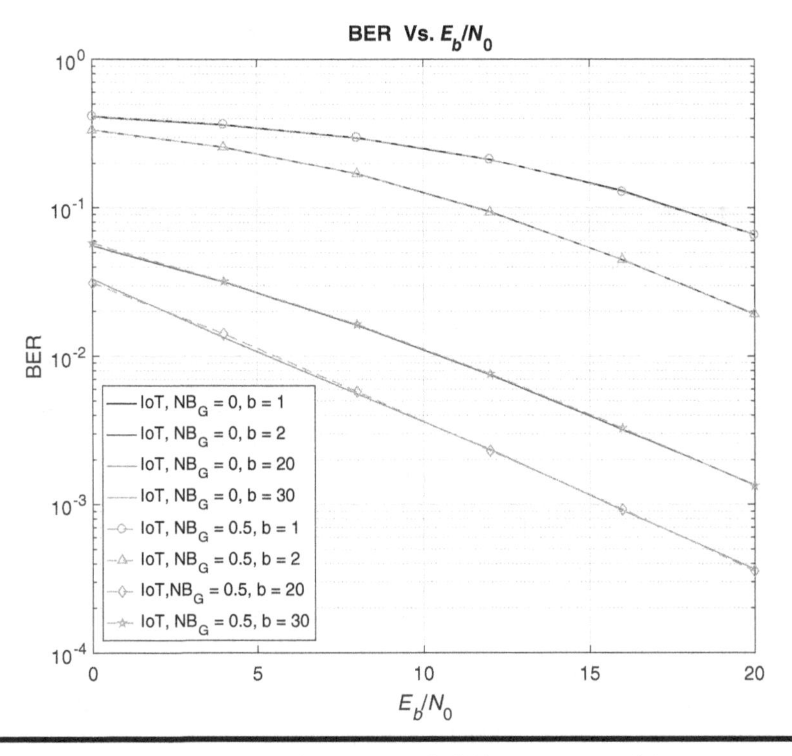

Figure 9.9 BER versus E_b/N_0 of NB-IoT with different $N_{B,G}$ and b.

9.5 Conclusions

In this chapter, a comprehensive uplink coexistence system model for NB-IoT and LTE was constructed for the purpose of investigating the potential influence caused by the mismatched sampling rate. By considering an arbitrary NB-IoT sample length, the mathematical expressions of the received signals were derived for both services. It is revealed that the LTE signal suffers from the interference generated by the NB-IoT device, and the power of interference is a closed-form function of the NB-IoT sample length as well as the guard band between the two services. For the NB-IoT signal, although it is free of interference from LTE UE, the misaligned sampling rate introduces distortion into its desired signal. At last, simulations were performed to verify the effectiveness of the proposed system model and derivations.

The design in this chapter is explicitly for in-band and guard-band NB-IoT systems. However, it can be considered as a guide to support the network slicing and service multiplexing in future 5G network. In addition, the ideas presented in this chapter provide us with a new train of thought on sample duration optimization.

References

1. Ericsson mobility report on the pulse of the network society. Technical report, Telefonaktiebolaget L. M. Ericsson, June 2016.
2. NarrowBand IoT wide range of opportunities. Technical report, Huawei Technologies Co., Ltd., February 2016.
3. 3GPP RAN4. Evolved Universal Terrestrial Radio Access (E-UTRA) NB-IoT technical report for BS and UE radio transmission and reception. Technical report (TR) 36.802, 3rd Generation Partnership Project (3GPP), June 2016. Version 13.0.0.
4. Ayesha Ijaz, Lei Zhang, Maxime Grau, Abdelrahim Mohamed, Serdar Vural, Atta U. Quddus, Muhammad Ali Imran, Chuan Heng Foh, and Rahim Tafazolli. Enabling massive IoT in 5G and beyond systems: PHY radio frame design considerations. *IEEE Access*, 4: 3322–3339, 2016.
5. Bowen Yang, Lei Zhang, Deli Qiao, Guodong Zhao, and Muhammad Imran. NarrowBand Internet of Things (NB-IoT) and LTE systems co-existence analysis. In *IEEE GLOBECOM*, pp. 1–6, 2018.
6. Lei Zhang, Ayesha Ijaz, Pei Xiao, and Rahim Tafazolli. Channel equalization and interference analysis for uplink Narrowband Internet of Things (NB-IoT). *IEEE Communications Letters*, 21(10): 2206–2209, 2017.
7. Nan Jiang, Yansha Deng, Massimo Condoluci, Weisi Guo, Arumugam Nallanathan, and Mischa Dohler. RACH preamble repetition in NB-IoT network. *IEEE Communications Letters*, 22(6): 1244–1247, 2018.
8. Xingqin Lin, Ansuman Adhikary, and Y.-P. Eric Wang. Random access preamble design and detection for 3GPP narrowband IoT systems. *IEEE Wireless Communications Letters*, 5(6): 640–643, 2016.
9. Changsheng Yu, Li Yu, Yuan Wu, Yanfei He, and Qun Lu. Uplink scheduling and link adaptation for narrowband internet of things systems. *IEEE Access*, 5: 1724–1734, 2017.
10. Lei Zhang, Ayesha Ijaz, Juquan Mao, Pei Xiao, and Rahim Tafazolli. Multi-service signal multiplexing and isolation for physical-layer network slicing (PNS). In *Vehicular Technology Conference (VTC-Fall), 2017 IEEE 86th*, pp. 1–6. IEEE, 2017.
11. Lei Zhang, Ayesha Ijaz, Pei Xiao, and Rahim Tafazolli. Multi-service system: An enabler of flexible 5G air interface. *IEEE Communications Magazine*, 55(10): 152–159, 2017.
12. Lei Zhang, Ayesha Ijaz, Pei Xiao, Atta Quddus, and Rahim Tafazolli. Subband filtered multi-carrier systems for multi-service wireless communications. *IEEE Transactions on Wireless Communications*, 16(3): 1893–1907, 2017.
13. Xiaoying Zhang, Lei Zhang, Pei Xiao, Dongtang Ma, Jibo Wei, and Yu Xin. Mixed numerologies interference analysis and inter-numerology interference cancellation for windowed ofdm systems. *IEEE Transactions on Vehicular Technology*, 67(8): 7047-7061, 2018.

14. Lei Zhang, Pei Xiao, Adnan Zafar, Atta ul Quddus, and Rahim Tafazolli. FBMC system: An insight into doubly dispersive channel impact. *IEEE Transactions on Vehicular Technology*, 66(5): 3942–3956, 2017.
15. Gerhard Fettweis, Marco Krondorf, and Steffen Bittner. GFDM—Generalized frequency division multiplexing. In *Vehicular Technology Conference, 2009. VTC Spring 2009. IEEE 69th*, pp. 1–4. IEEE, 2009.
16. Lei Zhang, Ayesha Ijaz, Pei Xiao, Mehdi M. Molu, and Rahim Tafazolli. Filtered OFDM systems, algorithms, and performance analysis for 5G and beyond. *IEEE Transactions on Communications*, 66(3): 1205–1218, 2018.
17. Lei Zhang, Pei Xiao, and Atta Quddus. Cyclic prefix-based universal filtered multicarrier system and performance analysis. *IEEE Signal Processing Letters*, 23(9): 1197–1201, 2016.
18. 3GPP RAN4. Evolved Universal Terrestrial Radio Access (E-UTRA) physical channels and modulation. Technical report (TR) 36.803, 3rd Generation Partnership Project (3GPP), April 2016. Version 1.1.0.

Chapter 10

Internet of Things Wireless Spectrum Sharing for Radio Access

Akhil Gupta and Raabia Kausar

Contents

10.1 Introduction

Fifth generation (5G) wireless technology is the up and coming age of portable interchanges innovation and is intended to give (in comparison with 4G) more noteworthy limit, quicker information speeds, and offer low idleness and high unwavering quality while empowering imaginative new administrations crosswise over various industry segments. The primary influx of 5G business items is required to be accessible in 2020 albeit some "pre-5G" arrangements are as of now expected in 2018. 5G innovation models are currently being worked on and will incorporate both a development of existing (4G) and new radio (NR) advancements (5G NR).

5G administrations and applications can be assembled into three distinctive classes:

- Enhanced mobile broadband
- Massive machine-type communications
- Ultra-reliable and low latency communications

Talking about the monstrous machine compose correspondence: The Internet-of-Things (IoT)—where sensors, actuators, shopper hardware apparatuses, road lighting and so on remotely interface with the web and each other. This is as of now occurring on existing 4G systems and the innovation is being utilized as a part of everything from brilliant homes to wearables. 5G should help the development of IoT administrations and applications and enhance communication between various stages and in addition empowering the vision of 50 billion gadgets getting to be associated by 2030. Conceivable future applications incorporate ongoing wellbeing observing of patients; streamlining of road lighting to suit the climate or movement; natural checking and keen agribusiness. Information security and protection issues should be viewed as given gigantic measures of information could be exchanged over an open system. We take note of that numerous IoT administrations are now being offered or will be offered in the following couple of years over existing and developed 4G systems, for example, utilizing narrowband IoT (NB-IoT), long-term evolution–machine (LTE-M), or NB-LTE-M innovations. 5G, around there, is probably going to kick-in by around 2025 where we hope to see the blast of new IoT administrations for which the advancement of LTE cannot give the required versatility prerequisites.

10.2 Radio Access Technologies for 5G

Unlike previous generations, where a new radio access technology replaced the old one, 5G will integrate different radio technologies. Some of these will be the evolution of already existing radio access technologies while some will be new. Different service classes could rely on different radio interfaces. Evolutions of the latest version of the 4G radio interface (LTE-Advanced Pro) are likely to be used to provide a coverage layer via macro cells. A new cellular radio interface (being developed in Third Generation Partnership Project [3GPP] under the name "New Radio" or "NR") operating at frequencies up to 50 GHz will be used to provide very high data rates, ultra-low latencies and to serve a very large number of devices via a large number of small cells. Low-cost, low-battery consumption IoT services are likely to be delivered initially using evolved 4G technologies, as described in the "Introduction" section, with a migration to 5G by 2025.

Wireless fidelity (Wi-Fi), evolutions will also play an important role for consumers, in particular to provide 5G services within homes or offices. In addition, it is expected that satellite technologies will play a role in 5G, in particular for wide area coverage in IoT application space (e.g., tracking of goods and vehicles), and also as a mechanism to offload broadcast and multicast linear TV traffic from 5G cellular networks [1].

10.3 Spectrum Sharing in 5G

While over 6 GHz, expansive pieces of range are required to wind up accessible for 5G frameworks, the measure of range in the sub-GHz and beneath 6 GHz run is undeniably restricted. The sub-6 GHz band is required to help critical uses of 5G, such a machine-type correspondences because of astounding proliferation and indoor infiltration attributes while the main influx of 5G versatile correspondence frameworks are relied upon to be sent in the 3.6 GHz recurrence go; here, in conjunction with the utilization of massive multiple input–multiple output (MIMO) and full-dimensional MIMO (FD-MIMO) advances, 100 Mbps+ information rates could be upheld while additionally keeping cell-sizes adequately expansive for practical organization. It is, along these lines of incredible significance to investigate choices for sharing of these valuable segments of 5G range. Because of the nature of administration, necessities of the 5G utilize cases that are required to be upheld, a critical choice for sharing of these groups is the development of Licensed Shared Access (LSA) [2]. In this approach, authorized clients, called LSA licensees, can get to underutilized authorized range on an elite premise, consequently getting a charge out of unsurprising quality of service (QoS), when it is not being utilized by the occupant, henceforth shielding it from unsafe obstruction.

10.3.1 Sharing mm-wave Spectrum

Millimeter-wave (mm-wave) interchanges have developed as a key problematic innovation for both cell systems (5G and past) [3] and remote Local Area Networks (802.11ad and past). While range accessibility is constrained in customary groups beneath 6 GHz, mm-wave frequencies offer request of greatness more prominent transmission capacities. Furthermore, mm-wave correspondence is regularly portrayed by transmissions with exceptionally tight bars, empowering further picks up from directional disengagement between mobiles. This mix of huge transfer speed and spatial degrees of opportunity make it workable for mm-wave to meet a portion of the boldest 5G necessities, including higher pinnacle per-client information rate, high activity thickness and low dormancy. The utilization of mm-wave groups for 5G display a number novel highlights not present at bring down frequencies:

- *Beamforming requirement:* A typical normal for all frameworks working in mm-wave frequencies is that pillar shaping is obligatory to make up for the fundamentally higher path loss in these frequencies, for example, the IEEE 802.11ad standard backings up to four transmitter reception apparatuses, four recipient receiving wires, and 128 divisions. Beam shaping is obligatory in 802.11ad, and both transmitter-side and beneficiary side beamforming are bolstered. Moreover, particular of beamforming for 5G are relied upon to be concluded by 3GPP, as a feature of 5G NR work thing. Thusly, shafts give a typical new measurement to sharing of range among various access innovations.

- *Potential for "infinite" spatial reuse:* Remote correspondences frameworks as of now depend on spatial sharing of range and the whole idea of cell interchanges depends on spatial re-utilization of radio range. In mm-wave frameworks with the utilization of both transmit-side and get side beamforming, spatial range reutilize can be pushed considerably encourage towards one-measurement, with the impression of impedance from every transmission interface winding up near a line, instead of a territory. In the admired instance of ultra-slender bars this would, subsequently, empowers vast spatial reuse of range.

10.3.2 Sharing with Satellite Services

Fixed satellite service (FSS) is the official grouping for geostationary correspondences satellites that give, for example, communicate feeds to TV channels, radio stations and communicate systems. The FSS uplink (from FSS to satellite) is designated in the band from 27.5 to 30 GHz, which is adjoining the 24.25–27.5 GHz band distinguished for 5G. In this way, there could be potential issues with sharing in the vicinity of 5G and FSS because of nearby channel impedance. A few intellectual procedures can be connected to enhance the 5G-FSS conjunction in the vicinity of 5G and FSS because of contiguous channel impedance. A few psychological procedures can be connected to enhance the 5G-FSS conjunction The concurrence amongst FSSs and portable cell base stations (BSs) in the mm-wave groups have been the subject of just few, and mostly hypothetical, considers. Imperative new parameters that should be considered are the manner by which the impedance levels could be diminished by abusing various reception apparatus arrangements by 5G mm-wave frameworks and also examining the total obstruction coming about because of monstrous sending of 5G frameworks on uplink FSS. The examinations in [4,5], performed in the most dire outcome imaginable of co-channel sharing, have demonstrated that, because of the utilization of shaft shaping innovation joined with the generally short-scope of interchanges in mm-wave frequencies, spatial sharing is substantially more achievable than on account of International Mobile Telecommunications (IMT)–propelled frameworks. Specifically, even in this more awful case situation the required insurance remove around FSS are significantly littler (~1 km rather than several kilometers) than those prescribed beforehand. Besides, by utilizing coordination among numerous 5G BS additionally picks up in range sharing can be accomplished. These examinations additionally show that

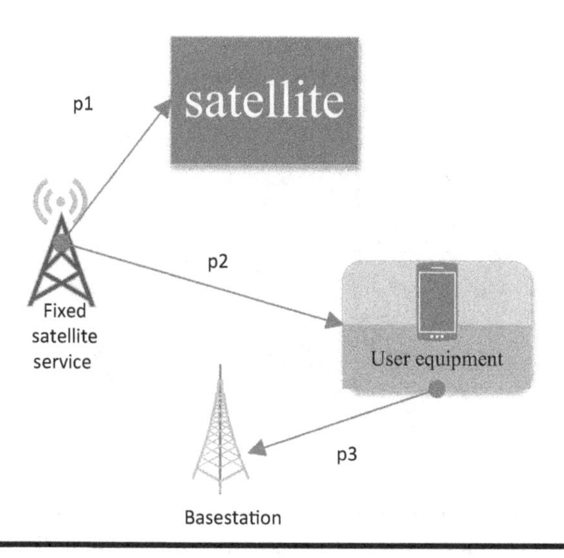

Figure 10.1 Impact of fixed satellite service uplink transmission on the coverage of a mm-wave 5G network in the worse-case co-channel sharing of spectrum in an interference scenario. Abbreviations: p1, Primary Link; p2, Interference Link; p3, Secondary Link. (From Nekovee, M. and Richard R., 5G spectrum sharing, *arXiv preprint arXiv:1708.03772*, 2017.)

nearness of profoundly directional FSS transmission can cause blackout in the scope of 5G mm-wave organize. In any case, due to the profoundly directional FSS transmission, the blackout locale is very much restricted (as is appeared in Figure 10.1) and its effect could be relieved utilizing a blend of invalid shaping at 5G UEs and participation by numerous BS to support flag qualities at the casualty UE.

10.3.3 Sharing the Unlicensed Millimeter-Wave Spectrum

The present pattern in cell correspondence is to use both the authorized and unlicensed range at the same time for expanding accessible framework transfer speed. In this specific circumstance, LTE in unlicensed range, alluded to as LTE-U, is proposed to empower portable administrators to offload information activity onto unlicensed frequencies all the more productively and adequately, and gives elite and consistent client encounter. Coordination of unlicensed groups is additionally considered as one of the key empowering agents for 5G cell frameworks. Be that as it may, not at all like the regular activity in authorized groups, where working BS have selective access to range and in this way can arrange by trading of motioning to relieve common obstruction, such a multi-standard and multi-administrator range sharing situation (as appeared in Figure 10.2) forces critical difficulties on concurrence regarding impedance alleviation. Authorized Assisted Access (LAA) with tune in before talk (LBT) convention has been proposed for the present

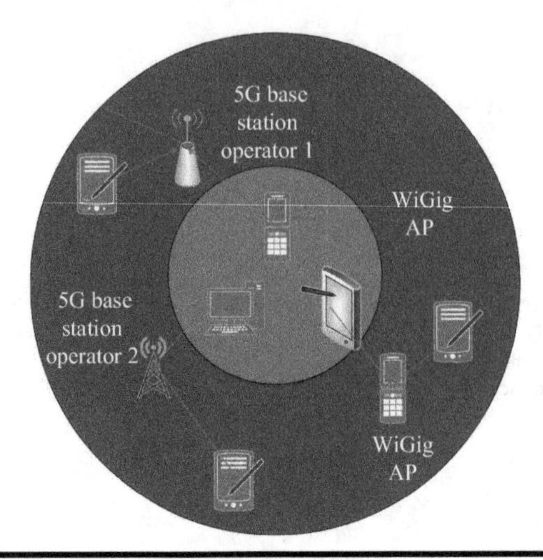

Figure 10.2 Multi-standards and multi-operator sharing of unlicensed mm-wave bands in future 5G systems, with a beam collision interference scenario.

conjunction system of LTE-U. On account of mm-wave unlicensed sharing, a noteworthy issue is that the utilization of exceedingly directional reception apparatuses as one of the key empowering influences for 5G systems ends up risky for the present concurrence components where omni-directional receiving wires were for the most part accepted. For instance, as appeared in Figure 10.2 transmission by an alternate adjacent 5G BS or Wireless Gigabit (WiGig) passage access points (APs) may not be identified because of the tight pillar that has been utilized, bringing about "bar impact" which can cause significantly more over the top obstruction than in traditional frameworks.

We take note of that such pillar impact obstruction situations can likewise happen in solely utilized mm-wave range too. Nonetheless, in such situations, unified asset allotment calculations from 4G can be stretched out to incorporate shaft planning among numerous base stations to maintain a strategic distance from over the top obstruction. On account of unlicensed sharing of mm-wave range, brought together coordination is not conceivable, and novel systems should be created. Work toward this path has just as of late been begun as a feature of another investigation thing in 3GPP 5G-NR which is relied upon to be finished in 2018 [6,7]. Different system for sharing are being proposed, including dispersed and self-sorted out component for pillar coordination [6], and approaches in light of range pooling [8].

10.4 Device Positioning

It is based in collecting data from users in order to arrange statistical data enabling optimization of marketing activities and thereby increase revenue and improve operations efficiency. In addition to marketing, users' data collection is useful for several other applications such as security or elderly-care. Users' data collection requires monitoring systems characterized by limited (or lack of) user activity, round-the-clock operation and tools for analyzing collected data according to the customer expectations. The customer (beneficiary) of the system will be its administrator, or companies interested in having information of the behavior of potential users. It is assumed that the users, i.e., the owners of devices (e.g., smartphone), are not constrained to install any new software in their own devices.

The users' data collection in marketing may innovate/improve operations such as to control the frequency of visits to a shopping center, to differentiate users with regard to visiting purpose or purchase, to identify users from the population on the basis of technical data, to monitor staff and comparison with revenues, to make heating maps (human activity) into shopping centers, and many other potential functionalities. From the technological point of view, the system could be based on users' hardware or be fitted with a special device that communicates with the users' devices. In this chapter, we propose an implementation of a system that aims to monitor the current location of user devices located in buildings (e.g., shopping mall) without the explicit awareness from the users. This means, in turn, that the positioning process will be carried out mainly based on infrastructure held by administrator and there is no possibility to install a dedicated software on localized devices. These assumptions mean that the proposed solution should be based on measurements of the power of the received signal strength (RSS) radio signal under different technologies, most frequently used by the users. At the same time the measurement of RSS is performed during normal operation of monitored devices (e.g., during update of the list of active access Wi-Fi access points) and does not use additional features requiring support from the application layer. The use of location technologies which requiring support (interaction) from the localized device is inefficient in the most scenarios, since such an approach causes a significant reduction in the number of monitored devices (only belong to users who have consciously made the appropriate configuration). At the control plane [Media Access Control [MAC] and routing layers], the proposed platform groups together the considered technologies into an open and programmable platform, which is easily adaptable to concrete requirements from the administrators of the equipment positioning system. Therefore, the full support of each radio technology will be implemented on the basis of the software without the need for dedicated hardware resources. In this way, it will be possible to use the protocol defined by the IEEE 802.15.4 standard instead of a closed expensive ZigBee solutions, which leads to reduction of the costs of the products. The undoubted benefit of using a single hardware platform and software level technology support (eliminating expensive hardware acceleration) is much lower cost of transmission and reception. In result, much more antennas can be deployed in a given area, which significantly improves the accuracy of positioning.

10.4.1 Data Collection System

The main prerequisite of the framework is that it ought to take into account observing/finding the best number of individuals moving in the checked regions, along these lines it is attractive to use for this reason radio arrangements executed in various advances that can be produced into the gadgets claimed by the clients. Specifically, these innovations include Wi-Fi, Bluetooth, radio frequency identification (RFID), different frameworks in view of IEEE 802.15.4. It ought to be noticed that in spite of the fact that there are answers for finding objects inside the rooms, which utilize in excess of one innovation, there are no arrangements that offer help for the greater part of the above advances. Current arrangements generally utilize close to two advances for finding the clients' gadgets: one of which is normally Wi-Fi innovation, and the second one, contingent upon the approach, might be one of the accompanying: Inertial Measurement Unit (IMU) [9] Bluetooth [10], ultrasound [11]. Parallel observing of gadgets actualizing a few advances may, from one perspective, fundamentally increment the quantity of checked gadgets (clients) and, then again, it builds the unpredictability and cost of the framework. Since, most of the previously mentioned advancements work in a similar recurrence and extended 2.4 GHz band. So the impedance between the frameworks requires less precision.

So as to neglect cross-innovation obstructions, we propose to utilize basic piece of the radio (reception apparatus framework) for these advances. This is conceivable by the way that the previously mentioned innovations work in a similar recurrence band, and additionally the framework will be utilized just with the end goal of area of clients (no other information will be sent that require high data transfer capacity). Regular radio access is accomplished by isolating the capacities, which are in charge of the transmission on the radio channel and the control capacities (counting higher layers). Thus, the control capacities (counting MAC and system layers) might be completely programmable on programming based stage, and additionally it is conceivable to utilize virtualization advancements taking benefit of its advantages (e.g., particularity). Figure 10.3 demonstrates the design of the proposed framework where the innovation controller will be produced on virtualization programming stage. The proposed arrangement includes that the radio part will on the other hand serve every last one of the previously mentioned advancements (time division will be connected). As per [12], the conjunction of various remote gadget relies upon three elements: recurrence, area in space and time. The individual radio systems will have the capacity to work on the off chance that it contrasts in no less than one of the above variables. For our situation, since we utilize similar radio wires (which implies a similar recurrence and position), the conjunction of various innovations can be executed just with time division. The two primary favorable circumstances of this arrangement is the absence of obstruction between gadgets from various advancements (in light of the fact that at the specific time, radio wire set plays out the elements of just a single innovation) and essentially bring down cost of usage contrasted and parallel establishments for every one of the advances

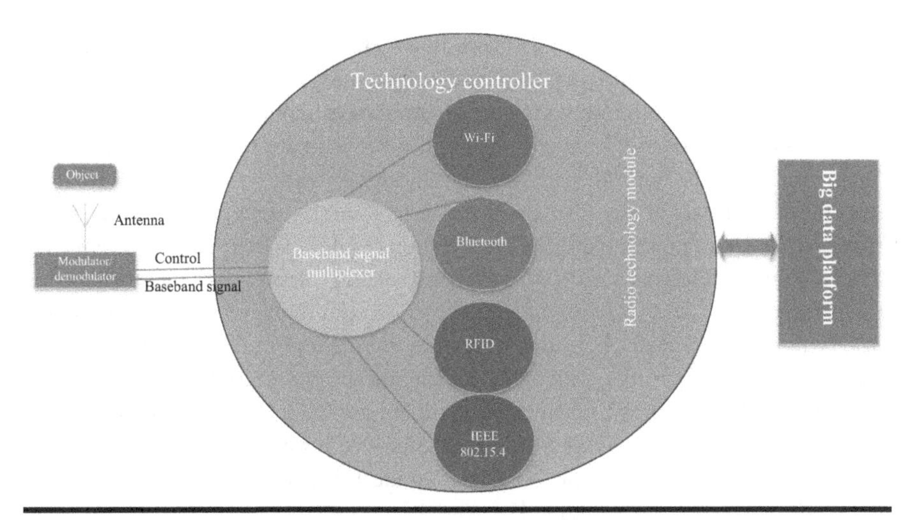

Figure 10.3 Proposed system architecture.

handled. From the perspective of the creating framework, the most imperative favorable position of the proposed arrangement is the likelihood of giving more exact estimations of the articles. This is because of two components: right off the bat, more noteworthy precision is the consequence of an absence of twists coming about because of the conjunction of radio innovations in the network, which are working in a similar recurrence band. Besides, a much lower cost of radio system segments takes into account sending of substantially more receiving wires in the zone, which essentially builds the precision of situating. It is worth to focus on the authenticity of the utilization of RFID innovation. For the larger part of utilizations, this innovation is utilized to recognize objects in light of RFID Tag for short separations (up to many centimeters). Clearly, these arrangements, because of the little range and the distinctive scope of frequencies utilized are not appropriate from the perspective of our proposed limitation framework. In any case, progressed RFID frameworks working in the unlicensed recurrence band and utilizing Active Tags (with their own capacity supply) accomplish a few meters go, which is reasonable with the end goal of the framework in a few fields as, e.g., representatives observing.

10.4.2 Computational Advanced Algorithm

Notwithstanding the utilization of cutting edge calculations, practically speaking there are various confinements that essentially influence the exactness of situating. Concentrates in [13] demonstrated that the RSS esteem, and accordingly the situating exactness, fundamentally relies upon the introduction (turn) of the estimating gadget. This is because of the radio flag abnormalities, which causes that the deliberate power relies upon the course of introduction of the receiving wire, segments of the reflected radio flag and the vicinity of the client's body, which because of the high water content in the human body assimilates a piece of radio signs [13–15]. The previously mentioned connections imply that the estimations of situating stage in pragmatic applications quite often happen in a domain unique in relation to the estimations of alignment stage. It ought to likewise be noticed that typically contraption with very surprising attributes are utilized as a part of alignment and situating stages. At long last, in various pragmatic applications, there are shifting proliferation conditions (e.g., customer remaining alongside someone else or a bed with merchandise can generously smother the flag from the particular AP), and there might be obstruction with different frameworks working in a similar frequency band. This all outcomes in botches in indicating area in encase spaces. In this manner, we underscore the significance of utilizing multi-innovation multi-radio wire condition for constraining situating mistake. In our framework, multi-innovation multi-receiving wire framework joined with fingerprinting strategy allows to get adequately exact outcomes. For creating ideal situating calculations we propose to exploit best works on, including [14,16–19].

10.4.3 Wireless Access

A radio access is proposed that isolates physical layer from MAC and system layers, comparably as it happens in Radio Access Networks in 5G systems. The primary contrast is that our stage requires diverse tweak for various remote innovations. To achieve this goal, the radio wire ought to create multi-modulator that consistently sends casings of every last one of the advancements. Accordingly, the receiving wire sends innovation multiplex in time. The transmission and the gathering from the clients' gadgets ought to be synchronized. This implies when a given innovation sends signals and holds up reactions in at least one channels, alternate advances are crippled (in the radio access). 802.15.4 Standard tweaks the flag with counterbalance quadrature stage move keying (offset quadrature phase-shift keying [O-QPSK]). Likewise Bluetooth innovation bases on 802.15.4 for the physical layer execution, Wi-Fi innovation regulates the flag by methods for differential binary PSK (DBPSK) for 1 megabit for every second information rate flag, and differential quadrature PSK (DQPSK) for 2 Mbps information rate flag. Nonetheless, different augmentations of 802.11 standard make utilization of different tweaks and coding components (e.g., 802.11b included Complementary Code Keying for 5.5 and 11 Mbps rates). For higher rates (standard 802.11 g), Wi-Fi conveyed symmetrical recurrence division multiplexing (OFDM), where the accessible radio band is partitioned into various sub-channels and the last chip succession is isolated and encoded between the radio sub-channels. The transmitter encodes the bit streams on the 64 subcarriers utilizing binary phase shift keying (BPSK), QPSK, or one of two levels of quadrature amplitude modulation (16, or 64-QAM). Since the adjustment and in addition the recurrence extend (2.4, 5 or 60 GHz) are distinctive for various Wi-Fi guidelines, the Controller should actualize diverse Wi-Fi hubs for various benchmarks for the situation when gadgets utilizing diverse models are believed to be utilized in the specific area.

Finally, dynamic RFID utilizes double sideband regulation. The receiving wire should execute all these modulators and get data about which modulator must transmit the bit stream got from the controller. Every hub of the Technology Controller (see Figure 10.3) maps the casing into a surge of bits that will be specifically regulated by the reception apparatus. For instance, the 802.15.4 innovation makes the supposed baseband chip grouping, which is the aftereffect of the change from image to chip (source: Standard IEEE Std 802.15.4™-2011). Every four bits of the crude information stream are changed over in 32 chips, so the baseband chip arrangement is eight times longer than the 802.15.4 edge made by the controller. In this way, the connection between the controller and the receiving wire ought to guarantee high limit (no less than 1 Gbps). Additionally, issues with synchronization may show up for high bitrates and long piece streams. Along these lines, straightforward persuade link amongst controller and receiving wire cannot be sufficient if the quantity of clients situated in the reception apparatus filtering region is high. The hubs into the Controller send data about the fundamental regulation for each piece stream in parallel to the baseband chip succession by utilizing the control interface amongst receiving wire and controller. The receiving wire does not know about higher layers functionalities (e.g., MAC) which stay under the control of the innovation hubs into the controller. The other way (i.e., for the correspondence between clients' gadgets and controller), the radio wire ought to have the capacity to determine the demodulation utilized for getting the bit stream and pass this data to the controller (by and by utilizing the control connect). This data is utilized at the host of the controller with a specific end goal to deliver the bit stream to the comparing hub. The radio wire performs activities consistently, so when it sends signals from one innovation, the reception apparatus holds up until the point that getting the reactions frames the gadgets. In the wake of completing the process of sending and getting activities, the reception apparatus changes to another innovation and requests another piece stream from another innovation hub. All the correspondence for synchronizing receiving wire and controller ought to be sent by the control connect.

References

1. 5G PPP, White Paper on 5G and Media & Entertainment (2016).
2. Matinmikko, M., Okkonen, H., Palola, M., Yrjola, S., Ahokangas, P., Mustonen, M., Spectrum sharing using licensed shared access: The concept and its workflow for LTE-advanced networks. *IEEE Wireless Communications Magazine* 72–79 (2014).
3. Roh, W., Seol, J.-Y., Park, J., Lee, B., Lee, J., Kim, Y., Cho, J., Cheun, K., Aryanfar, F., Millimeter-wave beamforming as an enabling technology for 5G cellular communications: Theoretical feasibility and prototype results. *IEEE Communications Magazine* 52(2): 106–113 (2014).
4. Guidolin, F., Nekovee, M., Investigating spectrum sharing between 5G millimeter wave networks and fixed satellite systems. *IEEE Globecom Workshops* 1–7 (2015).

5. Nekovee, M., Qi, Y., Wang, Y., Self-organized beam scheduling as enabler for coexistence in 5G unlicensed bands. *IEIC Transactions on Communication* (2017).
6. 3G PPP TSG RAN Meeting #75, RP-170828, Study on NR-based Access to Unlicensed Spectrum (2017).
7. Boccardi, F., Shokri-Ghadikolaei, H., Fodor, G., Erkip, E., Fischione, C., Kountouris, M., Popovski, P., Zorzi, M., Spectrum pooling in mm-wave networks: Opportunities, challenges and enablers. *IEEE Communications Magazine* 54(11): 33–39 (2016).
8. Nekovee, M., Richard R., 5G spectrum sharing. *arXiv preprint arXiv:1708.03772* (2017).
9. Laoudias, C., Larkou, G., Zeinalipour-Yazti, D., Panayiotou, C.G. (University of Cyprus), Li, C.-L., Tsai, Y.-K., Cywee Corporation Ltd: Accurate multi-sensor localization on android devices. Microsoft Indoor Localization Competition (2014).
10. Dentamaro, V., Colucci, D., Ambrosini, P., Nextome: Indoor positioning and navigation system. Microsoft Indoor Localization Competition (2014).
11. Jiangy, Z., Xiy, W., Li, X.-Y., Zhaoy, J., Hany, J., HiLoc: A TDoA-fingerprint hybrid indoor localization system. Technical report, Microsoft Indoor Localization Competition. 5G White Paper (2014). https://www.ngmn.org/, Accessed June 2, 2015.
12. LaSorte, N., Rajab, S., Refai, H., Experimental assessment of wireless coexistence for 802.15.4 in the presence of 802.11 g/n. In: IEEEEMC'12, pp. 473–479 (2012).
13. Seco, F., Jimenez, A., Prieto, C., Roa, J., Koutsou, K., A survey of mathematical methods for indoor localization. In: *IEEE International Symposium on Intelligent Signal Processing* 9–14 (2009).
14. Vaupel, T., Seitz, J., Kiefer, F., Haimerl, S., Thielecke, J., Wi-Fi positioning: System considerations and device calibration. In: *2010 International Conference on Indoor Positioning and Indoor Navigation* (IPIN) (2010).
15. Ferris, B., Hahnel, D., Fox, D., Gaussian processes for signal strength-based location estimation. In: Sukhatme, G.S., Schaal, S., Burgard, W., Fox, D. (eds.) *Robotics: Science and Systems, Sukhatme.* The MIT Press, Cambridge, UK (2006).
16. Microsoft indoor localization competition.
17. Reimann, R., Bestmann, A., Ernst, M., Locating technology for AAL applications with direction finding and distance measurement by narrow bandwidth phase analysis. In: Chessa S., Knauth S. (eds) *Evaluating AAL Systems Through Competitive Benchmarking.* EvAAL 2012. Communications in Computer and Information Science, vol. 362. Springer, Berlin, Germany (2013).
18. Knauth, S. (eds.), *Evaluating AAL Systems through Competitive Benchmarking. Communications in Computer and Information Science,* vol. 362, pp. 52–62. Springer, Berlin, Germany (2013).
19. Batalla, J. M., Mavromoustakis, C. X., Mastorakis, G., Sienkiewicz, K., On the track of 5G radio access network for IoT wireless spectrum sharing in device positioning applications. *Internet of Things (IoT) in 5G Mobile Technologies.* Springer, Cham 25–35 (2016).

Chapter 11

Random Access Modeling for the Cellular-Based Massive Internet of Things

Nan Jiang, Yansha Deng, Arumugam Nallanathan, and Lei Zhang

Contents

11.1 Introduction

The massive Internet of Things (mIoT) has provided an auspicious opportunity to build powerful and ubiquitous connections that face a plethora of new challenges, where cellular-based networks are potential solutions due to their high scalability, reliability, and efficiency. In view of this, a new radio access technology was developed by the Third Generation Partnership Project (3GPP) called the narrowband IoT (NB-IoT) that provides reliable connections among inexpensive IoT devices with extended coverage and low power. The NB-IoT is built from existing cellular networks to minimize development effort consumption, but which are also incorporated into fifth-generation New Radio (5G NR) to fulfill the requirements of 5G low-power wide-area (LPWA) use cases [1]. As these use cases favor delay-tolerant uplink data traffic with small size, the key target of NB-IoT design is to deal with the sporadic uplink transmissions of massive IoT devices, raising the challenge of establishing connections between IoT devices and base stations (BSs), namely, Random Access CHannel (RACH) procedure.

IoT devices perform the RACH procedure to request channel resources for uplink transmission in the cellular-based mIoT network, where the massive mIoT traffic imposes enormous load at the Radio Access Network (RAN) level. To improve the quality of service and reduce power consumption of IoT devices, the efficient RACH procedure is required to enhance the success RACH performance. Two types of RACH exist for IoT devices accessing the network: (1) the contention-free RACH for delayed-constrained access requests (e.g, handover), where the BS distributes one of the reserved dedicated preamble to a device and then the device uses its dedicated preamble to initiate a contention-free RACH; and (2) the contention-based RACH for delay-tolerant access requests (e.g, data transmission), where an IoT device randomly chooses a preamble from non-dedicated preambles to transmit to its associated BS [2]. The contention-based RACH is favored by the mIoT network for the initial association to the network, the transmission resources request, and the connection reestablishment during failure, due to the massiveness, unpredictability and delay-tolerance characteristics of mIoT traffic [3], such that most works have analyzed its scalability characteristics in supporting massive concurrent access requests [4–8].

The contention-based RACH is based on ALOHA-type access (i.e., request access in the first available opportunity), where an IoT device randomly selects a non-dedicated preamble (i.e., an orthogonal pseudo code, such as Zadoff-Chu sequence) transmitting to its associated BS via Physical Random Access CHannel (PRACH) in the first step of RACH [2]. Because a single preamble provides a single RACH opportunity, preambles contention among IoT devices represents their competition of uplink channel resources. When competing simultaneously, IoT devices choosing the same preamble bring mutual interference and collision risks in preamble detection, resulting in performance degradation in terms of high RA failure probability [3,9,10]. Because the number of IoT devices is expected to increase to more than 30,000 per cell and such IoT devices may request access simultaneously for their small-sized data packets uplink transmission [1,11,12], improving the contention-based RACH mechanisms is one of key challenges for cellular-based mIoT networks [3,10,12].

This chapter will review the state-of-the-art modeling approaches that analyze the RACH procedure in the general cellular-based mIoT and NB-IoT networks, especially with regard to interactions between static properties of physical radio channels and dynamic properties of queue evolving in each IoT device [13–17]. The development of these modeling approaches technically and chronologically occurred in the follow order: (1) a basic spatial-temporal mathematical framework for analyzing the RACH in traffic-aware cellular-based mIoT networks was first proposed in [13,14]; (2) based on this traffic-aware spatio-temporal model, an advanced model jointly considering both the signal-to-interference-plus-noise ratio (SINR) outage and collision events to analyze RACH success probability, queue length, and time delay has been represented in [15,16]; and (3) based on [13–16], a specific framework to analyze the RACH in the NB-IoT system was developed in [17], which evaluated how the repetition mechanism fulfills the requirement of improving RACH reliability in NB-IoT networks. The chapter is organized as follows:

- Overview of the RACH procedure and introduce RACH modeling challenges
- Description of the spatial-temporal RACH model
- Analytical results of RACH success probability in the spatial domain for general cellular-based mIoT networks and NB-IoT networks
- Analytical results of queue evolution considering the RACH success probability in the temporal domain for different RACH schemes
- Conclusion

11.2 Random Access Procedure and Modeling Challenges

11.2.1 Random Access Procedure

The contention-based RACH procedure is shown in Figure 11.1 and consists of four steps. In step 1, each IoT device transmits a randomly selected preamble (Msg1) from the specific preamble pool via the dedicated PRACH. The number of available preambles is denoted as ξ, where each preamble has an equal probability ($1/\xi$) to

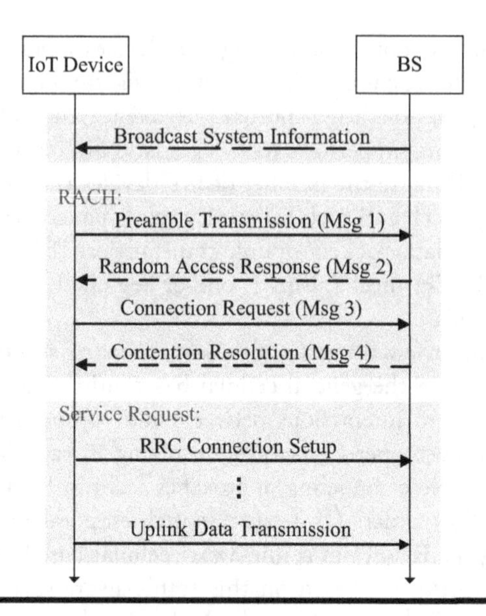

Figure 11.1 Contention-based RACH procedure.

be selected by an IoT device. In step 2, the BS acknowledges each IoT device via a random access response (RAR; Msg2) to identify if the preambles were detected. In step 3, each IoT device confirms its RAR by looking up the index of the preamble and sending the connection setup request message (Msg3) via a dedicated uplink channel in NPUSCH. In step 4, the BS transmits Radio Resource Control (RRC) connection setup messages (Msg4) to each IoT device [2].

We consider the RACH procedure a failure because (1) the BS cannot decode a received preamble due to low Signal-to-Interference-and-Noise Ratio (SINR) in step 1, namely, SINR outage, and (2) a collision occurs when two or more IoT devices select the same preamble in step 1, and the BS cannot decode Msg3.

In the cellular-based mIoT network, a huge number of IoT devices are expected to request access frequently, such that network congestion may occur due to mass concurrent data and signaling transmission [12]. This network congestion can cause severe SINR outage and collision and thus result in a great number of packets accumulating in buffers, which may cause unexpected delays. A possible solution is to restrict the access attempts in each IoT device according to some RACH control mechanism. However, the efficiency of such RACH control mechanisms need to be studied because overly restricting access requests also creates unacceptable delay as well as leading to low channel resource utilization. Here, we study the following three schemes:

- **Baseline scheme**: Each IoT device immediately attempts RACH when a packet is in the buffer. The baseline (BL) scheme is the simplest scheme without any control of traffic. Due to RACH attempts not being alleviated at the

IoT devices, the BL scheme can contribute to relatively faster buffer flushing in non-overloaded network scenarios. However, once the network is overloaded, high delays and service unavailability appear due to mass simultaneous access requests.

- **Access class barring scheme**: Each non-empty IoT device draws a random number, $q \in [0,1]$, and attempts to RACH only when $q \leq P_{ACB}$. Here P_{ACB} is the access class barring (ACB) factor specified by the BS according to the network condition [2,12]. An ACB scheme is a basic congestion control method that reduces RACH attempts from the side of IoT devices based on the ACB factor. A suitable ACB factor can maintain allowable access in a reasonable density and assure a relatively high data transmission rate when the network is overloaded.
- **Back-off scheme**: Each non-empty IoT device transmits packets the same as the BL scheme, when a packet is in the buffer. However, when RACH fails, the IoT device automatically defers the RACH reattempt and waits for t_{BO} (i.e., the back-off [BO] factor specified by the BS) time slots until it trys again. The BO scheme is another basic congestion control method, where each IoT device can automatically alleviate congestion and requires less of a control message from the BS than does the ACB scheme [10].

11.2.2 RACH Modeling Challenges

The contention-based RACH has been widely studied in conventional cellular networks, where the most critical point of this issue concerns modeling and analyzing time-varying queues and RACH schemes in the Media Access Control (MAC) layer [18,19]. Recently, a number of studies have been launched to discuss whether the contention-based RACH of long-term evolution (LTE) is suitable for mIoT and how to evolve cellular systems to provide efficient access for mIoT networks [3,7,20,21]. However, in [7,18–19], the collision events were considered as the main outage condition, and the preamble transmission failure that was impacted by the physical channel propagation characteristics was simplified. Generally speaking, in the large-scale cellular-based mIoT network, the physical layer characteristics can strongly influence the performance of RACH success because the SINR received at the BS can be severely degraded by the mutual interference generated from massive IoT devices. In this scenario, the random positions of the transmitters make accurate modeling and analysis of this interference even more complicated.

Stochastic geometry has been regarded as a powerful tool to model and analyze mutual interference between transceivers in the wireless networks, such as conventional cellular networks [22–24], wireless sensor networks [25], cognitive radio networks [26,27], and heterogenous cellular networks [28–30]. However, there are two aspects that limit the application of conventional stochastic geometry analysis to the RACH analysis of the cellular-based mIoT networks: (1) conventional stochastic geometry works focused on analyzing normal uplink and downlink data transmission channel, where the intra-cell interference is not considered because of the ideal assumption that each orthogonal sub-channel is not reused in a cell, whereas massive IoT devices in a cell may randomly choose and transmit the same preamble using the same sub-channel; and (2) these conventional stochastic geometry works only modeled the spatial distribution of transceivers and ignored the interactions between static properties of physical layer network and the dynamic properties of queue evolving in each transmitter due to the assumptions of backlogged network with saturated queues [31–33].

To model these aforementioned interactions, recent investigations have studied the stability of spatially spread interacting queues in the network based on stochastic geometry and queuing theory [30–33]. The work in [31] is the first paper applying the stochastic geometry and queuing theory to analyze the performance of RACH in distributed networks where each transmitter is composed of an infinite buffer and its location is changed following a high mobility random walk. The work in [32] investigated the stable packet arrival rate region of a discrete-time slotted RACH network where the transceivers were static and distributed as independent Poisson point processes (PPPs). The work in [30] analyzed the delay in the heterogeneous cellular networks with spatio-temporal random arrival of traffic where the traffic of each device was modeled by a marked Poisson process and the statistics of such traffic with different offloading policies are compared. In [33], the authors modeled the randomness in the locations of IoT devices and BSs via PPPs, and leveraged the discrete-time Markov chain to model the queue and protocol states of each IoT device. However, the model was limited in capturing the dynamic preamble success probability during the time evolution, such that it could only derive the analytical result during the steady state, and this result could not be verified by simulations.

11.3 System Model

11.3.1 Physical Layer Description

We consider an uplink model for a cellular-based mIoT network that consists of BSs and IoT devices, which are spatially distributed in $\mathbb{R}^2$ following two independent homogeneous Poisson point process (HPPP), Φ_B and Φ_D, with intensities λ_B and λ_D, respectively. Each IoT device associates to its geographically closest BS and thus forms a Voronoi tesselation, where the BSs are uniformly distributed in a Voronoi cell. Path-loss attenuation is defined as $r^{-\alpha}$, where r is the propagation distance and α is the path-loss exponent. We considered an identically distributed Rayleigh fading channel, where the channel power gain, h, is assumed to be an exponentially distributed random variable with a unit mean. According to [1,34], the transmitted power of IoT devices determined by the full path-loss inversion power control, where each IoT device maintains the received signal power in the BS equal to the same threshold, ρ, by compensating for its own path-loss.

Note that the RACH procedure in NB-IoT networks is advanced on the basis of the general cellular network system. In the following, we focus on introducing the RACH model in NB-IoT networks, and then briefly introduce the RACH model in the general cellular-based mIoT system. To mitigate the SINR outage in NB-IoT networks, a repetition scheme was designed in that the IoT device continuously transmited the same preamble for a dedicated number of times equal to the repetition value N_r. Each CE group can configure its specific repetition value from the set that $\mathcal{R}$ belongs to {1, 2, 4, 8, 16, 32, 64, 128}. The repetition values of three Coverage Enhancement (CE) groups follow the rule of $N_{r,0} < N_{r,1} < N_{r,2}$, and $N_{r,0}$ can only configure from the subset {1, 2}.

Single tone, signal frequency-hopping algorithms[1] have been designed for the narrowband physical random access channel (NPRACH) to obtain the time and frequency diversity gain. A preamble consists of four symbol groups as a basic unit, which is transmitted via four different sub-carriers with the fixed-size frequency hopping [1]. During the preamble transmission, the pseudo-random hopping is used for different repetitions, decided on the basis of the current repetition time and the narrowband physical cell indentification (NCellID) [1]. The details of frequency hopping can be found in [17,35]. Note that the collision cannot be relieved via the frequency hopping because all sub-carriers in the hopping are already determined by NCellID (i.e., the start sub-carrier has a one-to-one correspondence with all of the following sub-carriers).

In the uplink of NB-IoT, the narrowband physical uplink shared channels (NPUSCHs) are used for data transmission and the NPRACHs are used for preamble transmission. In more detail, 180 kHz of spectrum is occupied with a 3.75-kHz tone spacing (i.e., spans are over 48 sub-carriers) or a 15-kHz tone spacing (i.e., spans are over 12 sub-carriers), where the NPRACH only supports a 3.75-kHz tone spacing. To fulfill the coverage requirements of different IoT devices, the NB-IoT network can configure up to three repetition values from the set {1, 2, 4, 8, 16, 32, 64, 128} in a cell and allows the flexible configuration of NPRACH resources [36]. In this model, we consider a single repetition value to provide fundamental insights due to repetition, where the related resources assignment of NPRACHs only takes place in the beginning of a transmission time interval (TTI), as shown in Figure 11.2.

Recalling the repetition scheme, an active IoT device repeats a same preamble N_τ times (i.e., the dedicated repetition value). In each repetition, a preamble consists of four symbol groups, where the first preamble symbol group is transmitted via a sub-carrier determined by pseudo-random hopping (i.e., the hopping depends

[1] Signal frequency hopping algorithms can also facilitate the time-of-arrival estimation [35].

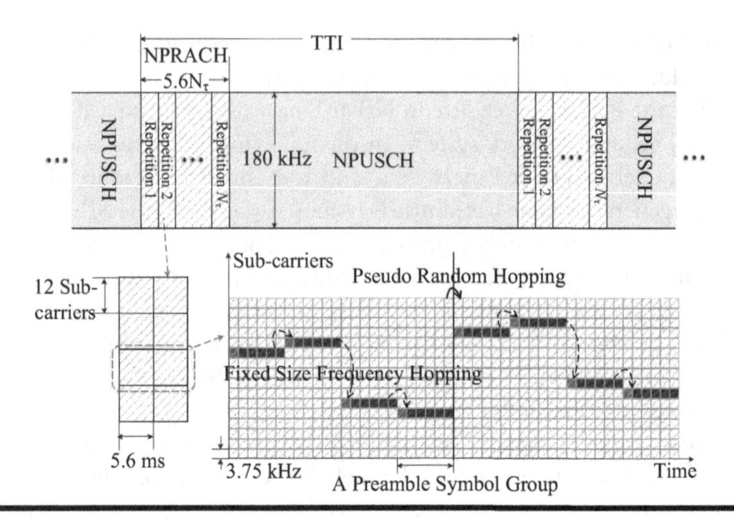

Figure 11.2 Structure of NPUSCH and NPRACH.

on the current repetition time and the NCellID), and the following three preamble symbol groups are transmitted via sub-carriers determined by the fixed-size frequency hopping [34]. In other words, if two or more IoT devices chose the same first sub-carrier in a single RACH opportunity, the following sub-carriers (i.e., in the same RACH opportunity) would be same because these two hopping algorithms lead to one-to-one correspondences between the first sub-carrier and the following sub-carriers.

We first formulate the SINR outage condition. In each TTI, a preamble transmission success occurs if any repetition succeeds, and in a single repetition, a preamble is successfully received at the associated BS if all four received SINRs are above the SINR threshold, γ_{th}. Overall, the preamble transmission success probability under N_τ repetitions conditioning on n number of intra-cell interfering IoT devices is expressed as

$$p_S(N_\tau, n) = 1 - \prod_{n_\tau=1}^{N_\tau} \left(1 - \mathbb{P}\{\theta_{n_\tau} \mid Z_D = n\}\right), \tag{11.1}$$

where $Z_D = \left\| \mathcal{Z}_{\text{intra}} \right\|$ is the number of active intra-cell interfering IoT devices, and

$$\theta_{n_\tau} \Delta = \begin{Bmatrix} SINR_{n_\tau,1} \geq \gamma_{th}, SINR_{n_\tau,2} \geq \gamma_{th}, \\ SINR_{n_\tau,3} \geq \gamma_{th}, SINR_{n_\tau,4} \geq \gamma_{th} \end{Bmatrix}. \tag{11.2}$$

In Eq. (11.2), γ_{th} is the SINR threshold, and $SINR_{n_\tau,1}$ is the received SINR of the first symbol group in the n_τth repetition. The BSs successfully decode a preamble when the received SINR is above the threshold. Based on Slivnyak's theorem [37], we formulated the SINR of a typical BS located at the origin in the mth time slot as

$$\mathrm{SINR} = \rho h_o \Big/ \underbrace{\sum_{u_j \in \mathcal{Z}_{\mathrm{in}}} \mathbf{1}_{\{N^m_{\mathrm{New}_j} + N^m_{\mathrm{Cum}_j} > 0\}} \mathbf{1}_{\{UR\}} \rho h_j}_{\mathcal{I}_{\mathrm{intra}}} + \underbrace{\sum_{u_i \in \mathcal{Z}_{\mathrm{out}}} \mathbf{1}_{\{N^m_{\mathrm{New}_i} + N^m_{\mathrm{Cum}_i} > 0\}} \mathbf{1}_{\{UR\}} P_i h_i \|u_i\|^{-\alpha}}_{\mathcal{I}_{\mathrm{inter}}} + \sigma^2, \quad (11.3)$$

where ρ is the full path-loss inversion power control threshold, h_o is the channel power gain from the typical IoT device to its associated BS, $\mathcal{Z}_{\mathrm{in}}$ is the set of intra-cell interfering IoT devices, $\mathcal{Z}_{\mathrm{out}}$ is the set of inter-cell interfering IoT devices, $N^m_{\mathrm{New}_j}$ is the numbers of new arrived packets in the mth time slot of jth interfering IoT device, $N^m_{\mathrm{Cum}_j}$ is the numbers of accumulated packets in the mth time slot of jth interfering IoT device, $\|\cdot\|$ is the Euclidean norm, u_i is the distance between the ith inter-cell IoT device and the typical BS, $P_i = \rho r_i^{\alpha}$ is the actual transmit power of the ith inter-cell IoT device with the distance r_i from its associated BS, σ^2 is the noise power, and $\mathcal{I}_{\mathrm{intra}}$ and $\mathcal{I}_{\mathrm{inter}}$ are the aggregate intra-cell and inter-cell interferences, respectively.

In Eq. (11.3), $\mathbf{1}_{\{\cdot\}}$ is the indicator function that takes the value 1 if the statement $\mathbf{1}_{\{\cdot\}}$ is true, and zero otherwise. Whether an IoT device generates interference depends on two conditions: (1) $\mathbf{1}_{\{N^m_{\mathrm{New}} + N^m_{\mathrm{Cum}_i} > 0\}}$, which means that an IoT device is able to generate interference only when its buffer is non-empty; and (2) $\mathbf{1}_{\{UR\}}$, which means that an IoT device is able to generate interference only when the IoT device does not defer its access attempt due to the RACH scheme. In addition, once the two conditions are satisfied, we say that the IoT device is active.

Mathematically, the non-empty probability of each IoT device can be treated using the thinning process. We assume that the non-empty probability, $\mathcal{T}^m$, and the non-restrict probability, $\mathcal{R}^m$, of each IoT device in the mth time slot are defined as

$$\begin{aligned} \mathcal{T}^m &= \mathbb{P}\{N^m_{\mathrm{New}} + N^m_{\mathrm{Cum}} > 0\}, \text{ and} \\ \mathcal{R}^m &= \mathbb{P}\{unrestricted\} \end{aligned} \quad (11.4)$$

where the non-restrict probability $\mathcal{R}^m$ depends on the RACH schemes, which will be discussed next.

In Eq. (11.3), $\mathbf{1}_{\{UR\}}$ presents that an IoT device generates interference only when its RACH attempt is not restricted by the RACH scheme (such as in the ACB scheme, generating $q > \mathrm{P}_{ACB}$ leads to $\mathbf{1}_{\{UR\}} = 0$), and $\mathbf{1}_{\{N^m_{\mathrm{New}_i} + N^m_{\mathrm{Cum}_i} > 0\}}$ presents that only an IoT device with non-empty buffer generating interference. The queue status of an IoT device is jointly populated by the new arrival packets (i.e., according to Poisson arrival process Λ_{New}) and the accumulated packets in the previous time slots.

Then, we formulate the RACH success probability under both SINR outage and collision conditions. Recalling that a collision occurs if a BS receives multiple preambles from the same set of sub-carriers at the same time. Consequently, a RACH succeeds when the preamble is successfully received in the BS and no collision occurs, which is presented as

$$\mathcal{P}_{N_\tau} = \sum_{n=0}^{\infty}\left(\mathbb{P}\{Z_D = n\}\, p_{S,0}(N_\tau, n)\prod_{l=1}^{n}\left(1 - p_{S,l}(N_\tau, n)\right)\right), \tag{11.5}$$

where $p_{S,0}(N_\tau, n)$ is the preamble transmission success probability of the typical IoT device, $p_{S,l}(N_\tau, n)$ is the preamble transmission success probability of the lth interfering IoT device located in the typical cell, and $\mathbb{P}\{Z_D = n\}$ is the probability of n number of interfering IoT devices located in the typical cell.

In the general cellular-based mIoT networks, the RACH success probability is still defined based on Eq. (11.5), but the repetition scheme is not considered, and a preamble only consists of one symbol group. Consequently, the preamble transmission success probability given in Eq. (11.1) is modified to

$$p_{S,CB}(1, n) = \mathbb{P}\{\text{SINR} \geq \gamma_{th} \mid Z_D = n\}, \tag{11.6}$$

where SINR is as given in Eq. (11.3).

11.3.2 Media Access Control Layer Description

We consider a time-slotted mIoT network, where the NPRACH (i.e., PRACH in general cellular-based mIoT networks) happens at the beginning of a TTI within a small time interval, τ_c, and the least time of a TTI is a gap interval duration, τ_g, for data transmission via NPUSCH, as shown in Figure 11.1. Here, the active IoT device represents that an IoT device is with non-empty buffers (i.e., $N_{New}^{m} + N_{Cum}^{m} > 0$, where N_{New}^{m} is the number of newly arrived packets, and N_{Cum}^{m} is the number of accumulated packets) without access restriction, which will be detailed in the following section.

Without loss of generality, we assume the size of buffer in each IoT device is infinite, and none of the packets will be dropped off. At the beginning of the NPRACH in the mth time slot, each IoT device checks its buffer status to determine whether it needs to attempt RACH as shown in the Figure 11.1. In detail, the buffer status (i.e., queuing packets) are determined by the new arrived packets, and the accumulated packets that unsuccessfully departs (i.e., unsuccessfully RACH attempts or never been scheduled) before the last time slot.

Once a RACH succeeds, the IoT device will transmit the corresponding data sequences with the scheduled uplink channel resources. Here, we interchangeably use packet to represent the data sequences. In each device, the packets line in a queue waiting to be transmitted, where each packet has the same priority, and the BSs are unaware of the queue status of their associated IoT devices. It is assumed that the BS will only schedule uplink channel resources for the head-of-line packet and each IoT deivce transmits packets via a first-come-first-serve packet scheduling scheme—the basic and the most simplest packet scheduling scheme, where all packets are treated equally by placing them at the end of the queue once they arrive [38].

We model the new arrived packets (N^m_{New}) in the mth time slot at each IoT device as independent Poisson arrival process, Λ^m_{New} with the same intensity $\varepsilon^m_{\text{New}}$ as [5,39,40] (i.e., these new packets are actually arrived within the $(m-1)$th time slot, but they are first considered in the mth time slot due to the slotted-ALOHA behavior). Therefore, the number of new arrival packets N^m_{New} in a specific time slot (i.e., within the time duration $\tau_c + \tau_g$) is described by the Poisson distribution with $N^m_{\text{New}} \sim \text{Pois}(\mu^m_{\text{New}})$, where $\mu^m_{\text{New}} = (\tau_c + \tau_g)\varepsilon^m_{\text{New}}$. The accumulated packets (N^m_{Cum}) at each IoT device is evolved following transmission condition over time, which is described in Table 11.1. Specifically, a packet is removed from the buffer once the RACH succeeds, otherwise, this packet will be still in the first place of the queue, and the IoT device will try to request channel resources for the packet in the next available RACH. Note that the data transmission after a successful RACH can be easily extended following the analysis of preamble transmission success probability in RACH. Because the main focus of this chapter is to analyze the contention-based RACH in the mIoT network, we assume that the actual intended packet transmission is always successful if the corresponding RACH succeeds.

Table 11.1 Packets Evolution in the Typical IoT Device

Slot	*Success*	*Failure*
1st	$N^1_{Cum} = 0$	$N^1_{Cum} = 0$
2nd	$N^2_{Cum} = N^1_{New} - 1$	$N^2_{Cum} = N^1_{New}$
3rd	$N^3_{Cum} = N^2_{Cum} + N^2_{New} - 1$	$N^3_{Cum} = N^2_{Cum} + N^2_{New}$
⋮	⋮	⋮
*m*th	$N^m_{Cum} = N^{m-1}_{Cum} + N^{m-1}_{New} - 1$	$N^m_{Cum} = N^{m-1}_{Cum} + N^{m-1}_{New}$

11.4 RACH Success Probability in the Spatial Domain

In this section, we provide a general single time slot analytical model for RACH success probability in general cellular-based mIoT networks and in NB-IoT networks, respectively.

11.4.1 General Cellular-Based mIoT Networks

In general cellular-based mIoT networks, the repetition scheme is not considered, and a preamble consists of only one symbol group. Recall that the RACH success probability is performed on a BS associating with a randomly chosen active IoT device, which refers to the preamble being successfully transmitted to the associated BS (i.e., received SINR is greater than the SINR threshold defined in Eq. [11.6]) and no collision occurs (i.e., no other IoT devices successfully transmits a same preamble to the typical BS simultaneously). In this general case, the RACH success probability given in Eq. (11.5) is simplified to

$$\mathcal{P}_{CB}^{m} = \sum_{n=0}^{\infty} \left\{ \underbrace{\mathbb{P}[Z_D = n]}_{\text{I}} \underbrace{\mathbb{P}[\text{SINR}_o \geq \gamma_{th} \mid Z_D = n]}_{\text{II}} \underbrace{\left(\prod_{i=1}^{n} \mathbb{P}[\text{SINR}_i < \gamma_{th} \mid Z_D = n]\right)}_{\text{III}} \right\}, \quad (11.7)$$

where γ_{th} is the SINR threshold, Z_D is the number of intra-cell interfering IoT devices (i.e., transmitting the same preamble as the typical IoT device simultaneously), SINR_o and SINR_i are the received SINRs of the preamble from the typical and the ith interfering IoT device following from Eq. (11.3), I in Eq. (11.7) is the probability of Z_D number of interfering IoT devices located in the typical BS, II in Eq. (11.7) represents the preamble transmission success probability that the typical IoT device successfully transmits the preamble to the associated BS conditioning on $Z_D = n$, and III in Eq. (11.7) represents the preamble transmission failure probability that the preambles transmitting from other n intra-cell interfering IoT devices are not successfully received by the BS conditioning on $Z_D = n$.

We assume $\hat{Z}_{in}$ denotes the number of active IoT device in a specific Voronoi cell, and let $Z_D = |\hat{Z}_{in}| - 1$ denotes the number of active interfering IoT devices in such cell, where the Laplace Transform of aggregate intra-cell interference is conditioned on Z_D. The Probability Density Function (PDF) of the number of active interfering IoT devices in a Voronoi cell has been derived by the Monte Carlo method in [41] and conditioned on a randomly chosen IoT device in its cell, the PMF of the number of interfering intra-cell IoT devices in that cell Z_D is expressed as [14, Eq. (11.5)]

$$\mathbb{P}\{Z_D = n\} = \frac{c^{(c+1)}\Gamma(n+c+1)\left(\frac{\mathcal{T}^m \mathcal{R}^m \lambda_{Dp}}{\lambda_B}\right)^n}{\Gamma(c+1)\Gamma(n+1)\left(\frac{\mathcal{T}^m \mathcal{R}^m \lambda_{Dp}}{\lambda_B} + c\right)^{n+c+1}}, \quad (11.8)$$

where $c = 3.575$ is a constant related to the approximate Probability Mass function (PMF) of the PPP Voronoi cell [41], $\Gamma(\cdot)$ is the gamma function, λ_{Dp} is the density of IoT devices using the same preamble.

Next, we derive the preamble transmission success probability presented in II of Eq. (11.7). According to the Slivnyak's Theorem [37], the locations of inter-cell IoT devices follow the Palm distribution of Φ_{Dp}, which is the same as the original Φ_{Dp}. The probability that the received SINR at the BS from a randomly chosen IoT device exceeds a certain threshold γ_{th} conditioning on the given number of interfering IoT devices in that cell n_1 is presented in following lemma.

Lemma 11.1: *The probability that the received SINR at the BS from a randomly chosen IoT device exceeds a certain threshold γ_{th} conditioning on a given number of interfering IoT devices in that cell n is expressed as [14, Eq. (11.7)]*

$$\mathbb{P}\left[\frac{\rho h_o}{\mathcal{I}_{\text{intra}} + \mathcal{I}_{\text{inter}} + \sigma^2} \geq \gamma_{th} \mid Z_D = n\right] = \mathbb{P}\left[h_o \geq \frac{\gamma_{th}}{\rho}(\mathcal{I}_{\text{intra}} + \mathcal{I}_{\text{inter}} + \sigma^2) \mid Z_D = n\right]$$

$$\overset{(a)}{=} \mathbb{E}\left[\exp\{\frac{\gamma_{th}}{\rho}(\mathcal{I}_{\text{intra}} + \mathcal{I}_{\text{inter}} + \sigma^2)\} \mid Z_D = n\right]$$

$$= \exp(-\frac{\gamma_{th}}{\rho}\sigma^2)\mathcal{L}_{\mathcal{I}_{\text{inter}}}(\frac{\gamma_{th}}{\rho})\mathcal{L}_{\mathcal{I}_{\text{intra}}}(\frac{\gamma_{th}}{\rho} \mid Z_D = n), \tag{11.9}$$

where the expectation in (a) is with respective to $\mathcal{I}_{\text{inter}}$ and $\mathcal{I}_{\text{intra}}$, $\mathcal{L}_{\mathcal{I}_{\text{intra}}}(\cdot)$ denotes the Laplace transform of the aggregate intra-cell interference $\mathcal{I}_{\text{intra}}$, and $\mathcal{L}_{\mathcal{I}_{\text{inter}}}(\cdot)$ denotes the Laplace transform of the aggregate inter-cell interference $\mathcal{I}_{\text{inter}}$. In Eq. (11.9), the Laplace Transform of $\mathcal{I}_{\text{inter}}$ and $\mathcal{I}_{\text{intra}}$ derived in [14, Eqs. (11.4) and (11.6)], are respectively given as

$$\mathcal{L}_{\mathcal{I}_{\text{inter}}}(\frac{\gamma_{th}}{\rho}) = \exp\left(-2(\gamma_{th})^{\frac{2}{\alpha}}\frac{\mathcal{T}^m\mathcal{R}^m\lambda_{Dp}}{\lambda_B}\int_{(\gamma_{th})^{\frac{-1}{\alpha}}}^{\infty}\frac{y}{1+y^{\alpha}}dy\right), \text{ and}$$

$$\mathcal{L}_{\mathcal{I}_{\text{intra}}}(\frac{\gamma_{th}}{\rho} \mid Z_D = n) = \frac{1}{(1+\gamma_{th})^n}, \tag{11.10}$$

where $\mathcal{T}^m$ and $\mathcal{R}^m$ are defined in Eq. (11.4). Recall that λ_{Dp} is the intensity of IoT devices using the same preamble.

Proof. See **Appendices A** and **B** in [14].

Substituting Eqs. (11.8) and (11.9) into Eq. (11.7), we derive the RACH success probability in the m th time slot $\mathcal{P}_{CB}^m$ in the following theorem.

Theorem 11.1: *In the depicted cellular-based mIoT network, the RACH success probability of a randomly chosen IoT device in the mth time slot is derived as*

$$\mathcal{P}_{CB}^{m} = \sum_{n=0}^{\infty} \left\{ \underbrace{\frac{c^{(c+1)}\Gamma(n+c+1)\left(\frac{\mathcal{T}^m \mathcal{R}^m \lambda_{Dp}}{\lambda_B}\right)^n}{\Gamma(c+1)\Gamma(n+1)\left(\frac{\mathcal{T}^m \mathcal{R}^m \lambda_{Dp}}{\lambda_B} + c\right)^{n+c+1}}}_{\text{I}} \right. \underbrace{\frac{\exp\left(-\frac{\gamma_{th}\sigma^2}{\rho} - 2(\gamma_{th})^{\frac{2}{\alpha}} \frac{\mathcal{T}^m \mathcal{R}^m \lambda_{Dp}}{\lambda_B} \int_{(\gamma_{th})^{\frac{-1}{\alpha}}}^{\infty} \frac{y}{1+y^{\alpha}} dy\right)}{(1+\gamma_{th})^n}}_{\text{II}} \left. \underbrace{\left(1 - \frac{\exp\left(-\frac{\gamma_{th}\sigma^2}{\rho} - 2(\gamma_{th})^{\frac{2}{\alpha}} \frac{T^m R^m \lambda_{Dp}}{\lambda_B} \int_{(\gamma_{th})^{\frac{-1}{\alpha}}}^{\infty} \frac{y}{1+y^{\alpha}} dy\right)}{(1+\gamma_{th})^n}\right)^n}_{\text{III}} \right\}. \tag{11.11}$$

In Eq. (11.11), it can be shown that the preamble transmission *success* probability of the typical IoT device is inversely proportional to the received SINR threshold, γ_{th}, and the preamble transmission *failure* probabilities of other interfering IoT devices are directly proportional to the received SINR threshold, γ_{th}, which leads to the fact that the non-collision probability (i.e., the probability of a successful transmission preamble does not collide with others) of the typical IoT devices is also directly proportional to the received SINR threshold, γ_{th}. Therefore, a trade-off between preamble transmission success probability and non-collision probability is observed. For illustration, the relationship among RACH success probability, the preamble tranmission success probability, and the non-collision probability are shown in Figure 11.3.

11.4.2 Narrowband Internet of Things Networks

In this subsection, we derive the RACH success probability of a randomly chosen IoT device in NB-IoT networks. Because each intra-cell–interfering IoT device has the same preamble transmission success probability, the RACH success probability presented in Eq. (11.5) can be simplified as

$$\mathcal{P}_{N_\tau}^{m} = \sum_{n=0}^{\infty} \mathbb{P}\{Z_D = n\} p_S(N_\tau, n)\left(1 - p_S(N_\tau, n)\right)^n, \tag{11.12}$$

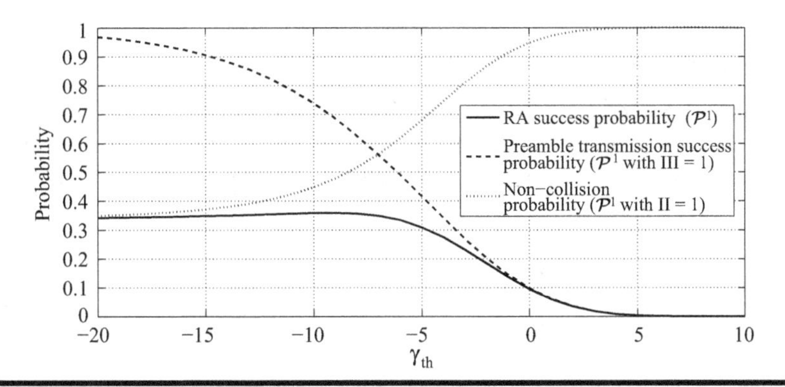

Figure 11.3 Comparing RACH success probability ($\mathcal{P}^1_{CB}$), preamble transmission success probability ($\mathcal{P}^1_{CB}$ with III = 1), and non-collision probability ($\mathcal{P}^1_{CB}$ with II = 1). The parameters are $\lambda_B = 10$ BS/km², $\lambda_{Dp} = 100$ IoT deivces/preamble/km², $\rho = -90$ dBm, $\sigma^2 = -90$ dBm, $\mathcal{R}^1 = 1$, and the new packets arrival rate $\mu^1_{\text{New}} = 0.1$ packets/time slot).

where $\mathbb{P}\{Z_D{=}n\}$ is the probability of the number of interfering IoT devices $Z_D = n$, and $p_S(N_\tau, n)$ is the preamble transmission success probability of an IoT device conditioning on $\{Z_D{=}n\}$. The probability mass function (PMF) of the number of interfering IoT devices, Z_D, is represented as Eq. (11.8).

Without loss of generality, we assume each IoT device remains spatially static during a TTI (i.e., we assume that the preamble format 0 is used, where a preamble repetition only takes 5.6 ms [34]), and thus the mutual interference among IoT devices is temporally correlated [42]. This temporal correlation complicates the derivation of the preamble transmission success probability, which is the main challenge of RACH success analysis. The preamble transmission success probability with N_τ repetitions $p_S(N_\tau, n)$ is derived in the following theorem.

Theorem 11.2: *The preamble transmission success probability of a randomly chosen IoT device with N_τ preamble repetitions $p_S(N_\tau, n)$ is expressed as*

$$\begin{aligned} p_S(N_\tau, n) &= 1 - \prod_{n_\tau=1}^{N_\tau} \left(1 - \mathbb{P}\{\theta_{n_\tau} \mid Z_D = n\}\right) \\ &= \sum_{n_\tau=1}^{N_\tau} (-1)^{n_\tau+1} \binom{N_\tau}{n_\tau} \mathbb{P}\{\theta_1, \cdots, \theta_{n_\tau} \mid Z_D = n\}, \end{aligned} \tag{11.13}$$

where $\binom{N_\tau}{n_\tau} = \frac{N_\tau!}{n_\tau!(N_\tau - n_\tau)!}$ *is the binomial coefficient, and* $\mathbb{P}\{\theta_1,\cdots,\theta_{n_\tau} \mid Z_D = n\}$ *is the probability that all of* $4 \times n_\tau$ *(i.e., a preamble consists of four symbol groups) time-correlated preamble symbol groups are successfully received in the BS. For ease of presentation, we assume* $m = 4 \times n_\tau$, *and* $\mathbb{P}\{\theta_1,\cdots,\theta_{n_\tau} \mid Z_D = n\}$ *is expressed as [17, Eq. (11.11)]*

$$\mathbb{P}\{\theta_1,\cdots,\theta_{n_\tau} \mid Z_D = n\} = \frac{\exp\left(-\frac{m\gamma_{th}\sigma^2}{\rho} - 2(\gamma_{th})^{\frac{2}{\alpha}} \frac{T^m R^m \lambda_{Dp}}{\lambda_B} \int_{(\gamma_{th})^{-\frac{1}{\alpha}}}^{\infty} \left[1 - \left(\frac{1}{1+y^{-\alpha}}\right)^m\right] y dy\right)}{(1+\gamma_{th})^{mn}}, \tag{11.14}$$

where α *is the path-loss parameter,* γ_{th} *is the received SINR threshold,* σ^2 *is the noise,* ρ *is the full path-loss inversion power control threshold, and* T^m *and* $\mathcal{R}^m$ *are as defined in Eq. (11.4).*

Proof. See proof of **Theorem 11.1** in [17].

For simplicity, we present a special case of RACH with two preamble repetitions, and the preamble transmission success probability $p_S(2,n)$ is expressed as

$$\begin{aligned} p_S(2,n) &= 1 - (1 - \mathbb{P}\{\theta_1 \mid Z_D = n\})(1 - \mathbb{P}\{\theta_2 \mid Z_D = n\}) \\ &\overset{(a)}{=} 2\mathbb{P}\{\theta_1 \mid Z_D = n\} - \mathbb{P}\{\theta_1,\theta_2 \mid Z_D = n\}, \end{aligned} \tag{11.15}$$

where (a) follows from $\mathbb{P}\{\theta_1 \mid Z_D = n\} = \mathbb{P}\{\theta_2 \mid Z\_D = n\}$. Substituting Eqs. (11.15) and (11.8) into Eq. (11.12), we obtain the RACH success probability when $N_\tau = 2$.

For illustration, we plot the RACH success probabilities of a randomly chosen IoT device in the light traffic scenario and the heavy traffic scenario shown in Figure 11.4a and b, respectively. The analytical curves of the RACH success probability are plotted using Eq. (11.12), which closely matches with simulation points that validate the accuracy of developed mathematical framework. We first observe that for all curves, the RACH success probability decreases with the increase of the density ratio between IoT devices and BSs (λ_D / λ_B), which is due to the following two reasons: (1) increasing the number of IoT devices generates more interference, leading to lower received SINR at the BS; and (2) increasing the number of IoT devices results in higher probability of collision. In both sub-figures, increasing the repetition value increases the RACH success probability, which is because it offers more opportunities to retransmit a preamble with the time and frequency diversity.

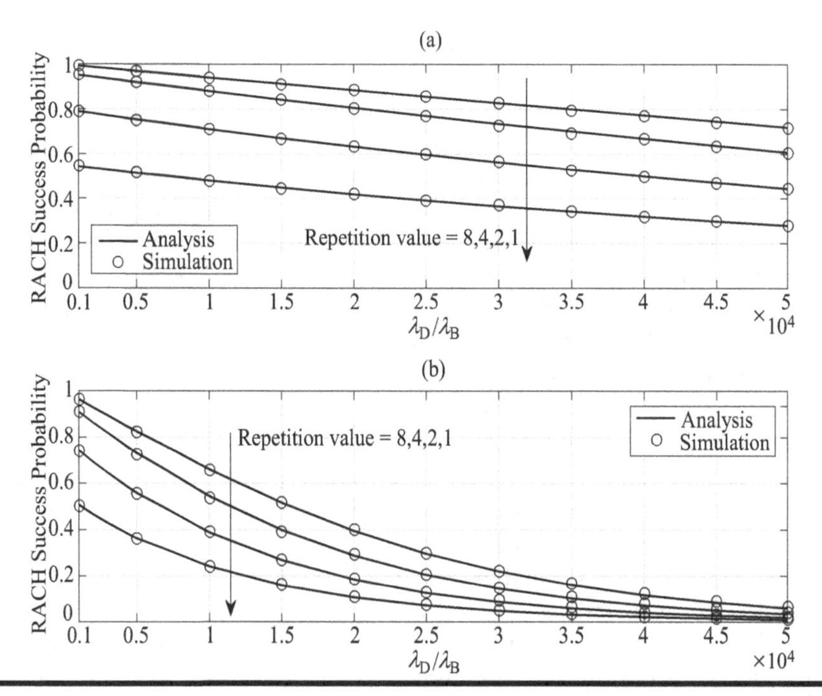

Figure 11.4 Comparing RA success probability with different repetition value in NB-IoT networks. We set $\lambda_B = 0.1$ BSs/km^2, $\gamma_{th} = 0$ dB, $\alpha = 4$, $\rho = -130$ dBm, the bandwidth of a sub-carrier is $BW = 3.75$ kHz, and thus the noise is $\sigma^2 = -174 + 10log_{10}(BW) = -138.3$ dBm. Different from coventional cellular networks with TTI = 40 ms [2], we set TTI in NB-IoT network as 640 ms following the defined Narrowband Physical Broadcast Channel (NPBCH) TTI [1]. The packet arrival periods of IoT devices are from a few minutes to several days [43], hence we assume two traffic scenarios, where (a) the light traffic scenario is 1 packet/h of each IoT device and (b) the heavy traffic scenario is 1 packet/10 min of each IoT device. Therefore, the active probabilities of each IoT device during 640 ms are $p_{a_l} = 640/3600000 = 0.00018$ and $p_{a_h} = 640/600000 = 0.0011$, respectively.

11.5 Queue Evolution Model in the Temporal Domain

In this section, we analyze the performance of the cellular-based mIoT network in each time slot with different schemes. Note that any cellular-based mIoT network (i.e., referring to either general cellular-based mIoT networks or NB-IoT networks in this chapter) can use the same model to capture the queue evolution of IoT devices. For the purpose of simplicity, we use $\mathcal{P}^m$ to represent the general RACH success probability in any network scenario, which represents $\mathcal{P}_{CB}^m$ in general cellular-based mIoT networks and represents $\mathcal{P}_{N_\tau}^m$ in NB-IoT networks. As mentioned in Eq. (11.4), the RACH success probability depends on the non-empty probability $\mathcal{T}^m$ and the non-restrict probability $\mathcal{R}^m$ of each IoT device, which raises the problem of how to study the queue status of each IoT device in each time slot.

The queue status and the preamble transmission are interdependent, and this imposes a causality problem. More specifically, the preamble transmission of a typical IoT device in the current time slot depends on the aggregate interference from those active IoT devices in that time slot, thus we need to know the current queue status, which is decided by the previous queue statuses, as well as the preamble transmission success probabilities of previous time slots. Recall that the evolution of queue status follows Table 11.1, where the accumulated packets come from the packets that are not successfully transmitted in the previous time slots.

Mathematically, to derive the RACH success probability of a randomly chosen IoT device in the mth time slot, we first derive the non-empty probability, $\mathcal{T}^m$, and the non-restrict probability, $\mathcal{R}^m$, of the IoT device, which are decided by $\mathcal{P}^{m-1}$, $\mathcal{T}^{m-1}$, and $\mathcal{R}^{m-1}$. As the number and locations of BSs and IoT devices are fixed all the time once they are deployed, the locations of active IoT devices are slightly correlated across time. However, this correlation has only very little impact on the distributions of active IoT devices, and thus we approximate the distributions of non-empty IoT devices following independent PPPs in each time slot.

In the rest of this section, we first describe the general analytical framework used to derive the non-empty probability, $\mathcal{T}^m$, in each time slot, and then delve into the analysis details of the non-restrict probability, $\mathcal{R}^m$, in each time slot for each RACH scheme.

11.5.1 Non-empty Probability $\mathcal{T}^m$

In the first time slot, the number of packets in an IoT device depends only on the new packets arrival process Λ_{New}^1, such that the non-empty probability of each IoT device, $\mathcal{T}^1$, in the first time slot is expressed as

$$\mathcal{T}^1 = \mathbb{P}\{N_{\text{New}}^1 > 0\} = 1 - e^{-\mu_{\text{New}}^1}, \tag{11.16}$$

where μ_{New}^1 is the intensity of new arrival packets. Note that the non-restrict probability in the first time slot, $\mathcal{R}^1 = 1$, with the BL scheme, and for other RACH schemes, $\mathcal{R}^1$ is determined by their transmission policies, which will be detailed in the following subsection. Substituting Eq. (11.16) and $\mathcal{R}^1$ into Eq. (11.7) (i.e., general cellular-based mIoT networks) or Eq. (11.12) (i.e., NB-IoT networks), we derive the RACH success probability of a randomly chosen IoT device in the first time slot, $\mathcal{P}^1$.

To derive the non-empty probability and the RACH success probability of a randomly chosen IoT device in the mth time slot, we can calculate the exact probability that IoT devices have each number of packets to be served, namely, probabilistic statistics approach. The related derivations have been given in the proof of Theorem 11.2 in [14], which will not be detailed here. However, as m increases, the complexity of these derivations exponentially increases, and thus they become more difficult to analyze. Because the new packet's arrival at each IoT device is modeled by an independent Poisson process, the packet's departure can be treated as an approximated thinning process (i.e., the thinning factor is a function relating to the RACH success probability, the non-empty probability, and the non-restrict probability) of the arrived packets. Therefore, after this thinning process in a specific time slot, the least number of the packets (i.e., the accumulated packets) at each IoT device can be approximated by a Poisson distribution with the same mean. As such, we approximate the number of accumulated packets in the mth time slot N^m_{Cum} by Poisson distribution Λ^m_{Cum} with the intensity μ^m_{Cum}.

Next, we derive the non-empty probability and the RACH success probability of a randomly chosen IoT device in the m th time slot in the following Theorem.

Theorem 11.3: *The accumulated packets number of an IoT device in any time slot should be approximately Poisson distributed. As such, we approximate the number of accumulated packets in the m th time slot N^m_{Cum} as the Poisson distribution Λ^m_{Cum} with intensity μ^m_{Cum}. The intensity of accumulated packets μ^m_{Cum} $(m > 1)$ in the mth time slot is derived as*

$$\mu^m_{\text{Cum}} = \mu^{m-1}_{\text{New}} + \mu^{m-1}_{\text{Cum}} - \mathcal{R}^{m-1}\mathcal{P}^{m-1}\left(1 - e^{-\mu^{m-1}_{\text{New}} - \mu^{m-1}_{\text{Cum}}}\right). \tag{11.17}$$

The non-empty probability of each IoT device in the mth time slot is derived as

$$\mathcal{T}^m = 1 - e^{-\mu^m_{\text{New}} - \mu^m_{\text{Cum}}}. \tag{11.18}$$

Substituting $\mathcal{T}^m$ and $\mathcal{R}^m$ into Eq. (11.7) (i.e., general cellular-based mIoT networks) or Eq. (11.12) (i.e., NB-IoT networks), we derive the RACH success probability of a randomly chosen IoT device in the mth time slot $\mathcal{P}^m$. Note that $\mathcal{R}^m = 1$ with the BL scheme, and for other RACH schemes, $\mathcal{R}^m$ are determined by their transmission policies, which will be detailed in the following subsection.[2]

Proof. See proof of **Theorem 11.2** in [14].

Figure 11.5 shows the Cumulative Distribution Functions (CDFs) of the number of accumulated packets via simulation, as well as calculating by the probabilistic statistics (i.e.,) and the Poisson approximation. We see the close match among the probabilistic statistics, Poisson approximation and the simulation results, which validates our approximation approach. More simulation results will be provided in the Section V to validate the Poisson approximation approach.

[2] With minor modification, this theorem can also be leveraged to study other traffic models, such as the time-limited uniform distribution and the time-limited beta distribution [12].

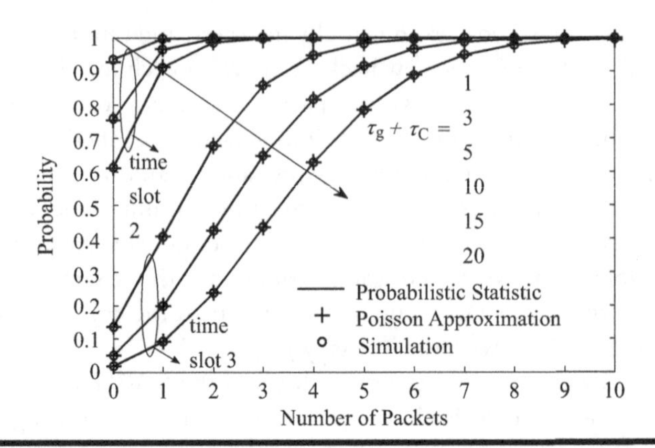

Figure 11.5 Comparing the CDFs of the number of accumulated packets between probabilistic statistics and Poisson approximation in the second and the third time slots. We present six scenarios with different RACH interval durations, where $(\tau_c + \tau_g) = 1,3,5,10,15$ and 20 ms. The simulation parameters are $\lambda_B = 10$ BS/km², $\lambda_{Dp} = 100$ IoT deivces/preamble/km², $\rho = -90$ dBm, $\sigma^2 = -90$ dBm, $\varepsilon_{\text{New}}^1 = \varepsilon_{\text{New}}^2 = \varepsilon_{\text{New}}^3 = 0.1$ packets/ms, and the BL scheme with $\mathcal{R}^m = 1$.

11.5.2 Load Control Schemes

Next, we illustrate the derivation of non-restrict probability, $\mathcal{R}^m$, for each scheme.

11.5.2.1 The Baseline Scheme

The BL scheme allows each IoT device to attempt RACH immediately when a packet is in the buffer, and thus the non-restrict probability is always equal to 1 in any time slot ($\mathcal{R}_{\text{BL}}^m = 1$).

11.5.2.2 The Access Class Barring Scheme

In the ACB scheme, the BS first broadcasts the ACB factor P_{ACB}, then each non-empty IoT device draws a random number $q \in [0,1]$, and attempts to RACH only when q is smaller than or equal to the ACB factor P_{ACB}. Therefore, the non-restrict probability is always equal to P_{ACB} in any time slot ($\mathcal{R}_{\text{ACB}}^m = P_{\text{ACB}}$).

11.5.2.3 The Back-Off Scheme

In the BO scheme, each IoT device defers its access and waits for t_{BO} time slots, when such IoT devices have failed to transmit a packet in the last time slot. The analysis of the non-restrict probability with the BO scheme $\mathcal{R}_{\text{BO}}^m$ is similar to the ACB scheme because the BO mechanism can be visualized as a group of IoT devices that are completely barred in a specific time slot. In the first time slot, none of IoT devices defer the access attempt, such that the transmission procedure is the same as for the BL scheme ($R_{\text{BL}}^{1} = 1$). After the first time slot, the BO mechanism starts to execute, an non-empty IoT device defers its access attempt if the BO is being triggered.

Due to the BO mechanism, only active IoT devices without RACH attempt failures in the last t_{BO} time slots can attempt to transmit a preamble, and only those IoT devices generate interference that determine the preamble transmission success probability in the mth time slot. The non-restrict probability with the BO scheme $\mathcal{R}_{\text{BO}}^{m}$ is derived as

$$\mathcal{R}_{\text{BO}}^{m} = \begin{cases} 1, & m = 1, \\ 1 - \left[\sum_{j=1}^{m-1} \underbrace{(1 - \mathcal{P}_{\text{BO}}^{j}) \mathcal{T}_{\text{BO}}^{j} \mathcal{R}_{\text{BO}}^{j}}_{(a)} \right] \mathcal{T}_{\text{BO}}^{m}, & (t_{\text{BO}} + 1) \geq m > 1, \\ 1 - \left[\sum_{j=m-t_{\text{BO}}}^{m-1} \underbrace{(1 - \mathcal{P}_{\text{BO}}^{j}) \mathcal{T}_{\text{BO}}^{j} \mathcal{R}_{\text{BO}}^{j}}_{(a)} \right] \mathcal{T}_{\text{BO}}^{m}, & m > t_{\text{BO}}, \end{cases} \tag{11.19}$$

where (a) is the probability that a randomly chosen IoT device fails to transmit a preamble in the jth time slot, and thus this IoT device would defer its RACH request in the mth time slot due to the BO mechanism.

11.5.3 Average Queue Length

The works on RACH has been mainly focused on minimizing the failure probabilities and the service delays [3,10]. The RACH success probability provides insights on the probability of access for a random IoT device in each time slot but does not evaluate the packet's accumulation status. Many previous works have indicated that the queue length is a good indication of network congestion [3,9,44]. The queue length refers to the number of packets that are waiting in the buffer to be transmitted [45].

Next, we evaluate the average queue length[3] $\mathbb{E}[Q^{m}]$, which denotes the average number of packets accumulated in the buffer in the mth time slot, which is measured by the mean average of the queue over all IoT devices in the network [45]. The average queue length of each packet over m time slots is derived as

$$\mathbb{E}[Q^{m}] = \mu_{\text{New}}^{m} + \mu_{\text{Cum}}^{m} - \mathcal{T}^{m}\mathcal{R}^{m}\mathcal{P}^{m}. \tag{11.20}$$

[3] The average queue length is looking at the average in space in a specific time slot.

For each RACH scheme, the network is considered stable if a randomly selected queue is finite, which requires the packets arrival rate to be less than the service rate. In other words, the stability only occurs when the queue distribution reaches a steady state. Therefore, the stability condition is related to the average queue length, which is given by

$$\lim_{m \to +\infty} (\mathbb{E}[Q^m] - [Q^{m-1}]) \approx 0. \tag{11.21}$$

11.5.4 Numerical Results

In this subsection, we validate the derived analytical results via independent simulations. We consider the general cellular-based mIoT networks, where the BSs and IoT devices are deployed via independent PPPs in a 400-km^2 area and each IoT device is associated with its closest BS and transmits with the channel inversion power control policy. Note that we simulate the real buffer at each IoT device to capture the packet's accumulated process evolved over time. In each time slot, IoT devices randomly move to a new position, and the active ones randomly choose a preamble for the current RACH attempt. In all figures of this section, "analytical" and "simulation" are abbreviated as "ana." and "sim.," respectively. We choose the same new packet's arrival rate for each time slot ($\mu_{\text{New}}^1 = \mu_{\text{New}}^2 = \cdots = \mu_{\text{New}}^m = 0.1$ packets/time slot), $\sigma^2 = -90$ dBm, $\rho = -90$ dBm, $\alpha = 4$, $\lambda_B = 10$ BS/km^2, $t_{\text{BO}} = 1$ for the BO scheme, and $\text{P}_{\text{ACB}} = 0.8$ for the ACB scheme.

Figures 11.6 and 11.7 plot the RACH success probability and the average queue length with three RACH schemes within the 30 time slots when $\gamma_{th} = -10$ and $\gamma_{th} = 0$dB, respectively. The density ratios between IoT devices transmitting the same preamble and BSs is set as $\lambda_{D_p}/\lambda_B = 1$. The close match between the analytical

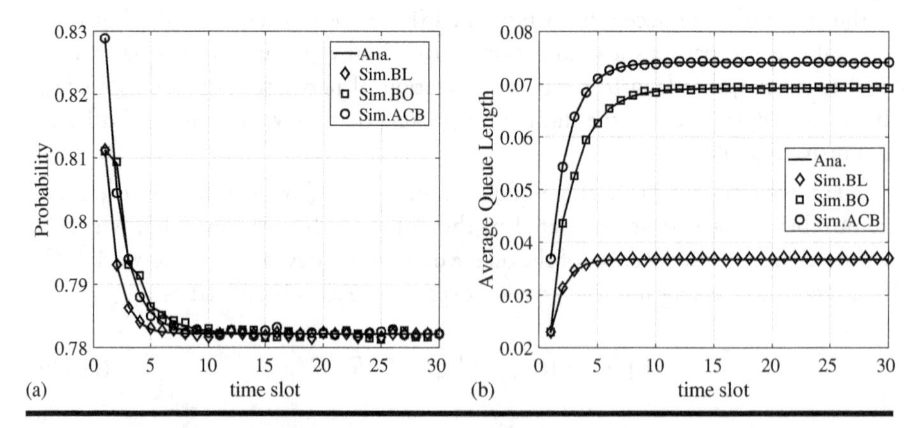

Figure 11.6 (a) The RA success probability and (b) the average queue length when $\gamma_{th} = -10$ dB.

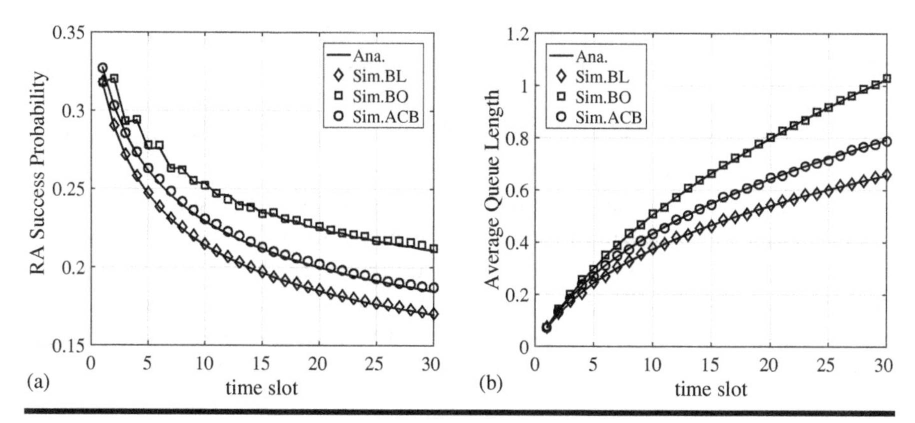

Figure 11.7 (a) The RA success probability and (b) the average queue length when γ_{th} = 0 dB.

curves and simulation points validates the accuracy of the developed spatio-temporal mathematical framework. We first observed that for all RACH schemes, the RACH success probabilities in Figure 11.6a outperformed those in Figure 11.7a. This is because of the lower SINR threshold leading to a higher preamble transmission success probability. The stability condition is given in Eq. (11.21). As can be seen in Figure 11.6b, all of the schemes can reach stability. The average queue lengths follow ACB>BO>BL, which sheds light on the buffer-flushing capability of each scheme in this network condition. In Figure 11.7b, we observe that no one scheme can reach stability because the average number of accumulated packets in each IoT device is much bigger than the serving rate.

Interestingly, in both Figures 11.6a and 11.7a, the RACH success probabilities follow the performance BO>ACB>BL, which is because of the higher probability of an RACH attempt being deferred in the IoT device site leads to less interference and collision probability. In detail, the RACH success probabilities are lower than 50%, leading to more than half of the IoT devices deferring their RACH attempts in the BO and ACB scheme, but the ACB scheme leads to only about 20% deferring their RACH attempts (i.e., P_{ACB} = 0.8). Therefore, the probabilities of deferring RACH attempt follows BO>ACB, which leads to the BO scheme outperforming the other two schemes.

11.6 Conclusion

The mIoT networks are expected to support sporadic uplink transmissions of massive IoT devices, where their key challenge is to efficiently establish connection between IoT devices and BSs referring to the RACH procedure. In this chapter, we reviewed the RACH procedure in cellular-based mIoT networks and introduced a spatio-temporal mathematical model to analyze the RACH of general cellular-based mIoT networks and NB-IoT networks. We provided analytical results of RACH success probability by taking into account the SINR outage problem as well as the collision problem. We then provided a Poisson approximation approach to model the queue evolution, which effectively captures the interdependence between the queue status and the RACH successes in each time slot. We finally extended the proposed queue evolution model to study the RACH success probability and the average queue length with the ACB and the BO schemes. Our numerical results show that the ACB and BO schemes outperform the BL scheme in terms of the RACH success probability due to their traffic control mechanism. We also show that the BL scheme outperforms the ACB and BO schemes in terms of the average queue length due to its higher buffer-flushing capability.

References

1. J. Schlienz and D. Raddino. Narrowband internet of things whitepaper. *IEEE Microw. Mag.*, 8(1): 76–82, 2016.
2. Erik Dahlman, Stefan Parkvall, and Johan Skold. *4G: LTE/LTE-Advanced for Mobile Broadband.* Academic Press, San Diego, CA, 2013.
3. Andres Laya, Luis Alonso, and Jesus Alonso-Zarate. Is the random access channel of LTE and LTE-A suitable for M2M communications? A survey of alternatives. *IEEE Commun. Surv. Tutor.*, 16(1): 4–16, 2014.
4. Zahra Alavikia and Abdorasoul Ghasemi. A multiple power level random access method for M2M communications in LTE-A network. *Trans. Emerg. Telecommun. Technol.*, 28: e3137, 2016.
5. Antonis G. Gotsis, Athanasios S. Lioumpas, and Angeliki Alexiou. Evolution of packet scheduling for machine-type communications over LTE: Algorithmic design and performance analysis. In *IEEE Global Communications Conference (GLOBECOM)*, pp. 1620–1625. IEEE, December 2012.
6. Israel Leyva-Mayorga, Luis Tello-Oquendo, Vicent Pla, Jorge Martinez-Bauset, and Vicente Casares-Giner. Performance analysis of access class barring for handling massive M2M traffic in LTE-A networks. In *International Conference on Communications (ICC)*, pp. 1–6. IEEE, 2016.
7. Guan-Yu Lin, Shi-Rong Chang, and Hung-Yu Wei. Estimation and adaptation for bursty LTE random access. *IEEE Trans. Veh. Technol.*, 65(4): 2560–2577, 2016.
8. Kan Zheng, Fanglong Hu, Wenbo Wang, Wei Xiang, and Mischa Dohler. Radio resource allocation in LTE-advanced cellular networks with M2M communications. *IEEE Commun. Mag.*, 50(7), 2012.
9. Adlen Ksentini, Yassine Hadjadj-Aoul, and Tarik Taleb. Cellular-based machine-to-machine: Overload control. *IEEE Netw.*, 26(6): 54–60, 2012.
10. Monowar Hasan, Ekram Hossain, and Dusit Niyato. Random access for machine-to-machine communication in LTE-advanced networks: issues and approaches. *IEEE Commun. Mag.*, 51(6): 86–93, 2013.
11. Evaluation of RACH congestion solutions. *3GPP, ZTE, Stockholm, Sweden, TSG RAN WG2 R2 103742*, June 2010.
12. Study on RAN improvements for machine-type communications. *3GPP, Sophia, Antipolis, France, TR 37.868 V11.0.0*, September 2011.

13. Nan Jiang, Yansha Deng, Xin Kang, and Arumugam Nallanathan. A new spatio-temporal model for random access in massive iot networks. In *IEEE Global Communications Conference (GLOBECOM)*, Singapore, December 2017.
14. Nan Jiang, Yansha Deng, Xin Kang, and Arumugam Nallanathan. Random access analysis for massive IoT networks under a new spatio-temporal model: A stochastic geometry approach. *IEEE Trans. Commun.*, 66(11): 5788–5803, 2018.
15. Nan Jiang, Yansha Deng, Arumugam Nallanathan, Xin Kang, and Tony Q. S. Quek. Collision analysis of mIot network with power ramping scheme. In *IEEE International Conference on Communications (ICC)*, Kansas City, MO, pp. 1–7, May 2018.
16. Nan Jiang, Yansha Deng, Arumugam Nallanathan, Xin Kang, and Tony Q. S. Quek. Analyzing random access collisions in massive IoT networks. *IEEE Trans. Wireless Commun.*, 17(10): 6853–6870, 2018.
17. Nan Jiang, Yansha Deng, Massimo Condoluci, Weisi Guo, Arumugam Nallanathan, and Mischa Dohler. RACH preamble repetition in NB-IoT network. *IEEE Commun. Lett.*, 22(6): 1244–1247, 2018.
18. Ping Zhou, Honglin Hu, Haifeng Wang, and Hsiao-hwa Chen. An efficient random access scheme for OFDMA systems with implicit message transmission. *IEEE Trans. Wireless Commun.*, 7(7), 2008.
19. Chia-Hung Wei, Giuseppe Bianchi, and Ray-Guang Cheng. Modeling and analysis of random access channels with bursty arrivals in OFDMA wireless networks. *IEEE Trans. Wireless Commun.*, 14(4): 1940–1953, 2015.
20. Sreekanth Dama, Thomas Valerrian Pasca, Vanlin Sathya, and Kiran Kuchi. A novel RACH mechanism for dense cellular-IoT deployments. In *Wireless Communications and Networking Conference (WCNC), 2016 IEEE*, pp. 1–6. IEEE, April 2016.
21. Prashant Wali and Debabrata Das. Optimal time-spatial randomization techniques for energy efficient IoT access in LTE-Advanced. *IEEE Trans. Veh. Technol.*, 66(8): 7346–7359, 2017.
22. Thomas D. Novlan, Harpreet S. Dhillon, and Jeffrey G. Andrews. Analytical modeling of uplink cellular networks. *IEEE Trans. Wireless Commun.*, 12(6): 2669–2679, 2013.
23. Jeffrey G. Andrews, Francois Baccelli, and Radha Krishna Ganti. A tractable approach to coverage and rate in cellular networks. *IEEE Trans. Commun.*, 59(11): 3122–3134, 2011.
24. Seung Min Yu and Seong-Lyun Kim. Downlink capacity and base station density in cellular networks. In *Proceedings of the International Symposium on Modeling and Optimization in Mobile, Ad Hoc & Wireless Network (WiOpt)*, pp. 119–124, May 2013.
25. Yansha Deng, Lifeng Wang, Maged Elkashlan, Arumugam Nallanathan, and Ranjan K Mallik. Physical layer security in three-tier wireless sensor networks: A stochastic geometry approach. *IEEE Trans. Inf. Forensics Secur.*, 11(6): 1128–1138, 2016.
26. Yansha Deng, Lifeng Wang, Syed Ali Raza Zaidi, Jinhong Yuan, and Maged Elkashlan. Artificial-noise aided secure transmission in large scale spectrum sharing networks. *IEEE Trans. Commun.*, 64(5): 2116–2129, 2016.
27. Hesham ElSawy, Ekram Hossain, and Martin Haenggi. Stochastic geometry for modeling, analysis, and design of multi-tier and cognitive cellular wireless networks: A survey. *IEEECommun. Surv. Tutor.*, 15(3): 996–1019, 2013.
28. Yansha Deng, Lifeng Wang, Maged Elkashlan, Marco Di Renzo, and Jinhong Yuan. Modeling and analysis of wireless power transfer in heterogeneous cellular networks. *IEEE Trans. Commun.*, 64(12): 5290–5303, 2016.

29. Hesham ElSawy and Ekram Hossain. On stochastic geometry modeling of cellular uplink transmission with truncated channel inversion power control. *IEEE Trans. Wireless Commun.*, 13(8): 4454–4469, 2014.
30. Yi Zhong, Tony Q. S. Quek, and Xiaohu Ge. Heterogeneous cellular networks with spatio-temporal traffic: Delay analysis and scheduling. *IEEE J. Sel. Areas Commun.*, 35(6): 1373–1386, 2017.
31. Kostas Stamatiou and Martin Haenggi. Random-access Poisson networks: Stability and delay. *IEEE Commun. Lett.*, 14(11): 1035–1037, 2010.
32. Yi Zhong, Martin Haenggi, Tony Q. S. Quek, and Wenyi Zhang. On the stability of static Poisson networks under random access. *IEEE Trans. Commun.*, 64(7): 2985–2998, 2016.
33. Mohammad Gharbieh, Hesham ElSawy, Ahmed Bader, and Mohamed-Slim Alouini. Spatiotemporal stochastic modeling of IoT enabled cellular networks: Scalability and stability analysis. *IEEE Trans. Commun.*, 65(8): 3585–3600, 2017.
34. Evolved universal terrestrial radio access (E-UTRA): Physical channels and modulation. *3GPP, TS 36.211 v.13.2.0, Release 13*, 2016.
35. Xingqin Lin, Ansuman Adhikary, and Y.-P. Eric Wang. Random access preamble design and detection for 3GPP narrowband IoT systems. *IEEE Wireless Commun. Lett.*, 5(6): 640–643, 2016.
36. Y.-P. Eric Wang, Xingqin Lin, Ansuman Adhikary, Asbjorn Grovlen, Yutao Sui, Yufei Blankenship, Johan Bergman, and Hazhir S. Razaghi. A primer on 3GPP narrowband internet of things. *IEEE Commun. Mag.*, 55(3): 117–123, 2017.
37. Martin Haenggi. *Stochastic Geometry for Wireless Networks.* Cambridge University Press, New York, 2012.
38. Gordon Gow and Richard Smith. *Mobile and Wireless Communications: An Introduction.* McGraw-Hill Education (UK), 2006.
39. Li Chen, Wenwen Chen, Xin Zhang, and Dacheng Yang. Analysis and simulation for spectrum aggregation in LTE-advanced system. In *70th Vehicle Technology Conference* (*VTC Fall*), pp. 1–6. IEEE, 2009.
40. Kaijie Zhou, Navid Nikaein, and Thrasyvoulos Spyropoulos. LTE/LTE-A discontinuous reception modeling for machine type communications. *IEEE Commun. Lett.*, 2(1): 102–105, 2013.
41. Járai-Szabó Ferenc and Zoltán Néda. On the size distribution of Poisson Voronoi cells. *Phys. A: Stat. Mech. Appl.*, 385(2): 518–526, 2007.
42. Radha Krishna Ganti and Martin Haenggi. Spatial and temporal correlation of the interference in ALOHA ad hoc networks. *IEEE Commun. Lett.*, 13(9), 2009.
43. Cellular system support for ultra-low complexity and low throughput Internet of Things (CIoT). *3GPP, Sophia, Antipolis, France, TR 45.820 V13.1.0*, November 2015.
44. Mingyu Chen, Xingzhe Fan, Manohar N. Murthi, T. Dilusha Wickramarathna, and Kamal Premaratne. Normalized queueing delay: Congestion control jointly utilizing delay and marking. *IEEE/ACM Trans. Netw.*, 17(2): 618–631, 2009.
45. Attahiru Sule Alfa. *Queueing Theory for Telecommunications: Discrete Time Modelling of a Single Node System*. Springer Science & Business Media, 2010.

PRIVACY AND SECURITY ISSUES

Chapter 12

Privacy and Security Issues in the 5G-Enabled Internet of Things

Liyuan Liu and Meng Han

Contents

12.1 Introduction

The Internet of Things (IoT) is an ecosystem of interrelated physical objects such as computing devices, machines, vehicles that are provided with unique identifiers and enable to communicate, compute and coordinate in anytime, anyplace. It can provide computing devices, machines, people and real-time business access the state of things, transfer data over a network. It does not require human-to-human or human-to-computer interaction, but instead it increases and encourages machine-to-machine (M2M) communication [62]. IoT is explosive growth in recent years and will lead the foreseeable future of technology. Cisco estimated in 2011 that the number of smart objects would rise to 50 billion, and their market value would reach $267 billion, by 2020 [17,53]. The IoT has great potential to enable a broad range of novel services and applications and to shape the future in many industries.

With the increasing number of IoT devices, the traditional 4G wireless technology cannot meet the efficiency, speed, and latency requirements of the IoT. Therefore, fifth-generation (5G) cellular wireless has been designed to satisfy the demands of future IoT devices and it can connect an unprecedented significantly greater number of devices. 5G will be a paradigm shift with very high carrier frequencies with massive bandwidth, extreme base station and device densities, and an unprecedented number of antennas [3]. According to the latest Ericsson's Annual Mobility Report [16], 5G will be able to transmit data about 10 times faster than 4G long-term evolution (LTE) and will have 550 million subscriptions by 2020. This means 5G could revolutionize the IoT.

Compared with previous wireless technology generations such as 1G, 2G, 3G and 4G, 5G has the advantage of not only fast downloads but also the unique combination of high-speed connectivity and low latency with being more effective and efficient. 5G will support IoT applications, especially for time-sensitive applications and remote service. For example, smart transportation—one of the most time-sensitive systems—can operate over a 5G network, use cameras and sensors connected to a shared network to monitor real-time traffic patterns, then change the traffic signals and increase the traffic flow: something beyond the capabilities of the 4G era. Smart grids connected to 5G can drive dynamic pricing, allowing citizens to choose where they buy their energy, while providing and distributing the extra capacity needed for charging electric vehicles to support their wide-scale adoption. In the 4G environment of computing over the past decade, the essential trend has been to provide hardware and systems software in the centralized data centers that provide the control, storage, and networking for Internet services. Its flexibility, automatic software updates, high ability for document control and so on have created a significant shift in computing services from traditional to an innovative methods. However, with the increasing number in influential end users, network edges, business-driven IT initiatives, and access devices in the 5G era, getting data processed from so many devices is becoming one of the main challenges of cloud computing. Cisco Systems predicts that cloud traffic is likely to rise nearly fourfold by 2020, increasing 14.1 zettabytes per year by 2020 [35].

Gartner predicts that 50% of enterprise-generated data will be created and processed beyond centralized cloud data centers via edge computing by the year 2022 [8]. Many possible changes and questions are raised: "Can the services be done closer to the end user and distributed computing?" "Can your smartphone become your data storage?" "Can your car monitor machine health, software updates and identify signs of time-sensitive maintenance issues in real-time?" "What if the smart edge devices process, store, and analysis the data of the time-sensitive applications on edge instead of on cloud?" Therefore, in the 5G-enabled IoT, how to process the high volumes of data at fast speed is the main focus. Computing, analysis, storage of suitable data in the end-user devices will be the most critical concerns in computing in the future.

Along with the IoT, 5G, distributed computing growth, and the vast amounts of data generated by IoT sensors, environmental monitors and personal devices, we need to rethink a problem: "Are there new security and privacy issues in the 5G-enabled IoT?" "Do the traditional security and privacy problems still exist, or do we need to deal with specific new problems?" Security and privacy problems are fundamental to every industry. The security and privacy threats will cause many issues such as data breaches. Data breaches happen daily and, in too many places, result in data loss including crucial, sensitive, and private personal, health, and financial information. The cost of security and privacy threats is extremely high. These threats are not only potentially damaging monetarily, but they also create other more pressing problems with consumer confidence, social trust, and personal safety. For example, in 2017, Equifax—one of the major credit reporting agencies—reported that a hacker had penetrated their network and leaked 143 million American citizens' sensitive information including names, social security numbers, birthdates, addresses, and even driver's license numbers. Facebook has also faced security and privacy problems since it was reported that Cambridge Analytica (a British consulting company) was accused of harvesting the data of up to 50 million Facebook users without permission. This incident not only caused a decrease in Facebook's stock price by $61 million to $476.4 billion, but it also caused other social media and technology company stock prices to fall [43]. In the traditional 4G-enabled IoT, using the cloud to store and compute will cause a series of security and privacy threats according to cloud-based IoT, including problems with identity privacy, location privacy, node compromise attacks, layer removing/adding attacks, and forward and backward security issues as well as semi-trusted and malicious cloud security problems. Compared with the 5G-enabled IoT, especially the edge-based IoT, except for a part of the traditional security and privacy issues that still exists, many new, specific security and privacy problems will emerge because of the characteristics of edge paradigms. For example, the IoT and 5G will add more data and generate more Internet-connected devices in more locations for remote and unmonitored devices, which can lead to additional security problems not seen in the 4G era. In addition, because the IoT has various use cases, most IoT devices don't have conventional hardware protocols and the software updates and security configuration, which have usually been needed during the lifecycle of a device are hard to implement.

In this chapter, we start with the preliminary definitions of 5G, the IoT, and some new distributed computing architectures; survey the existing research in the 5G, IoT, and edge computing area; point out the range of new issues in the 5G-enabled IoT; and discuss why it is difficult to address these new challenges with cloud-based computing and networking models. Then we introduce the modern architecture-edge computing of 5G-enabled IoT, consider the security and privacy issues particular to 5G-enabled IoT, and suggest future research opportunities. As shown in Table 12.1, this chapter provides a detailed survey of the security and privacy issues in the 5G-enabled IoT compared with previous research, illustrates and analyzes the state-of-the-art security and privacy in the edge paradigms, and points out the potential synergies among security mechanisms in 5G-enabled IoT environment.

This chapter is organized as follows. Section 12.2 introduces the most preliminary definitions of the IoT, 5G, cloud computing and edge paradigms, including their evolutions, use cases, importance, and standardization efforts, how they can cooperate with each other and change the technology. Section 12.3 presents security issues that affect the 5G-enabled IoT, especially for the edge paradigms and concludes with the different issues that target the 5G-enabled IoT and discusses the requirements and challenges of security mechanisms. Section 12.4 illustrates the privacy threats in the 5G-enabled IoT and discusses the current state of the art regarding privacy issues. Section 12.6 points out the challenges and opportunities in the 5G-enabled IoT environment and the potential research interests. This analysis does not just identify the challenges of security and privacy issues, it also shows how, in the new 5G-enabled IoT age, security and privacy mechanisms of other related fields can be incorporated into edge paradigms. Finally, conclusions are presented in Section 12.7.

Table 12.1 Contribution of Available Surveys

	[48]	*[46]*	*[40]*	*[1]*	*[9]*
Year of Publication	*2016*	*2015*	*2018*	*2017*	*2016*
IoT	Yes	Yes	Yes	Yes	Yes
5G	Yes	Yes	Yes	No	Yes
CC	No	Yes	Yes	Yes	Yes
FC	No	No	No	Yes	Yes
MEC	No	No	Yes	Yes	Yes
MCC	No	No	No	No	No
Security and Privacy	No	Yes	Yes	Yes	Yes

12.2 Preliminary Definition

We begin by introducing the preliminary definitions of 5G, IoT, cloud computing, edge paradigms, security, and privacy. We then compare the different edge paradigms with traditional cloud computing, and clarify what's the similarity and difference between security and privacy.

12.2.1 5G

Currently, advanced technologies and wireless communication networks are being pushed to their limits by the massive growth in demands for many wireless data services. High-throughput, enhancements in network capacity, low latency, ubiquitous connectivity, energy efficiency, high reliability, low-cost devices and quality of experience (QoE) are all part of the requirements that the next generation wireless needs [10]; 5G has emerged to meet these requirements. Since the early 1970s, when mobile wireless technology began its creation, revolution, and evolution, the first (1G), second (2G), third (3G), fourth (4G), and now the fifth (5G) generations have been continuously updated.

Compared with the original mobile radio-telephone system, 1G devices were lighter and cheaper. The first 1G telecommunication technology appeared in 1980. 1G was based on analogous techniques, and there were many different applications of 1G, including the paging system, cordless telephones, private mobile radios, and so on. The Advanced Mobile Phone System (AMPS), Nordic Mobile Telephone (NMT) and Total Access Communication System (TACS) were the preeminent systems used in 1G. Based on the digital system, 2G wireless systems were started in the 1980s and continued into the 1990s when the 2G system was completed and could transmit voice with digital signals. In addition, 2G networks had a more efficient spectrum and better data services. 2G used the global system for mobile communication (GSM), Digital AMPS, Personal Digital Communication (PDC), and Code Division Multiple Access (CDMA) technologies that enabled the various mobile networks to provide services such as voice, text and picture message transmission. Then, 3G made a lot of innovations and had a destructive effect all over the networks technology world and in many related fields since 2000. Compared with 2G standards, 3G was four times quicker, with speeds ranging from around 125 Kbps to 2 Mbps. In the 3G era, working with the Universal Mobile Telecommunications System (UMTS), High-Speed Packet Access (HSPA) and Evolved High-Speed Packet Access (HSPA+), for example, which belonged to the Third Generation Partnership Project (3GPP) family, mobile Internet access was developed along with fixed wireless Internet access, video calls, and mobile TV technologies. 4G started in 2010 and it was called "MAGIC" for its Mobile multimedia, Anywhere, Global mobility solutions over Integrated wireless and Customized services. Beyond providing the basic services of 3G, 4G offered much more mobile broadband Internet access using wireless modems between laptops and smartphones, mobile web access, Internet Protocol (IP) telephony, gaming services, high-definition mobile TV, video conferencing, three-dimensional television, and cloud computing. As an IP-based integrated system, 4G networks offered 100 Mbps for individuals on high mobility and 1 Gbps for those on low mobility. LTE, LTE Advanced, LTE Advanced Pro and Worldwide Interoperability for Microwave Access (WiMAX) technologies are widely used for 4G connectivity. 5G is the next wireless upgrade technology. 5G will go beyond 4G/IMT-Advanced standards and be better for user experience because it is faster and more efficient in its spectrum use. The primary objectives of the current research and development related to 5G are how to improve and support the M2M communication, how to connect billions of devices and sensors in the age of the IoT and, in the meantime, reduce cost, battery consumption, and latency compared with 4G [58]. Nevertheless, in the 5G networks, security and privacy are big issues. Figure 12.1 shows the evolution of the mobile communications [68]. It presents the time range from research to industry applications for each generation. 4G and 5G will be the main trends in the next 20 years. In 5G age, the arrival of the new mobile communications such as the smartphone, Google Home, SmartWatch, and autonomous vehicles will lead new possibilities for M2M communications across all market sectors. Table 12.2 presents the comparison between 1G, 2G, 3G, 4G and 5G from many different aspects [58]. Each generation has been faster, more reliable, and more functional than the previous generation. 5G will make wireless networks communications very fast, almost real-time, and IoT networks will enable M2M communications with embedded artificial intelligence (AI).

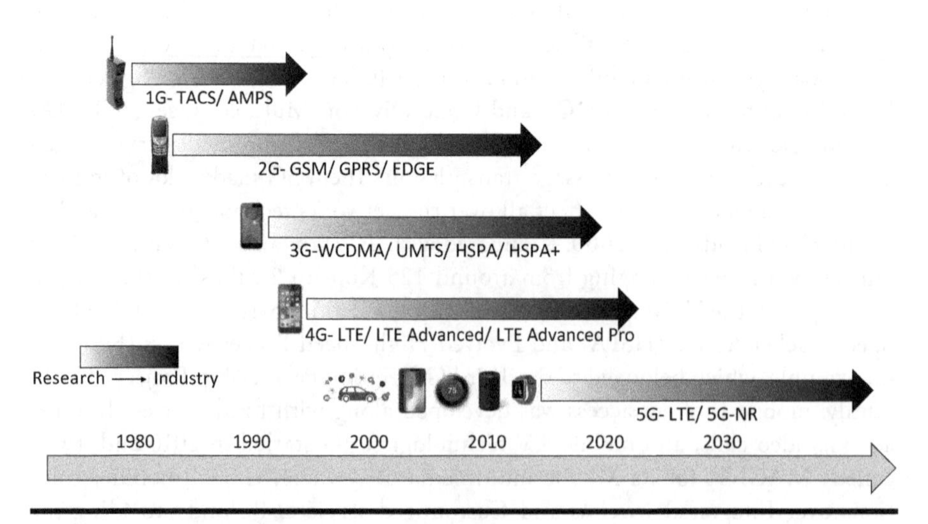

Figure 12.1 The evolution from 1G to 5G.

Table 12.2 Comparison 1G, 2G, 3G, 4G and 5G

	1G	*2G*	*3G*	*4G*	*5G*
Period	1980–1990	1990–2000	2000–2010	2010–2020	2020–2030
Data Rate	2 Kbps	14–64 Kbps	2 Mbps	200 Mbps	>1 Gbps
Frequency	30 KHz	1.8 GHz	1.6–2.0 GHz	2–8 GHz	3–300 GHz
Technology	Analog cellular (TACS)	Digital cellular (GSM)	Broadband width CDMA IP Technology	Unified IP LAN WAN WLAN PAN	4G+WWWW
Standards	AMPS TACS	GSM GPRS EDGE CDMA2000 1X	WCDMA HSPA HSPA+	LTE LTE Advanced LTE Advanced Pro WiMAX	LTE 5G-NR DECT-5G
Multiplexing	FDMA	TDMA/CDMA	CDMA	CDMA	CDMA
Switching	Circuit	Circuit	Partial packet	All Packet	All Packet
Core Network	PSTN	PSTN	Packet network	Internet	Internet
Handoff	Horizontal	Horizontal	Horizontal	Horizontal Vertical	Horizontal Vertical
Service	Voice-only telephone	Digital voice, short messages	Text messages, picture messages, MMS	Dynamic information access, various devices	Dynamic information access, IoT devices with AI capabilities

12.2.2 IoT

5G deployment and the increasing number of physical objects that are connected to the Internet provided support for the idea of the IoT [45]. There have been many definitions of the IoT, but the best definition would be that it is "an open and comprehensive network of intelligent objects that can auto-organize, share information, data, and resources, reacting and acting in the face of situations and changes in the environment." It was first introduced in 1999 and the purpose of the deployment of this technology is to promote radio frequency identification (RFID) technology. Table 12.3 displays the main significant incidents in IoT history and shows the details of each event [45]. From 1999 to 2018, the IoT has grown from concept to implementation with Internet-connected appliances

Table 12.3 Highlights of the History of the IoT

Year	*Organization or People*	*Detail*
1999	Kevin Ashton	The first time introduced IoT Helped to develop the Electronic Product Code
2000	LG	First Internet of refrigerator plans
2003–2004	US Department of Defense	RFID is deployed on a massive scale
2005	ITU	Published first report on the IoT topic
2008	EU	First IoT conference is held
2008	US National Intelligence Council	Listed the IoT as one of the six disruptive civil technologies with potential impacts on US interests out to 2025
2008–2009	Cisco	The IoT was born
2010	Chinese Government	Announced IoT is a strategic priority in the Five-Year-Plan
2011	Gartner	IoT listed in the "hype-cycle for emerging technologies"
2013	IDC	Reported Internet of Things would be a $8.9 trillion market in 2020
2014	Google	Announced to buy Nest
2017–2018	Microsoft, Dell, Amazon, etc.	Start establish edge computing portfolios instead of cloud computing

to all types of devices, from cloud-based to edge-based, and from common security and privacy issues to more personal issues.

Nowadays, with the development of the IoT, the IoT has various applications in the world. Figure 12.2 shows the different generic scenarios of IoT. The various IoT devices connect to each other in the real world, with the development of AI, IoT-embedded AI will help IoT devices process information and make autonomous decisions. Therefore, IoT can help the industry establish the widespread use of beneficial applications and services that are beyond our imagination [34]. The first widespread applications that will bring numerous benefits to the real world is the smart home. Smart home is an IoT technology that deals with the technological enrichment of the living environment to offer support to inhabitants and improve human's life quality [54]. The principal reason that the smart home idea has grown so quickly recently is that it can improve comfort, with high energy efficiency. There are many popular smart home IoT devices coming into our normal life. For example, you can talk to Amazon Echo and Google Home and they will control your Hue lights (Hue lights is the smart light product developed by Philips. It is a low-power, safe, and reliable technology that works even if the internet goes down.),

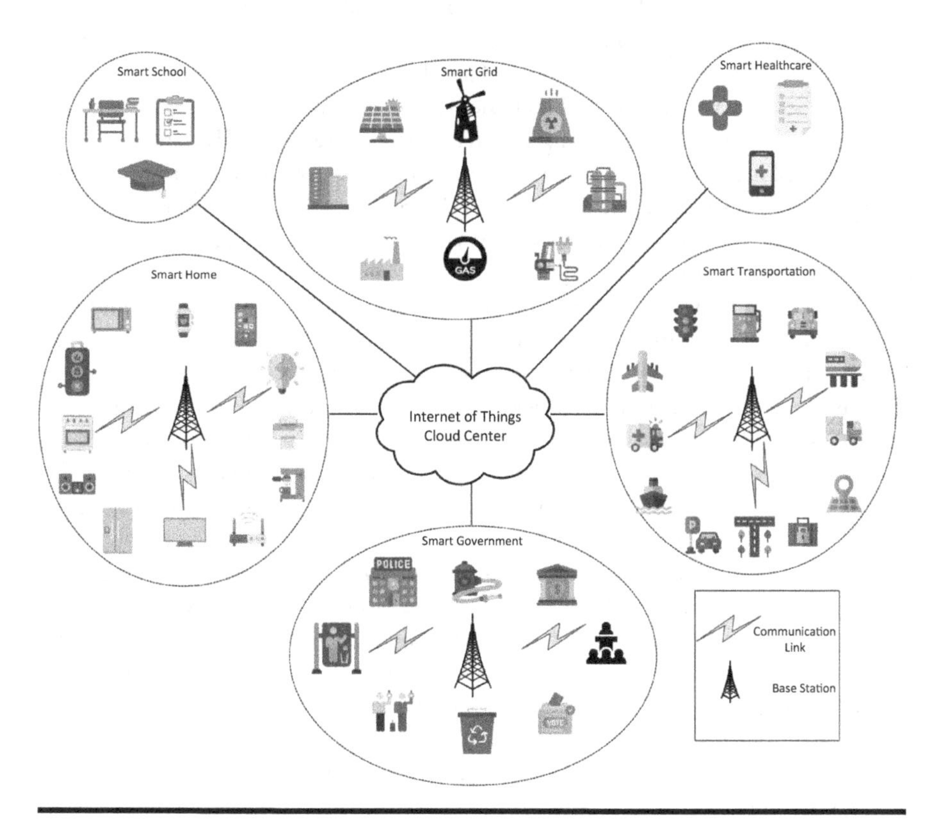

Figure 12.2 The scenarios of IoT.

play music, control appliances, and so on. The second large developing allocation is smart transportation. Smart transportation means the incorporation of advanced communications technologies into the transportation infrastructure and vehicles. The benefits of smart transportation are safety, mobility, and environmental sustainability. Much research has focused on the smart transportation IoT. Most researches have been centered on the vehicular networks. For example, how to protect the location privacy [20], how to make resource management more efficient [72] and how to solve the real time latency problem [52]. Smart grid, smart school, smart health care all are important services of the IoT. Embedding AI into the IoT will increase efficiency and provide beneficial services; this will also allow the IoT device to learn the user's behavior by itself to improve service quality and efficiency.

12.2.3 Cloud Computing and Edge Paradigms

In the past decade, cloud computing has been an essential part of the IoT. It uses some standard access mechanisms to multiple configurable resources such as applications, storage, networks, servers, and services. The first time cloud computing was mentioned to the world was in a Compaq internal document as early as 1996. Several years later, cloud computing has come into actuality. In 2006, Amazon introduced the created Elastic Compute Cloud (EC2) which can support Amazon Web Services. Two years later, Google delivered the Google App Engine. In the 2010s, various Smarter Computing foundations appeared after IBM released the IBM SmartCloud. Cloud computing is a critical component in these smarter computing frameworks [50]. There are many service models for cloud computing; three accessible models are Infrastructure as a Service (IaaS), Platform as a Service (PaaS), and Software as a Service (SaaS) [47].

Cloud computing has many advantages; one example is cost reduction. Minimal effort is required to manage the infrastructure, and users only need to pay when they need to process and store data in a private or public cloud. It can adapt to fluctuations according to demand, accelerate development work, and provide efficient computing [4,28]. However, with the soaring numbers of IoT devices, cloud computing face some challenges such as how to process and store the extensive data from billions of IoT devices. Regarding some time-sensitive applications, such as autonomous vehicles, can cloud computing handle the latency problem? In recent years, various innovative edge paradigms have appeared, such as fog computing (FC) [6], mobile edge computing (MEC) [29], and mobile cloud computing (MCC) [14].

These different distributed computing paradigms have a general concept, that unlike cloud computing, which pushes the resource into a centralized cloud, edge paradigms can provide computation and performance on end users, known as smart devices (edge devices). Table 12.4 shows the comparison between FC, MEC and MCC [15,51]. Because they are typically distributed paradigms and extensions of cloud computing, they have many common similarities such as low latency and jitter, high availability and scalability. Furthermore, edge paradigms refer to the computing

Table 12.4 Compare with FC, MEC and MCC

	FC	*MEC*	*MCC*
Support Organization	OpenFog Consortium	ETSI MEC	OpenFog Consortium
Service Provider	Telco operators	Private organization, individuals	Private organization, individuals
Hardware	Heterogeneous servers	Heterogeneous servers	Servers, user devices
Basic Architecture	Edge-fog-cloud	Edge-cloud	Edge-cloudlet-cloud
Context awareness	Medium	Medium	Low
Internode Communication	Yes	Partial	Partial
Geo-distributed	Yes		
Latency and Jitter	Yes		
Mobility Support	Yes		
Availability and Scalability	High		

infrastructure close to the sources of data and end user, allowing for lower latency and jitter compared with having to transmit all resources to the cloud. For example, for self-driving vehicles, if something happens on the road, it will be better to decide on the edge that avoids a collision. Also, there are some differences between these edge paradigms. First, the MEC deployment platform relies on mobile network infrastructure, such as 5G, the service provider can be only the Telecom operators. But for the FC and MCC, the providers can be any private entity or individuals. Therefore, for FC and MCC, the operators can closely incorporate with the third-party service providers to develop customized services for different business models. Second, FC and MCC push resources down to the local-area network-level of network architecture, computing data in a fog node or a cloudlet. The fog node and the cloudlet make up the interlayer between the edge and the cloud. MEC pushes the resource to an edge gateway or directly into the end-user devices. Third, in FC, the inter-fog nodes can totally communicate with each other compared with partial communication in MEC and MCC. Figure 12.3 displays an example of 5G-enabled IoT edge computing architecture. The IoT edge devices pushed the resources into edge data center through edge networks. The edge data centers are interconnected with the

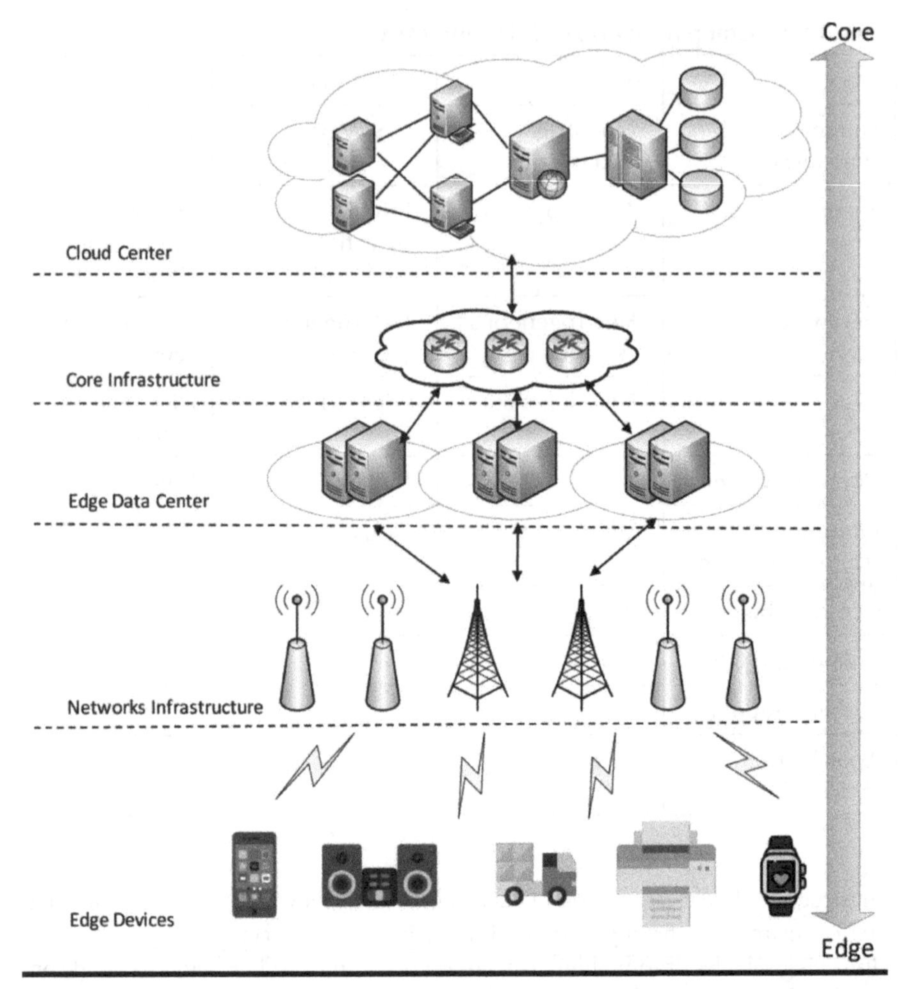

Figure 12.3 Architecture of edge computing in the 5G-enabled IoT.

cloud center via the network infrastructure. The core infrastructure allows for core network access, centralized cloud computing services and management capabilities for network edge devices. The cloud will compute and store the resources that the edge does not have the capacity, supporting the under layers.

12.2.4 Security and Privacy

Security and privacy are two big issues in the 5G-enabled IoT. The cost of security and privacy threats is prohibitive. It is not only considered potentially damaging monetarily, but it also causes other more pressing issues such as problems with consumer confidence, social trust, and personal safety. In this section, we mainly

Table 12.5 Difference between Security and Privacy

	Security	*Privacy*
Target	Against unauthorized access	Protect personal identifiable information
Data Type	All data	Private data
Program Focus	All information that an organization collected	The personal information such as name, address, SSN, etc.
Relationship	Security can be achieved without privacy	Privacy cannot be achieved without security

describe the difference between security and privacy. Security indicates protection against the unauthorized access of data. Privacy refers to protecting the personal identification information. For example, a credit report agency can protect the customers' personal information with security strategy, but the senior staff or data scientists may still have access to this private information. Table 12.5 displays the primary differences between security and privacy [7].

12.3 Security Threats in 5G-Enabled Internet of Things

The 5G-enabled IoT brings abundant benefits to the end users. However, it also carries some security and privacy challenges. Despite the common traditional security and privacy issues, the 5G-enabled IoT also has some unprecedented security and privacy challenges.

12.3.1 Architecture of 5G-Enabled Internet of Things

According to the existing research [32–34,57], we construct the new architecture of 5G-enabled IoT as Figure 12.2. The five layers from bottom to top are the recognition, connectivity/edge computing, support, application, and business layers.

Recognition layer: This layer is also known as the things layer, it is the foundation layer in 5G-enabled IoT. There are two main parts in this layer: IoT physical objects and hardware. The physical objects can be all the IoT devices such as smartphones, smart wearables, self-driving vehicles. The hardware in this layer includes sensors, controllers or any electronics hardware that obtain data from IoT devices. For example, RFID readers, barcode, microcontroller units, and so on.

The main task of this layer is to collect IoT information from physical devices and sensors, and to recognize and identify the IoT environment.

Connectivity/Edge Computing Layer: This layer task is to help connectivity and edge computing to define the communication protocols and networks. Use networks to transmit collected information from the first layer to the edge and cloud. In the 5G-enabled IoT, the underlying networks can be 4G, 5G, Wi-Fi, and so on. Another crucial research interest in this layer is how to decide the resource offload and resource allocation between the edge computing layer and the cloud layer.

Support Layer: This layer will offer the support platform for the upward and downward layers. It contains cloud computing powers, data analysis ability, and other intelligent computing ability. The cloud center has responsibility for the service management and processes the information that the edge computing layer cannot handle. In this layer, data analysis will be another vital part; it will support the AI embedded into IoT, making applications more intelligent. For example, for storing users' historical data, implementing AI algorithms can predict users' behavior to help 5G-enabled IoT devices make efficient and accurate decisions in the application layer.

Application Layer: This is a terminal layer that offers the various applications and service based on the information process from previous layers. The IoT devices can make personalized decisions according to the AI learning results. There are various applications in 5G-IoT, such as smart home, smart grid, smart health care, smart transportation, and smart wearables.

Business Layer: This layer manages all the IoT systems. It is required to create better business models that improve the IoT service quality. Much research has focused on how to design a better business model including ideas such as Real-Time Instrumentation return of investment (ROI), zero-capex IoT business models, and IoT data privacy and trust models [48].

12.3.2 Security Threats

In this section, we analyze the security threats in each layer of the 5G-enabled IoT and point out the security requirements and threats in the 5G-enabled IoT [18,51,57].

Recognition Layer:

1. Hardware security: The first essential threats in recognition layer is the hardware security. This layer consists of multiple sensors, controllers and different IoT devices.

   Unauthorized access and cloning the tag can occur in sensors and adversaries can reprogram the data. The malicious attackers can also control user-accessible IoT equipment and provide fake information. The wireless sensors such as RFIDs make it easy for the attacker to get confidential information such as passwords to eavesdrop on information.

2. Software security: It is important to understand that hardware security alone is not enough; most attacks are via software. Therefore, a clear understanding of the security threats of the 5G-enabled IoT is necessary. For any IoT device, integrity, authentication, and availability are factors that make software insecure. If the embedded AI system is maliciously attacked, IoT device information can be stolen or monitored and damage the software behavior. For most popular operating systems, such as IOS and Android, the most substantial security threat is a hacker attack. Malicious attackers on smart mobile devices can gain access to enterprise and personal data. The most likely malicious attacks include cryptographic attacks and code injection attacks.

Connectivity/Edge Computing Layer:

1. Network security: The main security threat in the 5G-enabled IoT is the denial of service (DoS) attack. This is a common attack through networks. In the edge computing scenario, the distributed denial-of-service (DDoS) attacks and wireless jamming are especially dangerous. The Man-in-the-Middle attack and the counterfeit attack also exist for this layer. Malicious attackers can control a section of the network and attack the network. The unauthorized attacker can even fake the identification and communicate as normal and obtain more IoT users' information. IP theft is also a danger for network security. IPv4 and IPv6 are different versions of Internet Protocol in the network [12]. IPv6, as the upgrade version of IPv4, has some advantages such as better multicast routing, a more simplistic header format, and no more network address translation. Even though they can provide seamless protection for applications, there are still have many security threats, such as the DDoS attack, the Man-in-the-Middle attack, and Packet sniffing.
2. Rogue node: In edge computing, after the end user pretends to connect to the edge node such as fog node, private cloudlets, edge devices, it gives a malicious attacker the chance to deploy fake gateway devices. The malicious attacker can manipulate users' requests, collect or tamper user data, and launch further attacks the same as the Man-in-the-Middle attack.
3. Edge data centers: Protecting the edge data center to avoid physical damage, and privilege escalation is necessary. Some edge paradigms' data center are managed by business organization, and preventing the attackers from accessing the data centers and damaging the devices is needed. In addition, to block external attackers, ongoing training of the security management individuals and maintenance of the data center professionals is necessary.

Support Layer:

1. Cloud computing: The main security threats in the cloud is the DoS attack, shared cloud computing services, system vulnerabilities, and the malicious or negligent insider. DDoS attacks against cloud platforms are incredible.

They will shut down the cloud system and denied the services [61]. Even though some security strategies can detect the DoS attack, with the increasing numbers of 5G-enabled IoT devices and the increases in high-speed 5G wireless, DoS attacks more likely since the previous protection methods cannot handle the high-volume data traffic. For some shared cloud services, they do not provide enough security between users and applications. Another security threat occurs when users share resources. System vulnerability can still exist in some complex cloud computing infrastructure, and if the attackers know the weakness, they can easily damage the cloud computing system. As with the edge data center, the cloud computing management team also need to prevent the internal security human-made security threats.

2. Data analysis: The key here is to prevent the data leakage threats that cause more severe privacy issues.

Application Layer:

1. Heterogeneous network: There are various 5G-enabled IoT applications in this application layer. Each different application domain has distinctive security threats. Moreover, these applications can share data through the heterogeneous network, which can easily cause several security issues: in particular, the DoS attack and malicious code injection. It is easy to attack and shut down the service if there is no secure authentication and key agreement. Also, it will cause data breaches and privacy problems, which we will explain in the next section.
2. Service management: The people who take the responsibility to maintain and manage the application infrastructures should be professional and educated, otherwise there will be other security issues.

Business Layer:

1. Data leakage: Owing to the requirements in the business model design, the designers will implement some APIs to help their modeling. In the meantime, the data have a significant exposure to leakage if the APIs lack security protection. The data can include many private pieces of information, whose loss will bring a severe cost.

12.4 Privacy Threats in the 5G-Enabled Internet of Things

With the deployment of 5G wireless and the rapidly growing number of IoT devices, the 5G-enabled IoT has more server privacy issues. Edge computing allows each physical object in the environment the ability to communication autonomously over 5G wireless or the Internet. In this section, we conclude some main privacy threats in the 5G-enabled IoT environment.

Broad Sensitive Information: The extremely high speed of 5G wireless lets more and more physical objects be IoT devices, including microwaves, robot vacuums, a cloth, even a small ring. These smart IoT equipment significantly improve human quality of life. In the meantime, widespread sensitive personal information will also be collected by these devices. Sometimes, the connected devices ask end users to input personal sensitive information such as a name, gender, age, ZIP code, or email address. Even though sometimes the data are just transmitted to a given end point, there is still a privacy risk. If the attacker integrates, collects and uses advanced data mining algorithms to analyze the fragmented data from multiple end points, sensitive personal information still can be ascertained [19]. For example, in a smart home, the refrigerator can embed a camera that monitors the grocery items and if they are spoiled. It appears to be only collecting grocery data; however, it can definitely use such historical data to predict the users' eating habits. As another example, the smart robot vacuum [5], when it works, it collects data to identify the locations of house walls and furniture. This helps them avoid crashing into the furniture, but it also can measure the square footage of home and create a map of the house and share it to the cloud.

Location Privacy: Location-based services (LBSs) are fundamental to many 5G-enabled IoT devices. Every smartphone and vehicle have built-in Global Positioning System (GPS). There are many phone Apps also based on LBS services such as Yelp, Lyft, Uber, and so on. The primary purpose of these services is to improve users' life quality and help them live convenience by, for example, recommending an adjoining restaurant, looking up reviews, and so on [56]. However, at the same time, the users' location data are at risk if their privacy has been attacked; attackers can use the location data to predict users' activity routine and their places of interest and life habits. Much research has focused on protecting location privacy information. One popular strategy is to employ well-known privacy metrics such as k-anonymity and differential privacy [27]. Nevertheless, most of these strategies can only provide privacy protection with a certain probability. When the attacker integrates the information from different devices and analyzes the hidden correlations of the data, location privacy can be easily destroyed [26].

Correlation Privacy: With more personal and identity information collected by 5G-enabled devices, the attackers will also collect more private information. After integrating and analyzing the correlations between fragmented data collected from many devices, the attacker will be able to identify the individual information by the specific time, location, and so on. For example, if an attack results in unauthorized access to a hospital patient's database, even if the information they get cannot identify the particular individual due to the encryption of the database, after integrating and analyzing multiple sources, the patients' information still can be determined.

12.5 Security and Privacy Threats in Specific Domain

Health care: Health care information is very sensitive. Health care information security requires significant attention both from providers and governments all over the world. In the digital era, most of the patients' information is stored in a database, which opens the patients' records to security and privacy risk. The first security threat is software security. Nowadays, there are more and more health and wellness programs can be finished on the device side. The patients can download the Apps in this mobile devices, then make an appointment, review the reports, and so on. Patients must be made aware of the risk of privacy data breaches and malicious attackers hacking the health information. The second issue is human. Whether it is malicious or ignorance, the information leakage by humans always has serious repercussions. It may be caused by employees stealing data or IoT devices, or breaching the data by accident. DDoS attacks are the third security issue. Health care is a regular target. As the smart health care industry continues to deliver new services, the security management leaders in health care must always see the big picture. It is also necessary to pay attention to threats to the supply chain of hospitals. Each vendor a hospital cooperates with can present a potential data breach incident. Make sure that security and privacy are required in each step in the supply chain transaction [60].

Smart Home: Confidentiality threats and authentication threats are general security and privacy threats in the smart home environment. The undesired personal sensitive information breach is known as a confidentiality threat. For example, the smart home control systems such as Nest can track the home temperature, electric usage, and the determine when the house is occupied. Moreover, lost passwords can also lead to unauthorized access threats. When the unauthorized attacker has access to the system, they may be able to reprogram the system and send fake requests to the things, such as releasing a smart lock and causing a more dangerous outcome [41]. In addition, in the smart home environment, most of the IoT devices collect private data. This can cause a series of privacy issues. For example, the AI-embedded IoT can predict the users' behavior, this kind of private information is also at risk of breaches.

Smart Grid: The first important thing in smart grid is privacy. Smart meters transmit the private customers' information to the utility company and service provider. This information can be used to infer customers' behavior and whether a house is occupied. Because in smart grid, several devices can manage the electricity supply and network demand at the same time, some attacks can enter the devices through the network. Most of the smart grid devices focus on a single specific functionality, which results in the devices lacking enough memory space or capability to deal with security threats [42]. M2M in control systems also can play a role in security issues. When one device is attacked, it may send the fake state to make another device act in an unwanted way. Many devices in smart grid use remote control, and they have a long lifecycle and require effective security software updates. Otherwise, when a DoS attack gains access to the equipment, the equipment will stop working.

Smart Logistics: In the supply chain IoT, there are three security and privacy issues. The first is data corruption that allows the attacker to access the system, send a fake request and cause the device to make the wrong decision. The second threat is in the maintenance of the equipment. The DoS attack can damage the equipment or facility if security protection is not updated or maintained. Data privacy is a common threat in smart logistics. The supply chain always contains users' private information that should be carefully protected [55].

12.6 Challenges and Opportunities

In the previous sections, we reviewed the security threats in each 5G-enabled IoT layer, illustrated the common privacy problems. In this section, we present the security and privacy challenges along with existing research and a discussion on the potential research areas. As Figure 12.4 shows, each layer has different security challenges security strategies [57]. The first recognition layer consists of all the IoT devices, sensors and controllers. To ensure the security with M2M, how to complete authentication and identification is a great challenge. Create efficiency and safety key agreement, and data encryption is another challenge to protect the sensor data. Trust management is also needed if the IoT devices are transferred from one user to another. Network security problems are the most critical security challenges in the second layer, the connectivity/edge computing layer. All the edge paradigm networks have a security challenge: protocols and mechanisms which have data confidentiality and integrality requirements. When two authenticated entities communicate with each other, make sure the data confidentiality and integrality is necessary. In addition, how to prevent unauthorized access to the edge networks and how to avoid a DoS attack are other challenges. In the third layer, multiple support architectures are important parts. Strong encryption algorithms and protocols are needed in this layer. In addition, for other support infrastructures, such as cloud infrastructures, secure virtualization mechanisms are necessary. In the next application layer, the first challenge will be how to design secure and efficiency applications. In the meantime, protecting the users' private information is necessary because the application platforms always contain sensitive information. To avoid authorized access to the heterogeneous network, better authentication and encryption algorithms are required in this layer. In addition, because the smart applications have management and are maintained by a third party, good training of the management team is essential. The last layer is the business layer, how to create a

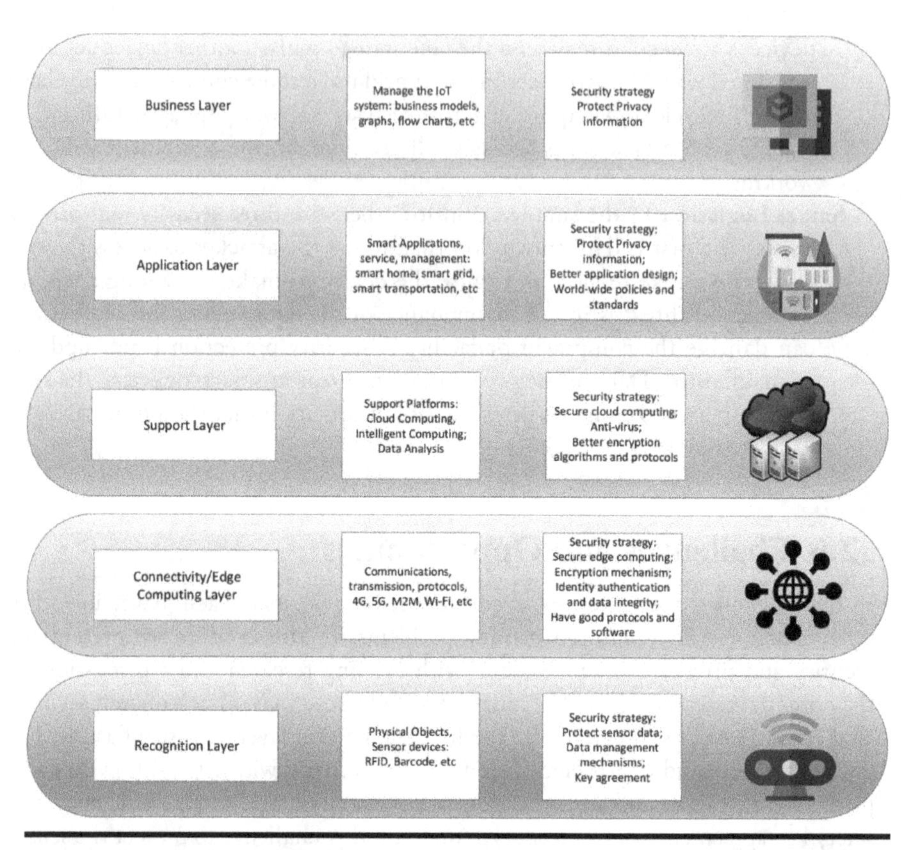

Figure 12.4 Architecture of the 5G-enabled IoT.

better business model and make sure the APIs are as secure as possible to protect the private information are the main challenges. Overall, we summarize the important security and privacy challenges and opportunities in 5G-enabled IoT environment.

12.6.1 Identity and Authentication

Efficiency and privacy authentication schemes are needed in the 5G-enabled IoT environment. Previously, the authentication protocols were designed based on the single-server environment, which are not suitable for the new architecture of 5G-enabled IoT environment–distributed services environment. Back in 2009, cloud computing came into our lives as new technology. With the huge growth of users and requests to share services, many researchers began to pay more attention to designing trust and security authentications between cloud users and the service. In addition, they found that most users, the past authentications such as SSL Authentication Protocol (SAP) were very complicated to use. Li et al. proposed a new identity-based authentication for cloud computing (IBACC), their authentication system is more efficient and lightweight than SAP [37]. Around 2015, more and more people found that much sensitive data had become digitalized, such as health care data. Therefore, many researchers construct more efficient authentication schemes to protect the E-health-care database. For example, Wu et al. and Jiang et al. all focused on proposing three-factor authentication protocols to defend against the different types of attacks [30,69]. After some researchers found a disadvantage to the single-server authentication schemes, some pursued the creation of inter-cloud identity management systems such as OpenID, SAML to provide Single-Sign-On (SSO) authentication. However, these authentication schemas relied on third parties, which may bring some new security threats. Therefore, how to design efficient and private authentication schemes will be a challenge in the distributed service environment.

There is some research focusing on this field. Lo et al. designed an efficient privacy-aware authentication scheme for MCC service. Their scheme allows mobile users to access multiple computing services from multiple providers using a single private key [59]. Recently, He et al. and Jiang et al. found some drawbacks of Lo's scheme and improved the scheme by using an identity-based signature scheme. Recall that IoT devices may transfer from one user to another. Because most IoT devices need to create a personal account and input the sensitive information, how to provide provides user anonymity with an efficient authentication should be another challenge. Li et al. [36], Yang et al. [71], Wu et al. [70] all focused on how to create an efficient authentication with user anonymity.

Even though there are many types of research in this domain, there are still some opportunities. The first opportunity in identity and authentication is synergies. We found most efficiency and privacy-aware authentication research are in the MCC environment. Finding the possible cooperation between MCC and other edge paradigms is necessary. The second potential research director is how to control the trade-off between security and privacy when designing an authentication protocol. For example, when using lightweight authentication, how to make sure the user-anonymity function still works well. For some battery-based IoT devices, how to save energy and protect the security at the same time is also an interesting topic.

12.6.2 Trust Management

In 5G-enabled IoT, especially for the different IoT devices, service providers and remote servers have existing trust issues. How to develop the trust relationship among device to device, device to user, user to server is a challenge. How to efficiently create the trust management framework is a most attractive research field for many researchers. Previously, trust management was developed for cloud computing environment to solve the trust issues between IoT users and the cloud server. At the beginning of trust management in cloud computing, Service Level Agreements (SLAs) were the foundation technique but they were not consistent among cloud providers. This left the user unable to identify a trustworthy cloud provider. Therefore, many trust management frameworks are proposed. Petri et al. proposed a trust model that manages the information and identifies the trust distribution [49]. Hammam et al. employed a trust management system to calculate the trust value for the node in the mobile ad hoc clouds environment. In their system, they combined the considerations of availability, neighbor's evaluation, and response quality and task completeness together to ensure efficiency [23]. In 2016, Zhu et al. implemented SLAs trust as the base and proposed an innovative trust model to integrate the objective and subjective trust in the MCC manner [13].

It is evident that most trust management researchers focused on centralized service. From 2016, more and more researchers begin to contribute to distributed computing service. However, trust management in distributed-based computing in 5G-enabled IoT will still be an open research direction in the future.

12.6.3 Encryption Method

Encryption has been a research interest for many researchers in many years. However, the traditional encryption methods such as Triple DSE (3DSE), Triple Data Encryption Algorithm (TDEA) have drawbacks in that the devices must know the identities of the information recipients and share credentials with them in advance. In the 5G-enabled IoT environment, in many scenarios, there are many unknown recipients, which means the traditional encryption algorithms cannot handle the complicated IoT system. Therefore, some encryption methods developed for the IoT environment have contributed many solutions. Attribute-Based Encryption (ABE) in one of the encryption algorithms consists of a key authority between the data sender and the recipient [66].

How to make the encryption scheme more secure and efficiency has become a challenge when designing encryption algorithms. Wu et al. combined two systems: the hierarchical identity-based encryption (HIBE) and the ciphertext attributed-based encryption (CP-ABE); and proposed an efficiency encryption scheme for users to the share the confidential data in the centralized cloud environment [64]. After 3 years. Li et al. introduced a novel encryption mechanism to protect the health care data in cloud servers. They leveraged the ABE techniques to encrypt each patient's private health record file [39]. With the development of edge paradigms, there are many extensive concerns of the edge paradigms encryption method. Alrawais et al. offered an efficient key exchange protocol based on CP-ABE and digital signature techniques in the fog computing environment and achieved more efficient performance on confidentiality, authentication, verifiability, and access control [2]. Jiang et al. also designed an encryption scheme based on CP-ABE for fog computing IoT [31]. Besides KP-ABE, there is another scheme in ABE: key-policy ABE (KP-ABE) which also can contribute to the big data security [67].

Unfortunately, most existing encryption mechanisms studies are mainly focused on cloud computing or fog computing environment, for other environments, such as MEC and MCC, there still have many opportunities. In addition, most of the research has focused more attention on the CP-ABE algorithms, but there be other encryption methods that can be incorporated with CP-ABE, such as fully homomorphic encryption (FHE), ciphertext policy attribute-based proxy re-encryption (CP-ABPRE) [21,22].

12.6.4 Access Control

Access control is used to ensure that unauthorized entities cannot access IoT devices and collect data. Previously, most research has focused on designing access control systems for cloud computing. Yu et al. created an access control system by exploiting techniques of many encryption schemes and building an efficient fine-grained data access control [73]. They also proposed another novel framework for access control to the health care domain within the cloud computing environment [38].

However, research into the access control mechanisms in edge computing is sparse. Implementing effective access control algorithms will be one challenge. Because of the enormous numbers of the edge devices in the 5G-enabled IoT, a new challenge to access control is how to efficiently find the access model and how to optimize the limited resources, especially for some battery-based devices.

12.6.5 Privacy

Because in the 5G-enabled IoT, the IoT architecture is complicated and distributed, the information that IoT devices collect is sensitive and private and more complex privacy mechanisms are needed now. In the past years, there have been many privacy protection mechanisms such as k-anonymity [24], differential privacy [25], quasi-identifier [74], and pseudonymization [11], and so on.

At the beginning of IoT privacy research, there were many privacy protection mechanisms focused on cloud computing. Wang et al. utilized and uniquely combined the public key-based homomorphic authenticator with random masking and then achieved the privacy-preserving public audit requirement [63]. Itani et al. presented the Privacy as a Service (PasS) security protocols for protection against the privacy threats in the cloud computing architecture. After edge paradigms emerged, Lu et al. presented the Lightweight Privacy-preserving Data Aggregation scheme for fog computing IoT [44]. Wang et al. proposed a privacy-preserving content-based publish/subscribe (PCP) with differential privacy in a fog computing context [65]. Therefore, more and more researchers are now paying attention to how to protect privacy by using efficient mechanisms.

However, there are also some challenges. Some IoT applications always use incentive mechanisms to collect data from users, such as crowdsourcing. The trade-off between incentive price the works privacy will be a fundamental research problem. Even though the better mechanisms can protect the privacy, it is hard to find one privacy protection mechanism that has no vulnerability. If the authorized attackers access the system and send a fake request, it will still cause the devices to take abnormal actions. Therefore, how to use AI-embedding IoT algorithms to predict the cyber attack or predict the users' behavior is essential.

12.7 Conclusion

With the development of 5G wireless and IoT devices, distributed-based computing is becoming an efficient and possible technological solution to handle the billions of the 5G-IoT devices. The new techniques provide us various conveniences and high quality of life. However, they also produce new security threats and put numerous users' private information at risk. We constructed the 5G-enabled IoT as five layers: recognition layer, connectivity/edge computing layer, support layer, application layer, and business layer. In each layer, we concluded the security and privacy threats and reviewed the challenges for 5G-enabled IoT. For future study, we suggest implementing the appropriate existing security and privacy strategies to the edge paradigms domain, incorporated with AI algorithms to improve the application service quality and defend against unauthorized access.

References

1. Arwa Alrawais, Abdulrahman Alhothaily, Chunqiang Hu, and Xiuzhen Cheng. Fog computing for the Internet of Things: Security and privacy issues. *IEEE Internet Computing*, 21(2): 34–42, 2017.
2. Arwa Alrawais, Abdulrahman Alhothaily, Chunqiang Hu, Xiaoshuang Xing, and Xi-uzhen Cheng. An attribute-based encryption scheme to secure fog communications. *IEEE Access*, 5: 9131–9138, 2017.
3. Jeffrey G. Andrews, Stefano Buzzi, Wan Choi, Stephen V. Hanly, Angel Lozano, Anthony C. K. Soong, and Jianzhong Charlie Zhang. What will 5G be? *IEEE Journal on Selected Areas in Communications*, 32(6): 1065–1082, 2014.
4. Michael Armbrust, Armando Fox, Rean Griffith, Anthony D. Joseph, Randy Katz, Andy Konwinski, Gunho Lee, David Patterson, Ariel Rabkin, Ion Stoica et al. A view of cloud computing. *Communications of the ACM*, 53(4): 50–58, 2010.
5. Maggie Astor. Your Roomba may be mapping your home, collecting data that could be Shared. https://www.nytimes.com/2017/07/25/technology/roomba-irobot-data-privacy.html, 2017.
6. Flavio Bonomi, Rodolfo Milito, Jiang Zhu, and Sateesh Addepalli. Fog computing and its role in the Internet of Things. In *Proceedings of the First Edition of the MCC Workshop on Mobile Cloud Computing*, pp. 13–16. ACM, 2012.
7. Sophia Cart. Difference Between Security and Privacy. http://isc2central. blogspot. com/2018/08/difference-between-security-and-privacy.html, 2018.
8. Antonio Celesti, Alina Buzachis, Antonino Galletta, Maria Fazio, and Massimo Villari. A nosql graph approach to manage IoTaaS in cloud/edge environments. In *2018 IEEE 6th International Conference on Future Internet of Things and Cloud (FiCloud)*, pp. 407–412. IEEE, 2018.

9. Mung Chiang and Tao Zhang. Fog and IoT: An overview of research opportunities. *IEEE Internet of Things Journal*, 3(6): 854–864, 2016.
10. Debabani Choudhury. 5G wireless and millimeter wave technology evolution: An overview. In *Microwave Symposium (IMS), 2015 IEEE MTT-S International*, pp. 1–4. IEEE, 2015.
11. F. Meyer De, G. Moor De, and L. Reed-Fourquet. Privacy protection through pseudonymisation in eHealth. *Studies in Health Technology and Informatics*, 141: 111–118, 2008.
12. Steve Deering and Robert Hinden. Internet protocol, version 6 (IPv6) specification. Technical report, 2017.
13. Qing Ding, Xiangzhou Chen, and Zhu Shi. A novel trust management framework for mobile cloud computing environment. In *Proceedings of the 2017: The 7th International Conference on Computer Engineering and Networks*, July 22–23. Shanghai, China, 2017.
14. Hoang T. Dinh, Chonho Lee, Dusit Niyato, and Ping Wang. A survey of mobile cloud computing: Architecture, applications, and approaches. *Wireless Communications and Mobile Computing*, 13(18): 1587–1611, 2013.
15. Koustabh Dolui and Soumya Kanti Datta. Comparison of edge computing implementations: Fog computing, cloudlet and mobile edge computing. In *Global Internet of Things Summit (GIoTS)*, 2017, pp. 1–6. IEEE, 2017.
16. A. B. Ericsson. Ericsson mobility report: On the pulse of the networked society. Ericsson, Sweden, Technical Report, EAB-14, 61078, 2015.
17. Dave Evans. The Internet of Things: How the next evolution of the internet is changing everything. *CISCO White Paper*, 1(2011): 1–11, 2011.
18. Muhammad Umar Farooq, Muhammad Waseem, Anjum Khairi, and Sadia Mazhar. A critical analysis on the security concerns of Internet of Things (IoT). *International Journal of Computer Applications*, 111(7), 2015.
19. Bree Fowler. Gifts That Snoop? The Internet of Things is wrapped in privacy concerns. https://www.consumerreports.org/internet-of-things/gifts-that-snoop-internet-of-things-privacy-concerns/, 2017.
20. Julien Freudiger, Maxim Raya, M'ark F'elegyh'azi, Panos Papadimitratos, and Jean-Pierre Hubaux. Mix-zones for location privacy in vehicular networks. In *ACM Workshop on Wireless Networking for Intelligent Transportation Systems (WiN-ITS)*, number LCA-CONF-2007-016, 2007.
21. Chunpeng Ge, Willy Susilo, Liming Fang, Jiandong Wang, and Yunqing Shi. A CCA-secure key-policy attribute-based proxy re-encryption in the adaptive corruption model for dropbox data sharing system. *Designs, Codes and Cryptography*, pp. 1–17, 2018.
22. Craig B. Gentry. Fully homomorphic encryption, July 25, 2017. US Patent 9,716,590.
23. Ahmed Hammam and Samah Senbel. A trust management system for ad-hoc mobile clouds. In *Computer Engineering & Systems (ICCES), 2013 8th International Conference on*, pp. 31–38. IEEE, 2013.
24. Meng Han, Zhuojun Duan, and Yingshu Li. Privacy issues for transportation cyber physical systems. In *Secure and Trustworthy Transportation Cyber-Physical Systems*, pp. 67–86. Springer, 2017.
25. Meng Han, Ji Li, Zhipeng Cai, and Qilong Han. Privacy reserved influence maximization in gps-enabled cyber-physical and online social networks. In *Big Data and Cloud Computing (BDCloud), Social Computing and Networking (SocialCom), Sustainable Computing and Communications (SustainCom)(BDCloud-SocialCom-SustainCom), 2016 IEEE International Conferences on*, pp. 284–292. IEEE, 2016.

26. Meng Han, Lei Li, Ying Xie, Jinbao Wang, Zhuojun Duan, Ji Li, and Mingyuan Yan. Cognitive approach for location privacy protection. *IEEE Access*, 6: 13466–13477, 2018.
27. Meng Han, Mingyuan Yan, Zhipeng Cai, Yingshu Li, Xingquan Cai, and Jiguo Yu. Influence maximization by probing partial communities in dynamic online social networks. *Transactions on Emerging Telecommunications Technologies*, 28(4): e3054, 2017.
28. Keiko Hashizume, David G. Rosado, Eduardo Fern´andez-Medina, and Eduardo B. Fernandez. An analysis of security issues for cloud computing. *Journal of Internet Services and Applications*, 4(1): 5, 2013.
29. Yun Chao Hu, Milan Patel, Dario Sabella, Nurit Sprecher, and Valerie Young. Mobile edge computing a key technology towards 5G. *ETSI White Paper*, 11(11): 1–16, 2015.
30. Qi Jiang, Muhammad Khurram Khan, Xiang Lu, Jianfeng Ma, and Debiao He. A privacy preserving three-factor authentication protocol for e-health clouds. *The Journal of Supercomputing*, 72(10): 3826–3849, 2016.
31. Yinhao Jiang, Willy Susilo, Yi Mu, and Fuchun Guo. Ciphertext-policy attribute-based encryption against key-delegation abuse in fog computing. *Future Generation Computer Systems*, 78: 720–729, 2018.
32. Qi Jing, Athanasios V. Vasilakos, Jiafu Wan, Jingwei Lu, and Dechao Qiu. Security of the Internet of Things: Perspectives and challenges. *Wireless Networks*, 20(8): 2481–2501, 2014.
33. Mike Kavis. Investor Guide to IoT Part 1-Understanding The Ecosystem. https://www.forbes.com/sites/mikekavis/2016/02/24/investors-guide-to-iot-part-1-understanding-the-ecosystem/ #195f511311a1, 2016.
34. Rafiullah Khan, Sarmad Ullah Khan, Rifaqat Zaheer, and Shahid Khan. Future internet: The Internet of Things architecture, possible applications and key challenges. In *Frontiers of Information Technology (FIT), 2012 10th International Conference on*, pp. 257–260. IEEE, 2012.
35. Mirosl Aw Klinkowski and Krzysztof Walkowiak. On the advantages of elastic optical networks for provisioning of cloud computing traffic. *IEEE Network*, 27(6): 44–51, 2013.
36. Chun-Ta Li, Tsu-Yang Wu, Chin-Ling Chen, Cheng-Chi Lee, and Chien-Ming Chen. An efficient user authentication and user anonymity scheme with provably security for iot-based medical care system. *Sensors*, 17(7): 1482, 2017.
37. Hongwei Li, Yuanshun Dai, Ling Tian, and Haomiao Yang. Identity-based authentication for cloud computing. In *IEEE International Conference on Cloud Computing*, pp. 157–166. Springer, 2009.
38. Ming Li, Shucheng Yu, Kui Ren, and Wenjing Lou. Securing personal health records in cloud computing: Patient-centric and fine-grained data access control in multi-owner settings. In *International Conference on Security and Privacy in Communication Systems*, pp. 89–106. Springer, 2010.
39. Ming Li, Shucheng Yu, Yao Zheng, Kui Ren, and Wenjing Lou. Scalable and secure sharing of personal health records in cloud computing using attribute-based encryption. *IEEE Transactions on Parallel and Distributed Systems*, 24(1): 131–143, 2013.
40. Shancang Li, Li Da Xu, and Shanshan Zhao. 5G Internet of Things: A survey. *Journal of Industrial Information Integration*, 10: 1-9, 2018.
41. Huichen Lin and Neil W. Bergmann. IoT privacy and security challenges for smart home environments. *Information*, 7(3): 44, 2016.

42. Jing Liu, Yang Xiao, Shuhui Li, Wei Liang, and C. L. Philip Chen. Cyber security and privacy issues in smart grids. *IEEE Communications Surveys & Tutorials*, 14(4): 981–997, 2012.
43. Liyuan Liu, Meng Han, Yan Wang, and Yiyun Zhou. Understanding data breach: A visualization aspect. In *International Conference on Wireless Algorithms, Systems, and Applications*, pp. 883–892. Springer, 2018.
44. Rongxing Lu, Kevin Heung, Arash Habibi Lashkari, and Ali A. Ghorbani. A lightweight privacy-preserving data aggregation scheme for fog computing-enhanced IoT. *IEEE Access*, 5: 3302–3312, 2017.
45. Somayya Madakam, R. Ramaswamy, and Siddharth Tripathi. Internet of Things (IoT): A literature review. *Journal of Computer and Communications*, 3(5): 164, 2015.
46. Rwan Mahmoud, Tasneem Yousuf, Fadi Aloul, and Imran Zualkernan. Internet of things (IoT) security: Current status, challenges and prospective measures. In *Internet Technology and Secured Transactions (ICITST), 2015 10th International Conference for*, pp. 336–341. IEEE, 2015.
47. Peter Mell, Tim Grance et al. The NIST definition of cloud computing. NIST Special Publications, 800, 2011.
48. Maria Rita Palattella, Mischa Dohler, Alfredo Grieco, Gianluca Rizzo, Johan Torsner, Thomas Engel, and Latif Ladid. Internet of Things in the 5G era: Enablers, architecture, and business models. *IEEE Journal on Selected Areas in Communications*, 34(3): 510–527, 2016.
49. Ioan Petri, Omer F. Rana, Yacine Rezgui, and Gheorghe Cosmin Silaghi. Trust modelling and analysis in peer-to-peer clouds. *International Journal of Cloud Computing*, 1(2–3): 221–239, 2012.
50. Ling Qian, Zhiguo Luo, Yujian Du, and Leitao Guo. Cloud computing: An overview. In *IEEE International Conference on Cloud Computing*, pp. 626–631. Springer, 2009.
51. Rodrigo Roman, Javier Lopez, and Masahiro Mambo. Mobile edge computing, Fog et al.: A survey and analysis of security threats and challenges. *Future Generation Computer Systems*, 78: 680–698, 2018.
52. Maheswaran Sathiamoorthy, Alexandros G. Dimakis, Bhaskar Krishnamachari, and Fan Bai. Distributed storage codes reduce latency in vehicular networks. *IEEE Transactions on Mobile Computing*, 13(9): 2016–2027, 2014.
53. Tweet Like Plus Share. 10 ways the Internet of Things (IoT) is redefining manufacturing.
54. Knud Erik Skouby and Per Lynggaard. Smart home and smart city solutions enabled by 5G, IoT, AAI and CoT services. In *Contemporary Computing and Informatics (IC3I), 2014 International Conference on*, pp. 874–878. IEEE, 2014.
55. Rob Stevens and Lenny Zeltser. IoT and Security in the Supply Chain: Making Smart Choices. https://www.inboundlogistics.com/cms/article/IoT-and-security-in-the-supply-chain-making-smart-choices/, 2018.
56. Gang Sun, Victor Chang, Muthu Ramachandran, Zhili Sun, Gangmin Li, Hongfang Yu, and Dan Liao. Efficient location privacy algorithm for Internet of Things (IoT) services and applications. *Journal of Network and Computer Applications*, 89: 3–13, 2017.
57. Hui Suo, Jiafu Wan, Caifeng Zou, and Jianqi Liu. Security in the Internet of Things: A review. In *Computer Science and Electronics Engineering (ICCSEE), 2012 International Conference on*, volume 3, pp. 648–651. IEEE, 2012.

58. Santosh M. Tondare, Sachin D. Panchal and Devidas T. Kushnure. Evolutionary steps from 1 g to 4.5 g. *International Journal of Advanced Research in Computer and Communication Engineering*, 3(4): 6163–6166, 2014.
59. Jia-Lun Tsai and Nai-Wei Lo. A privacy-aware authentication scheme for distributed mobile cloud computing services. *IEEE Systems Journal*, 9(3): 805–815, 2015.
60. UIC. Top 4 threats to healthcare security. https://healthinformatics.uic.edu/resources/articles/top-4-threats-to-healthcare-security/, 2013.
61. Gary Utley. 6 most common cloud computing security issues. https://www.cwps.com/blog/cloud-computing-security-issues, 2018.
62. Angel Leonardo Valdivieso Caraguay, Alberto Benito Peral, Lorena Isabel Barona Lopez, and Luis Javier Garcia Villalba. SDN: Evolution and opportunities in the development IoT applications. *International Journal of Distributed Sensor Networks*, 10(5): 735142, 2014.
63. Cong Wang, Qian Wang, Kui Ren, and Wenjing Lou. Privacy-preserving public auditing for data storage security in cloud computing. In *Infocom, 2010 Proceedings IEEE*, pp. 1–9. IEEE, 2010.
64. Guojun Wang, Qin Liu, and Jie Wu. Hierarchical attribute-based encryption for fine-grained access control in cloud storage services. In *Proceedings of the 17th ACM Conference on Computer and Communications Security*, pp. 735–737. ACM, 2010.
65. Qixu Wang, Dajiang Chen, Ning Zhang, Zhe Ding, and Zhiguang Qin. PCP: A privacy-preserving content-based publish–subscribe scheme with differential privacy in fog computing. *IEEE Access*, 5: 17962–17974, 2017.
66. Xinlei Wang, Jianqing Zhang, Eve M. Schooler, and Mihaela Ion. Performance evaluation of attribute-based encryption: Toward data privacy in the iot. In *Communications (ICC), 2014 IEEE International Conference on*, pp. 725–730. IEEE, 2014.
67. Zhiwei Wang, Cheng Cao, Nianhua Yang, and Victor Chang. Abe with improved auxiliary input for big data security. *Journal of Computer and System* Sciences, 89: 41–50, 2017.
68. Wordpress. Laying the foundations for 5G mobile. https://ytd2525.wordpress.com/2015/01/23/laying-the-foundations-for-5g-mobile/, 2015.
69. Fan Wu, Lili Xu, Saru Kumari, and Xiong Li. A novel and provably secure biometrics-based three-factor remote authentication scheme for mobile client–server networks. *Computers & Electrical Engineering*, 45: 274–285, 2015.
70. Libing Wu, Yubo Zhang, Yong Xie, Abdulhameed Alelaiw, and Jian Shen. An efficient and secure identity-based authentication and key agreement protocol with user anonymity for mobile devices. *Wireless Personal Communications*, 94(4): 3371–3387, 2017.
71. Xu Yang, Xinyi Huang, and Joseph K. Liu. Efficient handover authentication with user anonymity and untraceability for mobile cloud computing. *Future Generation Computer Systems*, 62: 190–195, 2016.
72. Rong Yu, Yan Zhang, Stein Gjessing, Wenlong Xia, and Kun Yang. Toward cloud-based vehicular networks with efficient resource management. *IEEE Network*, 27(5): 48–55, 2013.
73. Shucheng Yu, Cong Wang, Kui Ren, and Wenjing Lou. Achieving secure, scalable, and fine-grained data access control in cloud computing. In *Infocom, 2010 Proceedings IEEE*, pp. 1–9. IEEE, 2010.
74. Xuyun Zhang, Chang Liu, Surya Nepal, and Jinjun Chen. An efficient quasi-identifier index based approach for privacy preservation over incremental data sets on cloud. *Journal of Computer and System Sciences*, 79(5): 542–555, 2013.

Chapter 13

Privacy-Preserving Techniques for the 5G-Enabled Location-Based Services

Chen Wang, Ping Zhao, Haojun Huang, Rui Zhang, and Weixing Zhu

Contents

13.1 Introduction

Nowadays, with the convergence of the Internet and advanced communication technologies, the mobile Internet has become a reality. The advent of fifth generation (5G) wireless technology will further promote the rapid development of the Internet of Things (IoT) industry. Compared with the current 4G technology, the 5G technology will increase its peak rate by several tens of times and support access to more networks, which can better meet the massive access scenarios such as IoT [1]. The 5G delay is shorter, which makes it a great advantage in the field of vehicle network vehicle safety. The higher reliability of 5G enables home facilities such as home appliances with wireless communication capabilities in smart home applications to access smart terminals. It is foreseeable that the 5G will play a very important role and support in IoT era in the near future (Figure 13.1) [2].

In the meantime, IoT also arms location-based services (LBSs) with more possibilities. The application of the LBS system in IoT is mainly to realize the basic information and location information of the target object in real time by deploying sensing devices of various sizes to facilitate people's daily work and life. For example, in elderly apartments, LBS can get real-time information about the elderly, which greatly improves management efficiency and safety for the elderly. For upcoming 5G systems, positioning requirements are much stringent (less than 1 m accuracy

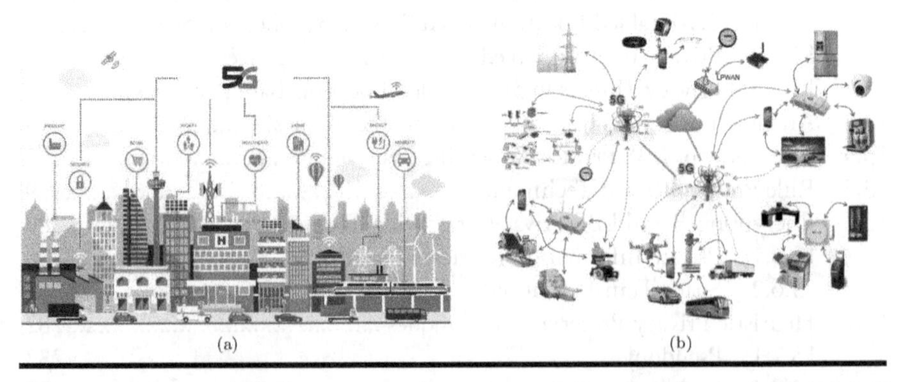

Figure 13.1 5G-enabled IoT scenarios. (a) 5G IoT networks and (b) 5G IoT applications.

for both indoor and outdoor users, including humans, devices, machines, vehicles, and so on). Consequently, significant improvements in positioning accuracy is achievable by appropriately redesigning positioning reference signals for 5G radio access technology. This will enable LBSs to continue enjoying growing popularity now and in near future [3–7]. The most popular LBSs include map applications (e.g., Google Maps), points of interest search (e.g., HERECITY), coupon or discount push (e.g., Shopkick), location-based game applications (e.g., MyTown), location-aware social networks (e.g., Foursquare), and check in applications (e.g., Gowalla), and so on, which have largely improved users' daily lives.

However, 5G-enabled LBSs have also raised severe privacy concerns [8–11]. When enjoying 5G-enabled LBSs, users have to provide their location information embedded in the LBS queries to the LBS server. Such location information often implies sensitive personal information, such as religious activities, social relationships, health conditions, and living habits. It is found that, the LBS server may either deliberately or inadvertently disclose the location information involved in LBS queries [12–15]. For example, the black hat "Peace" disclosed over 167 million users' information (along with their locations) from LinkedIn in 2013 [16] and about 360 million from MySpace in 2016 [17].

Alternatively, an adversary who has compromised the LBS server can also infer sensitive privacy information of the users. For example, Hoh et al. [18] and Krumm [19] showed that a driver's home location can be inferred from Global Positioning System (GPS) data collected on his/her vehicle even if the location data were pseudonymized or anonymized. In another study, Matsuo [20] exploited a user's indoor location data to infer a variety of personal information such as work role, smoker or not, coffee drinker or not, and even age. Moreover, Gruteser and Hoh [21,22] showed that individuals' tracks can be reassembled from completely anonymized GPS data from three or even five users by using multiple hypotheses tracking (MHT) [23].

Therefore, the location privacy disclosed in LBSs has raised severe privacy concerns of the LBS users. For instance (Figure 13.2), 90% of investigated users express concern about their location privacy that may be disclosed to advertisers, strangers, friends, and employers, 6% of investigated users do not care about their privacy, and 4% of investigated users are not sure whether their location privacy is disclosed in LBS [24]. In such grim situations, how to supervise LBSs while preserving user's location privacy has been a hot research topic for years [25,26].

Researchers have long been aware of the potential privacy risks associated with LBSs, and have proposed a number of promising schemes that can help users protect their privacy. In this chapter, we provide a comprehensive overview on the state of the art and the key fundamentals of LBS privacy-preserving techniques (LPPT). We classify the up-to-date techniques with respect to different privacy-preserving approaches, and point out the similarities and differences of all research efforts. To understand the trade-offs of each approach, we also present the advantages and disadvantages of each technique. Finally, we highlight

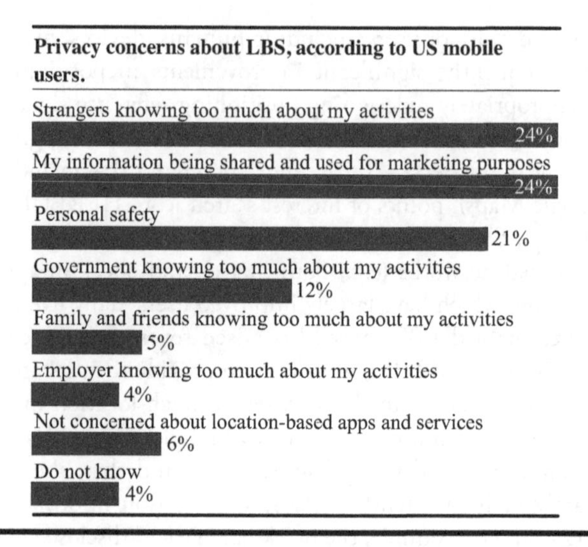

Figure 13.2 The privacy concerns about LBS, according to SACA's research on US mobile users.

the issues in existing solutions and new research directions to augment privacy protection in LBS systems.

The remainder of this chapter is organized as follows: Section 13.2 gives an overview of the context of 5G-enabled LBSs and the privacy threats in LBSs. Section 13.3 describes the general architecture LPPT. Section 13.4 presents the taxonomy of LPPT. Sections 13.5 through 13.8 review several typical works in each category of the existing LPPT. Section 13.9 discusses new challenges and future research topics. Finally, Section 13.10 concludes the paper.

13.2 Privacy Issues in 5G-Enabled Location-Based Services

13.2.1 5G-Enabled Location-Based Services Application Scenarios

LBSs have become popular over the last years due to the global adoption of smartphones and the worldwide availability of the GPS and other positioning methods. A typical LBS system consists of mobile users and the LBS server, as shown in Figure 13.3. In LBS, users send LBS queries to the LBS server, and the LBS server returns LBS results to mobile users. In general, there are two LBS application scenarios: single-query scenarios and continuous-query scenarios.

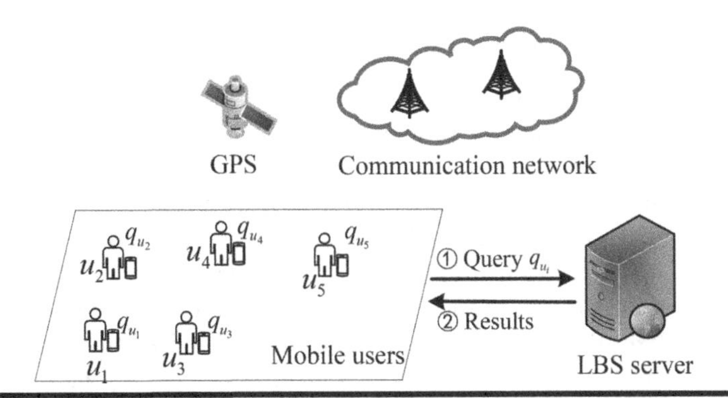

Figure 13.3 A typical architecture of 5G-enabled LBSs.

Single-query scenarios: In this scenario, users send only one or multiple queries containing users' location information to the LBS server. Upon receiving users' queries, the LBS server searches and returns personalized LBS (i.e., query results) to users, according to the users' current locations. For example, users can search POI (e.g., restaurant hotel) through HERECITY (Figure 13.4a), and MyTown provides location-based games (Figure 13.4b). Single-query scenarios include neighbor query (e.g., search hotels around users), Top-k query (e.g., search *k* nearest hotels), range query (e.g., search all hotels within a range), among others.

Continuous-query scenarios: In continuous-query scenarios, users continuously or periodically send LBS queries containing users' location information to the LBS server. The LBS server returns personalized LBS to users, according to the users' movement trajectories. For instance, Google Maps returns real-time traffic and navigation route to users, while users continuously provide Google Maps their locations (Figure 13.4c), and Gowalla provides a specific user with the locations of his friends around him when Foursquare acquires the knowledge of his continuous locations (Figure 13.4d).

Figure 13.4 Examples of (a,b) single-query applications and (c,d) continuous-query applications. (a) HERECITY, (b) MyTown, (c) Google Maps, and (d) Gowalla.

13.2.2 Privacy in 5G-Enabled Location-Based Services

Based on the two LBS application scenarios, the privacy in LBSs largely refers to both location privacy and trace privacy as described below.

Location privacy (unidentification) refers to users' location information contained in the LBS queries and other personal information (e.g., occupation, health, religion) that can be inferred from users' locations.

Specially, location privacy involves whether users can be precisely located, or whether other personal information can be deduced from the users' locations. Thus, protecting users' location privacy is to prevent attackers from getting users' exact locations and inferring other personal information from locations.

Trace privacy (unlinkability) refers to users' query content contained in LBS queries and other personal information inferred from query content.

When a query content is mapped to a specific user, the user' query privacy is breached because query content indicates the user's hobby or other personal information. Thus, query privacy involves whether a specific user's query content is identified by attackers, and whether other personal information can be deduced from query content.

13.2.3 Privacy Threats in 5G-Enabled Location-Based Services

Typically, the LBS server is assumed to be semi-honest (i.e., it provides LBSs to users while it attempts to disclose the users' location privacy), and wireless channels and mobile devices are assumed to be secure, which is the general privacy threat model in LBSs [27,28]. Therefore, location privacy in LBSs can be disclosed by mobile devices, wireless transmission, and the LBS server. First, users' personal information can be disclosed when mobile devices are attacked. Safeguarding mobile devices is a parallel topic and is out of the scope of this chapter; interested readers can refer to techniques [29,30]. Second, attackers may eavesdrop on users' queries during wireless transmission, which can be solved by Encryption and hashing mechanisms [31,32]. Third, the LBS server may be malicious or be attacked, and all users' location privacy will thus be breached.

13.3 Architecture of LBS Privacy-Preserving Techniques

The architecture of LPPT can be classified into centralized, distributed, and mixed architecture as shown in Figure 13.5.

The centralized architecture consists of mobile users, trusted server, and the LBS server (Figure 13.5a). The trusted server collects users' queries and protects users'

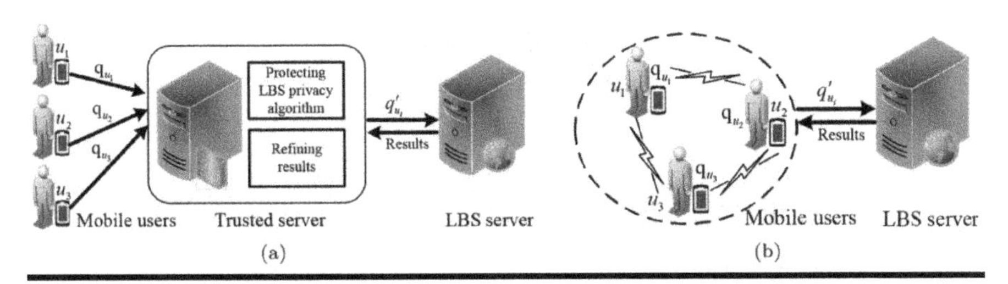

Figure 13.5 (a) The centralized architecture and (b) the distributed architecture of privacy preserving techniques.

privacy through privacy-preserving algorithms, and thereafter sends protected queries to the LBS server. Upon receiving the protected queries, the LBS server searches the results and sends them to the trusted server. The trusted server then forwards related results to users. However, the centralized LPPT suffer from two drawbacks: (1) It is nontrivial to deploy the trusted server in practice, and (2) when the trusted server is attacked, all users' privacy is breached.

The distributed architecture includes mobile users and the LBS server (Figure 13.5b). Users communicate with each other through single-hop and multi-hop communication to protect their privacy, or each user locally executes a privacy-preserving algorithm. Then protected LBS queries are sent to the LBS server. Finally, the LBS server returns the results to users. Although the distributed LPPT do not rely on the trusted server, they enlarge the overhead of mobile users. Furthermore, when some users taking part in the distributed privacy-preserving algorithms are malicious, then other trusted users' privacy can be disclosed. In addition, when users' locations are very sparse, it is time consuming to perform the distributed privacy-preserving algorithms.

The mixed architecture encapsulates both the centralized and distributed modules. So mixed privacy-preserving techniques can conserve both the overhead of mobile users and the trusted server. Unfortunately, complicated parameter setting limits the application of the mixed privacy-preserving techniques [33].

13.4 Taxonomy of LBS Privacy-Preserving Techniques

LBS privacy has attracted the attention of academia over the last decades, and we classify the privacy-preserving techniques as in Figure 13.6. At the beginning, LBS privacy-preserving techniques intuitively prevented potentially untrusted LBS server from accessing users' location information contained in LBS queries through rule protocol-based privacy-preserving techniques [34–37], for example, GeoPriv [38] and P3P [39]. But rule protocol-based privacy-preserving techniques cannot protect users' location privacy because the execution of rule-based protocols heavily rely on the supervision of economy, society, and government, and the

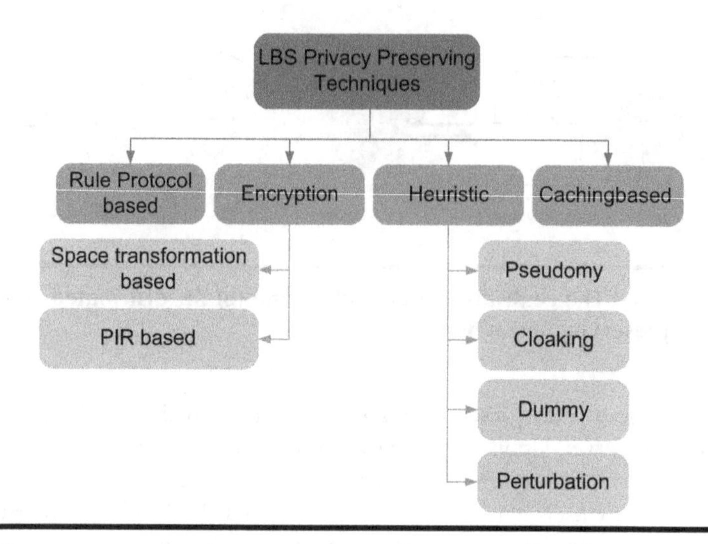

Figure 13.6 Taxonomy of LBS privacy-preserving techniques.

LBS server may often disclose users' location information to advertisement or analytic servers for the economic benefits [12–15]. Furthermore, rule protocol-based privacy-preserving techniques cannot weigh the privacy protection and quality of service (QoS) [34–37].

Another line of protecting privacy in LBSs is to encrypt the data so that the sensitive information carried by the data cannot be available to others. Though encryption techniques has been well established and powerful for privacy preserving, the common encryption techniques protect privacy at the sacrifice of the data usability, which is unsuitable to LBS. In LBS applications, the location information is the key to service, if it cannot be obtained by the server, then services cannot be provided, and this intrinsic conflict is exactly the difficulty of LBS privacy protection. Some cryptographic techniques like the private information retrieval (PIR) protocol [31,32] and space transformation [40,41] can be used in some applications of LBSs such as the Nearest Neighbor (NN) query and the shortest path calculation and achieve a high level of privacy and QoS. However, they suffer from disadvantages of complex deployment and high cost in computation and communication, making it necessary to design optimization algorithms.

Thus, the flow-up research work in LBS privacy preserving strives to design rigorous algorithms to protect LBS users' location privacy. Because the possibility of attackers getting a specific user's location and other personal information deduced from his location is proportional to the attackers' preknowledge about the user, it is nontrivial to compute this possibility. For example, some LPPT taking users' moving speed into consideration have to deal with the NP (Non-Polynomial) problems [42]. As a result, in this period, most if not all LPPT employ heuristic algorithms that assume that a specific user's location privacy is only related to the location information in the LBS query. The heuristic LPPT mainly include pseudonyms, cloaking, dummy and perturbation.

Pseudonym is used to prevent the LBS server from mapping a specific user's LBS queries to the user by replacing the user's identity information with pseudonyms [43]. So, Pseudonym mainly protects the query privacy of users. Thus, when a user applies for an LBS with Pseudonym, even though attackers can obtain the locations of the users, they cannot recognize the identity of the users corresponding to their locations directly. However, using only Pseudonym cannot provide enough protection level for query privacy. It is very easy for attackers to infer the true identity of users with some background knowledge. Even though frequently changing the pseudonym, attackers can still continuously track the user through an identity match attack model, thus invalidating the role of Pseudonym. So, a more effective technique called *Mix Zone* has been proposed. Levente et al. [44] divides the road network into the observed zone and the unobserved zone, which functions as a mix zone. The identity of vehicles in the observed zone are visible, whereas it is invisible in the unobserved zone, and vehicles mix and change pseudonyms within this zone, making it difficult for attackers to associate the pseudonyms used by a specific vehicle before entering and or after exiting this zone. So, the mix zone is more suitable for services that need to track a user's movement like Vehicle Ad hoc Networks (VANETs). Furthermore, it can be used in the continuous-query scenarios according to some modified version, which will be described in detail in Section 13.7.1. Somehow, mix zone can improve the effectiveness of changing a pseudonym, but weaknesses still exist: (1) mix zone is dependent on the third-party anonymity server, (2) mix zone fails to provide query service for users who locate in the mix zone such that the QoS of the system is degraded.

In 2003, Gruteser et al. introduced the notion of *k-anonymity* from relational database to the context of LBS and proposed spatiotemporal cloaking to achieve location k-anonymity.

k-anonymity is the most commonly used location privacy cloaking technique. Location k-anonymity works on the principle that generate a cloaking region that includes k users sharing some property of interest, and then querying the server for this cloaking region. The size of cloaking region and the number of users in this region, that is, k directly affects the efficiency and effectiveness of the privacy protection. So, many cloaking region generation methods that aim at generalizing the space have been proposed, including clique cloaking, Casper and Hilbert cloaking, and so on. On the other hand, aiming at the weaknesses of k-anonymity in protecting the sensitive attributes of queries, more privacy-preserving principles such as l-diversity, p-sensitivity and t-closeness have been proposed, all of which bear advantages and disadvantages, as will be discussed in Section 13.7.2.

Initially, the *spatial and temporal cloaking* [45] proposed by M. Gruteser et al. in 2003 relies heavily on the anonymizer and have a traceability problem, that is, an attacker can easily determine a rough trajectory through tracing a user's data for several minutes. In addition, in 2004, Hong et al. [46] proposed a method that users send a *landmark* instead of actual location to service providers. However, this method fails to provide accurate query results for users when a landmark is far from the user. Motivated by that prior work, Kido et al. proposed a dummy-based technique [47] in 2005 to address the traceability problem without the help of an anonymizer. It sends several fake locations and the actual location together to the server so that the LBS server cannot distinguish the user's location from the dummy ones. Compared with cloaking, the dummy-based technique leaks less location information about users. Service providers cannot get the precise location, but they can know the range of users who use the cloaking technique. While using a dummy-based technique, users can generate several distant fake locations such that even servers do not know the range. This method can guarantee the precision of query results, but it places extra burden on the server. Furthermore, dummy locations are useful only if they are sufficiently similar to the actual location. So, how to generate fake locations or traces which are sufficiently similar with the actual location is a hot problem in current research.

Mix zone, dummy and cloaking are designed on the basis of attackers' side information and are therefore susceptive to side information-based attacks (e.g., location tracking, restricted space attacks, observation attacks, location-dependent attacks [45,48,49]). What is more, when users require high-level privacy protection, they either send very coarse-grained location information or they cannot provide location information to the LBS server, resulting in deteriorated QoS.

In [50], the authors state that they believe that the abovementioned methods have failed to obtain a good trade-off between the desired level of privacy and the usefulness of the LBSs and therefore introduce the notion of differential privacy from statistic databases into LBSs. Differential privacy works on the principle that distort the sensitive data by adding noise to achieve the privacy preserving while keeping the statistic attributes of data. It is independent on the background knowledge that attackers have and provides a rigorous, quantitative representation and proof for privacy exposures. Presently, differential privacy is the most rigorous technique for LBS privacy preserving. However, it does not suit applications in which only a single user is involved.

In addition, there are other ways to achieve the privacy preserving the LBS from the perspective of reducing the interaction with servers. For example, *Cache* [51] uses the cache mechanism to cache part of data on local mobile phones. Users can query the local cache data with a request for LBS, which avoids disclosing location information to servers. Moreover, a collaborative framework *MobiCrowd* [52] proposes that users query neighbors for similar data first before query the LBS server. Many LBS privacy-preserving methods combine multiple techniques, making it difficult to simply classify them into one of these categories.

In summary, rule protocol-based privacy-preserving techniques can guarantee QoS, but can provide only a low level of privacy protection. Most heuristic privacy-preserving techniques can gain a better balance between QoS and privacy protection but are susceptive to side information-based attacks, and differential privacy suffers

Table 13.1 Comparison of Privacy-Preserving Techniques

Techniques		*Architecture*			*LBS Privacy*	
		Centralized	*Distributed*	*Mixed*	*Location*	*Trace*
Rule Protocol based			✓		✓	✓
Encryption			✓		✓	
Heuristic	Pseudonym	✓			✓	✓
	Cloaking	✓	✓		✓	✓
	Dummy		✓		✓	✓
	Perturbation		✓		✓	✓
Caching based			✓		✓	

from limit application scenarios; encryption-based privacy-preserving schemes can add protection privacy while rendering heavy computation and communication overhead as well as degrading the availability of the data. Their comparisons are shown in Table 13.1.

13.5 Rule Protocol-Based Techniques

Rule protocol–based LPPT prevent an untrusted LBS server from accessing, storing, and using users' location information through regulatory strategies. That is, the LBS server is required to obtain users authorization before accessing users' locations. For example, GeoPriv [38] defines a location object that encapsulates every user's location and an according regulatory strategy to prevent the user's location privacy being breached when his location information is created, stored and used. For another instance, W3C develops open standards to ensure the long-term growth of the Web that describes Web communication protocols (e.g., HTML and XHTML) and other blocks [39].

Because the privacy level of users' location information sometimes changes with the variation of locations, privacy policies should allow users to manage and publish location information according to their requirements. [53] proposed a privacy protection architecture for global location services where users can decide who will get, and to what extent they will get, the location and identity information. However, users are often reluctant to directly manage complex privacy policies, so they sometimes refuse to participate. In continuous-query scenarios, [54] proposed a method to protect privacy in continuous location-tracking applications. It claims that users' desired levels of privacy can be situation-dependent and adopts a context-aware application to enable automatic privacy policies decisions.

However, to preserve users' privacy through rule protocol–based privacy-preserving techniques, the LBS server should follow the regulatory strategies and be resilient to system attacks. Unfortunately, the LBS server is always untrusted, for example, according to [55], there are 15 of the 30 investigated Apps such as MySpace and Trapster, disclosing users' location data to advertisement or analytic servers. More severely, the black hat "Peace" disclosed over 167 million users information (including location information) from LinkedIn in 2013 [16] and about 360 million from MySpace in 2016 [17]. What is more, the LBS server is always vulnerable to system attack, even though it follows the regulatory strategies. For example, Heartbleed, a serious vulnerability in the popular OpenSSL cryptographic software library, allows stealing the information protected by the secure socket layer (SSL)/transmission layer security (TLS) encryption, under normal conditions [56].

In addition, in the era of Big Data and the 5G IoT, rule protocol–based privacy-preserving techniques cannot obtain the privacy-utility trade-off. Specially, in the era of Big Data and the IoT, data mining enables the LBS server to mine more utility of users' information contained in LBS queries. On the down side, there is high overhead for the LBS server to obtain users authorization before utilizing users' data. So rule protocol–based privacy-preserving techniques can only preserve users' privacy or data's utility.

13.6 Encryption-Based LBS Privacy-Preserving Techniques

Instead of cloaking users' true location inside a crowd or fake locations, encryption-based techniques aims at making it completely invisible to the LBS servers. Encryption-based privacy preserving techniques adopts decentralized architecture to achieve the tradeoff between the utility and quality of LBS that users wish to receive and the location privacy they are ready to compromise.

13.6.1 Private Information Retrieval Protocol

The PIR protocol [57] was first applied to outsourcing data in access networks that allow a user to retrieve an item in private from a server of a database without revealing which item is retrieved. According to the level of privacy protection, its implementation can be classified into information theoretic PIR protocol [57] and computational PIR protocol [58]. Computational PIR protocol ensures that an attacker cannot distinguish the access of user to different data items by reducing the complexity of theoretically insoluble or computationally infeasible problems. The information theoretic PIR protocol ensures privacy no matter how strong the computing capacity of attackers. However, due to the fact that the transmission cost of returned query results by information theoretic PIR protocol is too high, computational PIR protocol is more commonly used. Most techniques are expressed in a theoretical setting, where the database is an *n*-bit binary string *X*. The client wants to find the value of the *i*th bit of *X* (i.e., X_i). To preserve privacy, the client sends an encrypted request $q(i)$ to the server. The server responds with a value $r(X, q(i))$, which allows the client to compute X_i. Computational PRI employs cryptographic techniques and relies on the fact that it is computationally intractable for an attacker to find the value of *i* and given $q(i)$. Furthermore, the client can easily determine the value of X_i based on the server's response $r(X, q(i))$.

This approach has several advantages including that it does not reveal any spatial information, it is resistant against correlation attacks, and it does not require any trusted third party, among others. The approach is expensive in terms of communication and computation; some of the improvements and optimizations the authors have proposed include compressing the data sent by the server, having rectangular grids/matrices in addition to square ones, and using off-line data mining to avoid redundant computation. However, the server reveals large amounts of extra information, which results in exposing large portions of the server's data assets to the client.

13.6.2 Space Transformation

Space transformation makes over the information of location and queries to a different space using some encryption techniques like Space Filling Curves (SFC) and the one-way Hash function prior to transmitting it to the LBS and assessing the queries in this encrypted space. In this way, only mobile users can transform the transformed data into the original one. The data stored in the LBS server and the location information about the query sent by mobile users are encrypted, so the server can return correct query results without revealing the true location corresponding to the encryption location. For instance, HilCloak, a Hilbert curve-based location cloaking method, fist rotates the whole space at some angle, and builds a Hilbert curve with the key *H*, which is only available to mobile users and trusted entities in this rotated space. In the phase of preparing queries, trusted entities transform each point of interest (POI) into Hilbert value $H(p_i)$ and send this value to the server. When starting a query, user *q* sends $H(q)$ to the server and obtains a Hilbert value nearest to $H(q)$ from the server. HilCloak can ensure users' query privacy only once although it preserves the privacy of any location distribution. However, a standard Hilbert curve transforms POIs without considering the distribution characteristics of POIs, so it cannot achieve hierarchical access control due to identical granularity division.

Cryptographic techniques can achieve a good performance in privacy preserving, but this would degrade the quality of LBS services that mobile users can receive, especially in applications like POIs or navigation. Furthermore, they are costly in computation time and in the communication and resources needed.

13.7 Heuristic Privacy-Preserving Techniques

The success rate of obtaining users' location privacy is proportional to the side information attackers get. Thus, it is nontrivial to quantify the risk of users' location privacy being disclosed, for example, some privacy-preserving techniques considering users' moving speed have to deal with NP hard problems [42]. So most existing privacy-preserving techniques in this period are heuristic. These heuristic approaches can be generally classified into four categories: pseudonyms, dummy, cloaking and perturbation.

13.7.1 Pseudonym

Pseudonym employs the centralized architecture because the release, use and revocation, and so on need to be operated on a trusted server. It relies on replacing the users' identity with a pseudonym and changing pseudonym frequently or removing it directly to cut off the connection between the user's identity and his queries. Pseudonym is an object's identifying without containing the users' information that can be recognized by others rather than a real name. As a result, LBS queries cannot be connected with the users' identifying information. As such, the users' query privacy can be protected.

In general, to enhance the effectiveness of pseudonyms, it is necessary to combine some complicated encryption schemes. It is not enough to guarantee the privacy by employing only Pseudonym because attackers can obtain the location in a variety of ways (such as monitoring users' phone signal) and identify users with some public information. It can reduce the chance that an attacker will use the accumulated historical information to infer the users' identity or behavior by changing the pseudonym frequently.

However, attackers can still identify a user by associating all the pseudonyms connected to this user if the user's preference information is stored on the server. In addition, there is an association among the space–time information of users' movement, so the user's identity may still be tracked even if changing the pseudonym frequently. A further-developed framework that uses pseudonyms with mix zone was proposed and has been extensively studied; this framework is mainly employed in dynamic location or trace privacy preserving such as vehicle networks.

Mix zone was first proposed in [59] and is defined as a spatial region where applications cannot access any location information of the users therein. Thus, users in mix zone are indistinguishable by the LBS server. Furthermore, the LBS server still cannot identify users when they come out if they changed their pseudonym in the mix zone. For example, in Figure 13.7, u_1, u_2, u_3, and u_4 (assume pseudonym are id_1, id_2, id_3, and id_4) enter the mix zone $zone_1$ and change their pseudonym to id'_1, id'_2, id'_3, and id'_4. Then the LBS server cannot distinguish users after they

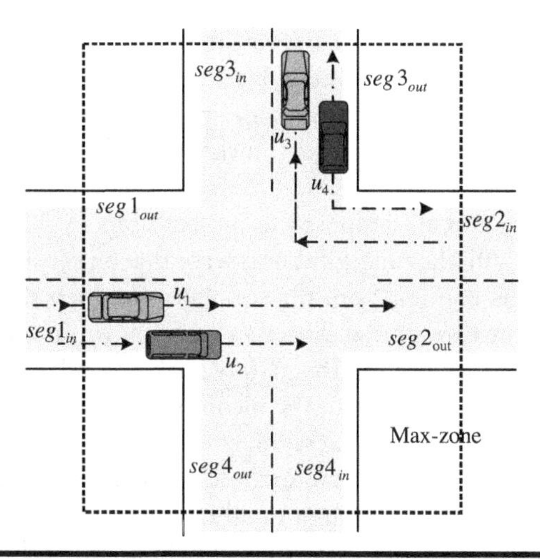

Figure 13.7 Illustration of the definition of max zones.

enter the mix zone. Thereafter, Beresford improved the Mix zone by optimizing the model and reducing the overhead in [60].

The earlier techniques about mix zone follow a naive refinement of the mix zone, using a rectangular or circular region. [22] proposed a perturbation algorithm that crosses paths in a region including no less than two users. [61] proposed a user-centric approach, where users can determine where and when to change their pseudonym. [62] proposed that users change their pseudonym only when they meet *k* users within a confusion radius in a similar direction. [63] concentrated on the self-interested users in mix zone and analyzes the non-cooperative behavior. Mix-zone is deployed by [64] to enable each user to pass through a mix-zone. [65] proposed encrypting users' information when users are in mix zone. However, such a naive refinement of the mix zone is vulnerable to timing attacks [66] and cannot guarantee users' privacy. For instance, in Figure 13.7, if the entering time of u_1, u_2, u_3, and u_4 differ by a large value, an attacker can infer the linking between Pseudonym and users.

To resist timing attacks, a time window is applied to a mix zone in the follow-up techniques. [67] proposed cryptographic "mix-zones" where the mix zone is located at an intersection, and the timing is determined by the delay of the intersection. [44] focused on investigating the effectiveness of changing pseudonyms in mix zone, considering the arriving and existing time. [68] proposed the optimal locations of mix zones, and cloaks a specific user u_i with users arriving at the mix zone within the time interval $|t_i - \tau_1,\ t_i + \tau_2|$, where τ_1 and τ_2 are small numbers, and t_i is the arriving time of u_i. The time window in [69] depends on users' arrival rate, and the mix zone is centered at the point at which it takes the same time for users to arrive. These types of work can protect users' privacy from timing attacks through the time window; for example, Figure 13.7 shows that users can enter and exist in the mix zone at a similar time, thereby confusing the LBS server. Unluckily, all this work is not resilient to transition attacks where attackers map the new pseudonym and the old one.

Thus, some techniques are proposed to protect users in the max zone against transition attacks. In [70], the time window is set so that on very outgoing segment *s*, there are enough users coming from the segments of which the transition probabilities to the segment *s* are similar. Specifically, as shown in Figure 13.8, assume a user u_i exists in the mix zone and that $T(u/y)$, $T(v/y)$, and $T(w/y)$ are transition probabilities that a user coming from the incoming segments *u*, *v*, and *w*, respectively, given he exists on the outgoing segment *y*. Assume $T(u/y)$ is much less than $T(v/y)$ and $T(w/y)$. Then attackers can exclude users entering through *u*. Thus, if there are enough users entering through *v* and *w*, attackers would be confused and cannot map u_i to a specific user in the mix zone. In addition, [71] proposed temporal, spatial and spatiotemporal delay-tolerant mix-zones that combing mix zone and spatiotemporal cloaking to resist transition attacks. However, all the work introduced above is based on a single mix zone and is susceptible to inferential attack when the attackers employ side information.

Thus, multiple mix zones are proposed to address this problem. [48] proposed optimal multiple mix zone placement to minimize the pairwise information correlation. Users in [72] use unlinkable pseudo-IDs to prevent attackers from linking users' locations to their identities, even when attackers have enough side information. [73] proposed a heuristic algorithm to select the optimal mix zones' locations by modeling it as a transportation problem. The users' moving velocity and direction are taken into consideration when locating multiple mix zones in [74]. However, in the era of Big Data and the IoT, mix zones cannot protect against side information-based inferential attack because attackers can obtain more side information from various applications.

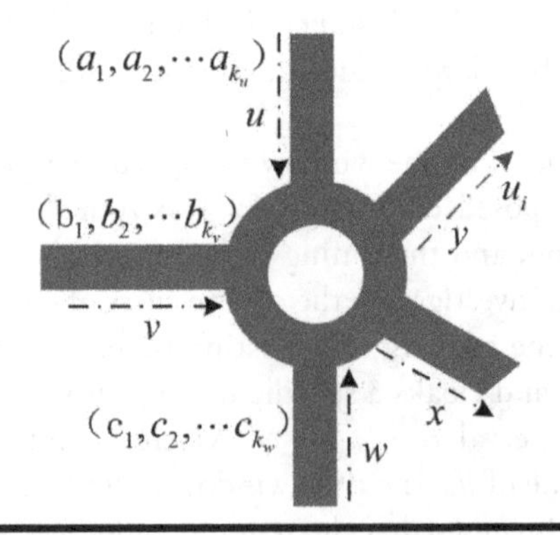

Figure 13.8 Illustration of the technique against transition attacks.

13.7.2 Cloaking

Cloaking includes spatial cloaking and temporal cloaking. Spatial cloaking is based on reducing the precision of users' actual location by sending a generalized region instead of a precise point to the server, whereas the temporal cloaking diminishes the precision in time value. The existence of delay introduced by temporal cloaking makes it so that it cannot work well in LBS, which requires high real time. So, cloaking techniques usually use spatial cloaking or spatiotemporal cloaking.

Interval cloaking is the first spatial-cloaking technique and was proposed by Gruteser and Grunwald in 2003 [45]. Inspired by [75], Gruteser and Grunwald employed *k-anonymous*, making a user's location information is indistinguishable from at least k - 1 other users, which is commonly achieved by forming a k-anonymizing spatial region containing the user and k - 1 additional users. Since that time, most cloaking-based techniques have been based on well-established *k-anonymous* and many mechanisms had been proposed to form the cloaking region.

A naive approach is to create a minimum bounding rectangle or circle that contains an anonymous set called Center cloak [76]. In [45], location information is represented by a tuple containing three intervals ($[x_1, x_2]$, $[y_1, y_2]$, $[t_1, t_2]$), which represent a user who appears in the geographic area $[x_1, x_2]$, $[y_1, y_2]$ during the time period $[t_1, t_2]$. The anonymity server organizes at least k users, including the target user present in this area during this time period. In this way, attackers cannot lock the target. Nevertheless, interval cloaking has some weaknesses: (1) the size of the anonymity set k is equal to all of users; (2) there is no limitation for the size of cloaking region; (3) its low spatial resolution will result in inaccurate query results and tons of unnecessary query results will be sent to users, degrading users' QoS and increasing the burden of communication and processing overhead; and (4) the extra delay introduced by temporal cloaking also decreases users' QoS.

Cloaking based on location k-anonymity has four main factors that reflect the privacy protection requirements. (1) k, the minimum number of users in the anonymity set required by the service requester, also known as degree of anonymity. (2) A_{min}, the minimum size of cloaking region. Setting A_{min} is to prevent the anonymity region from being too small in a populated area to easily expose the users' location. In extreme cases, k users gather in a position where the k-anonymity region is a specific location point. (3) A_{max}, the maximum size of cloaking region. If the anonymity region is too large, then the distribution of users would be very sparse and the server's cost of processing queries would be great. (4) T_{max}, the longest tolerable anonymous time, that is, the process of anonymization should be complete in T_{max} from the instant the user asks for anonymization. k and A_{min} are location-anonymous restrictions and refer to the minimum value of anonymous quality, whereas A_{max} and T_{max} are restrictions of service quality in LBS and refer to the worst QoS.

As previously mentioned, k-anonymity is the most important privacy cloaking technique (Figure 13.9).

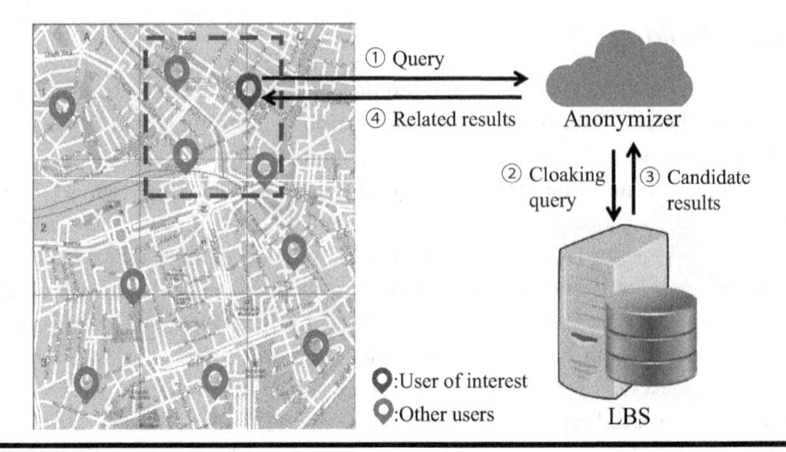

Figure 13.9 An illustration of cloaking-based privacy-preserving techniques.

Location *k*-anonymity cloaks a specific user's location into a region that includes no less than (*k* – 1) other users, so the location of every user in this region is distinguishable from others. However, *k*-anonymity also has its inherent shortcomings. So many improved versions of *k*-anonymity have been proposed, it is more difficult for attackers to get a user's sensitive information. At first, the *k*-anonymity-based privacy-preserving techniques focused on designing location *k*-anonymity. (1) Location *k*-anonymity cannot guarantee users' location privacy, for example, the region including no less than (*k* – 1) users is too small to disclose users' privacy in densely populated areas. Thus, [34] proposed to cloak more than (*k* – 1) users in a region the area of which is no less than A_{min}, through hierarchically decomposing the spatial space. (2) Users cannot be cloaked, or the cloaking region is too larger to increasing overhead and deteriorating the quantity of LBSs, in sparsely populated areas. To address this problem, [77] cloaked a specific user's location in a region that contains no less than (*k* – 1) historical locations of other users, and [78] proposed to cloak a specific user's location into dummy locations.

With further in-depth exploration about the *k*-anonymity-based privacy-preserving techniques, academia has gradually concentrated on the privacy level location *k*-anonymity provides, personalized requirements of users, the prior knowledge attackers obtained, and the trade-off between privacy and QoS, and the system architecture, and so on.

k-anonymity is vulnerable to two attacks: (1) the homogeneity attack, which leverages the case where all the values for a sensitive value within a set of *k* records are identical. In such cases, even though the data has been *k*-anonymized, the sensitive value for the set of *k* records may be exactly predicted; and (2) the background knowledge attack, which leverages an association between one or more quasi-identifier attributes with the sensitive attribute to reduce the set of possible values for the sensitive attribute.

An improved privacy principle *(α,k)-anonymity* [79] guarantees that the percentage of records that are related to any values of sensitive attributes in each equivalence class is less than α with the satisfaction of *k*-anonymity.

Machanavajjhala et al. [80] proposed *l*-diversity, which guarantees at least *l* "well-represented" values for sensitive attributes in each equivalence class such that the risk of linking a sensitive attribute to its issuer is less than $1/l$. However, *l*-diversity neglects the semantic similarity of sensitive attributes, which makes it is insufficient to defend against similarity attacks. A similarity attack refers to there being semantic similarity among sensitive attributes in an equivalent class although their values are different.

On the basis of *l*-diversity, [81] proposed the notion of *t-closeness,* which considers the distribution of sensitive attributes. Specifically, it requires that the distance between the distribution of a sensitive attribute in a equivalence class and the distribution of the attribute in the whole table is less than a threshold *t*.

Xiao et al. [82] aimed at preventing the sensitive attributes disclosure with consideration of the query diversity and semantic information and proposed *p-sensitivity*. It requires that in the query anonymity set of a user, the proportion of the number of sensitive queries account for this set is less than *p*, that is, the probability of any user in this corresponding cloaking region sending a sensitive query is less than *p*.

Xiao and Tao [83] proposed *m-invariance* to degrade the risk of privacy disclosure in republication of dynamic datasets. Assuming that $T_1^*, T_2^*, \ldots, Tn^*$ are a series of data released successively in dynamic context, the series of data achieve *m-invariance* if and only if two conditions are satisfied: (1) For each data released at time *i*, $Ti*$, there are at least *m* records in its every equivalence class and these records have different sensitive attributes. (2) If a record presents in several republication at different time, then the set formed by the value of sensitive attributes included in the equivalence class that releases this record every time must equal.

Cloaking techniques that are mainly based on *k*-anonymity protect query privacy through attackers' inability to distinguish the issuer from at least other $k - 1$ users, whereas *l*-diversity, *m*-invariance, *p*-sensitivity, and so on protect query privacy by obscuring the sensitive attributes within queries such that they increase the uncertainty of linking users with their sensitive attributes.

13.7.3 Dummy

Dummy-based privacy-preserving techniques always adopt distributed architecture. Specifically, each user locally generates dummy queries and sends both the dummy queries and his query to the LBS server. Then the LBS server returns all results corresponding to both dummy queries and real query to the user, and the user filters out his query results. As an example, in Figure 13.10, u_1 first generates three dummy users u_2, u_3, and u_4 (dummy users' queries are dummy queries, locations of dummy users are dummy locations), sends his real query q_1 and the dummy

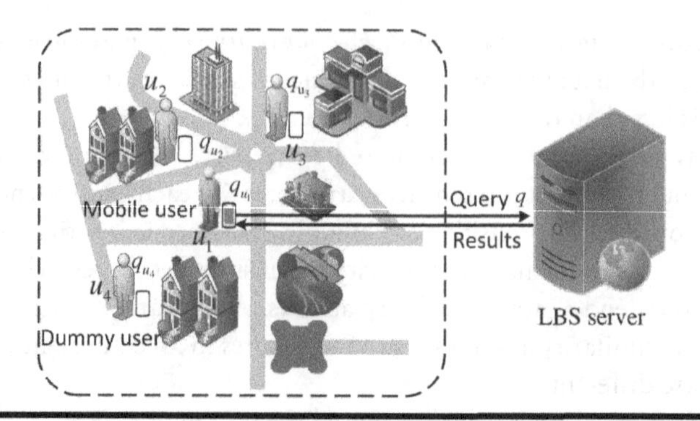

Figure 13.10 An illustration of dummy-based privacy-preserving techniques.

queries q_2, q_3, and q_4 to the LBS server, and filters out his query results upon receiving all query results from the LBS server.

At first, dummy-based privacy-preserving techniques generate dummy queries based on very simple heuristics to protect users' location privacy. [84] first proposed to generate dummy users according to dispersibility, density, and uniformity. [85] proposed to generate dummy queries based on a virtual grid or circle. [86] generated dummy queries through interpolation strategies. However, all these techniques [84,86] allow attackers to distinguish users' queries from dummies according to users' moving patterns.

In continuous query scenarios because attackers can continuously observe the location information of users, it is necessary to consider the correctness and continuity of the dummy locations. So, the key problem is how to measure the privacy protection level of generated dummy trajectories. [87] proposed three metrics: short-term disclosure (SD), long-term disclosure (LD) and distance deviation (DD). SD refers to the probability of successfully inferring each true location. LD is the probability of successfully identifying the true trajectory among all possible trajectories. DD is the average distance between the real trajectory and the dummy trajectory.

So, the follow-up research work takes restrictions in the real environment into consideration. Dummy users generated in [88] have consistent movement, and the anonymous area containing dummy locations and users' locations also satisfies users' privacy requirements. Dummy users are generated in [89] according to the users' future locations, and at the same time, service attribute values of dummy users are consistent with the query context. [90] considered a more realistic movement behavior that users always pause at some places of interest. But, these techniques are vulnerable to side information-based attacks [81,91] such as attacks that obtain a specific user's lifestyle details, which allows attackers to distinguish the user's location from dummies according to his lifestyles.

Thereafter, research work strove to propose dummy-based privacy-preserving techniques against side information-based attacks. [92,93] generated dummy users taking the attackers' own side information into consideration, and enhance privacy protection through choosing dummy locations from some locations that have the similar query probabilities with a specific user. However, all this work only captures the basic geographic features of the users' movement, ignoring the semantics of the users' movement, and deteriorate the utility of the users' location information [94].

Therefore, the follow-up techniques attempt to synthesize plausible location traces, in terms of geographic and semantic features. [94] generated fake, yet semantically real, traces (i.e., dummy traces) to protect users' location privacy while considering both geographic and semantic features. It first transformed traces into semantic space, then generated plausible location traces through sampling in semantic space. Nevertheless, it fails to capture the social relationships of mobile users, which would make the user suffer from de-anonymization attacks where social networks are used as the side-information.

However, in the era of Big Data and the IoT, dummy-based privacy-preserving techniques are not sufficient to protect users' privacy. Specifically dummy queries confuse attackers only when dummy queries are plausible to real queries. However, attackers can access more side information in the era of Big Data and the IoT, thus it is still difficult to synthesize plausible location traces considering attackers are exploiting more side information.

13.7.4 Perturbation

Perturbation is a straightforward method for privacy preserving. The main idea is to randomly perturb a user's location by adding predetermined noise distributions to coordinates, which makes a deviation between the true location and the location sent to servers. This method, in principle, derives from the "data distortion" in privacy-preserving data mining [95]. [19] generated random points by adding Gaussian noise to actual locations and sent the farthest point from the actual location to servers. However, selecting the farthest point will help attackers infer the actual location (i.e., where to start inferring). An improved method proposed by [96] employed a uniform distribution to remit this problem. However, this method is vulnerable to a noise-filtering attack. SpaceTwist [97] mainly aims at the issue of computing exact k nearest neighbor (kNN) query results. Instead of sending the true location to servers, SpaceTwist sends an anchor, a location near the true location to the final results, thereby protecting the location privacy. However, the farther that an anchor is from the true position, the higher the privacy protection level for users but the greater the cost because it needs to search more neighbors of anchors. Furthermore, it is unlikely to suffice for LBS because, with a predetermined noise distribution, the levels of privacy protection and LBS accuracy largely depend on the context, such as road and population density, around a user's location. For example, intuition suggests that, to achieve the same level of privacy and LBS accuracy, a user should (or could) deviate more from its real location in a rural area than in a downtown location.

Simply and intuitively adding noise to actual location: (1) degrades the user's experienced quality of LBS, (2) severely limits the availability of data in Big Data and the IoT, and (3) is vulnerable to noise filtering attacks. So, it is rarely studied now. Notably, there is a well-established noise-based perturbation technique called differential privacy, which is a privacy notion proposed by Dwork in 2006 [98] for the issues of privacy leakage in the context of statistical databases. In principle, differential privacy substantially transforms the exact query of a dataset into a distribution, making the probability of getting the same result for querying two adjacent datasets. Differential privacy is independent on background knowledge that attackers have and provides strict proof of privacy protection, which makes it extensively researched and applied in privacy protection of data release and data mining. [99] presented a synthetic data generation technique that can be used to publish statistic information about commuting patterns in a differentially private way. [100] used a quadtree spatial decomposition technique to ensure differential privacy in a database with location pattern-mining capabilities. So, differential privacy can be successfully applied in cases where aggregate information about several users is published.

There are some advantages for differential privacy: (1) Differing from the heuristic privacy-preserving techniques, differential privacy defines a very rigorous attack model and gives a rigorous, quantitative representation and proof of privacy exposures. It can reduce the risk of privacy disclosure while guaranteeing the usability of data. (2) It is free from the background information of the attackers, which greatly affects the privacy-preserving level provided by many heuristic methods. (3) Although it works on distorting data, the amount of noise added to data is independent of the data set size. As a result, for large data sets, by adding only a very small amount of noise, one can achieve high-level privacy protection. Accordingly, it raised a study upsurge because the concept of differential privacy was proposed and received a high-level of approval. However, this technique does not work well when only one user is involved and, in that scenario, risk of compromising the location information is high.

13.8 Caching-Based Techniques

The above three types of techniques to protect privacy are from the perspective of the sensitive attributes of queries submitted to the LBS server, whereas the viewpoint of the caching-based technique is to reduce the number of queries sent to the LBS server. Specifically, users' cache data obtained from historical queries to settle future queries. If fewer queries are sent to the server, then less sensitive information will be released, and hence there is less chance of exposing the user's privacy.

Amini et al. [51] introduced the cache system to preserve the user's privacy first. By pre-fetching the service data within a certain area before arriving at that area, mobile users can search the POIs locally rather than sending queries to an untrusted LBS server. However, mobile users need to store large volumes of service data for a large area due to the naive and heuristic caching method, which incurs high overhead in storage space and low-efficiency in answering queries. As an auxiliary method, [101] designed a distributed location privacy-preserving algorithm for a collaborative group, termed MobiCrowd. Users in MobiCrowd query neighbors for similar service data before sending a query to the LBS server. In this way, the probability of location exposing is decreased.

However, this scheme neglects the privacy-preserving requirement of users who need to send queries to the LBS server. In addition, it does not consider whether the cached data is able to meet users' queries, namely, the cache hit rate. Mobicache [102] combines caching and k-anonymity, and carefully designs a dummy selection scheme with consideration of improving the cache hit ratio. But it still simply tries to cache more data that has not been cached yet, and it does not consider the side information that an adversary may have. Thus, the adversary can infer the real location with the help of side information.

There are also several common problems with these caching-based solutions. They do not have an integrated privacy metric to measure the effect of caching on privacy, and their caching design is pretty straightforward but without consideration of some important factors that could affect the cache hit ratio. Authors in [103] proposed a caching-aware, dummy-based solution to protect location privacy in LBSs. This solution caches the service data obtained from both the real location and dummy locations of the current query to improve the cache hit ratio. And it shows the quantitative relation between caching and privacy. But there is still a risk of revealing the user's identity due to the first time communication with the LBS server.

From the level of users' QoS, there are three possible cases for cached data. (1) the cached data is able to answer the LBS query and is unexpired, (2) the cached data is able to answer the LBS query but is expired, and (3) the cached data cannot answer the LBS query. So it is necessary to consider several factors that could affect the cache hit ratio while protecting privacy: (1) the query probability of location, where caching locations with high query probability is of the priority; (2) normalized distance (the distance between the real location and dummy locations), where caching locations near to the real location is of the priority; and (3) data freshness, where caching locations close to expiration is of the priority. Synthesizing the above three factors, the dummies' contribution to cache can be calculated quantitatively.

13.9 Challenges and Future Opportunities

13.9.1 New Challenges in Big Data and the Internet of Things

In the era of Big Data and the IoT, the aforementioned four types of privacy-preserving techniques all face new challenges.

1. Rule protocol–based privacy-preserving techniques either constrain the utility of location information or disclose users' privacy in the era of Big Data and the IoT. In the era of Big Data and the IoT, data mining technology can excavate much new utility of large volumes of data. So the LBS server cannot

inform users of the new utility of the users' location data it used in advance, and it cannot bear the cost when it informs users and obtains users' authorization when it uses a new utility of users' location data. In addition, rule protocol–based privacy-preserving techniques cannot weigh QoS and privacy protection. In summary, rule protocol–based privacy-preserving techniques can only guarantee the utility of location information or users' privacy.

2. In the era of Big Data and the IoT, most if not all heuristic privacy-preserving techniques cannot preserve users' privacy because attackers can obtain large volumes and many kinds of data related to users' location information that can then be directly or indirectly used to infer the users' location privacy. First, attackers can get more data related to users' location information (e.g., social relationships, moving habits, hobbies, workplace) from various LBS Apps (e.g., Google Maps Foursquare Nearby Getyowza). Then it is more possible for attackers to infer users' location privacy when applying these side information. For example, users' locations cloaked by *k*-anonymity–based privacy-preserving techniques can be successfully de-anonymized when attackers obtain friendships from Facebook [104]. For another example, when attackers get users' moving habits from Google Maps, pseudonym and dummy-based privacy-preserving techniques suffer from moving mode-based attacks. Second, users' location privacy protected in one kind of LBS (e.g., Nearby) may be disclosed in another LBS (e.g., Check in).
3. In the era of Big Data and the IoT, the expanding users' location information is provided to the LBS server, making it impossible to apply encryption–based privacy-preserving schemes in practical systems. Specifically, first encryption-based privacy preserving techniques always heavily rely on additional hardware and complex algorithms and are resource-hungry so that they cannot efficiently deal with large volumes of users' data [105]. Second, the important features in the IoT, limited resources, constrains the application of encryption–based privacy-preserving techniques in Big Data and the IoT.

13.9.2 Future Research in Big Data and the Internet of Things

Facing those new challenges, the following are some open research problems that could be further investigated when it comes to location privacy preserving in LBSs.

- How to obtain a good balance between the level of privacy protection, the utility of location information, and the overhead
- How to systematically quantify LBS privacy and the risk of users' privacy being disclosed
- How to protect privacy when considering attackers' side information
- How to protect privacy for users with heterogeneous data

13.10 Conclusions

The increasing capabilities of position-determination technologies in mobile and hand-held devices facilitate the widespread use of 5G-Enabled LBSs. The emerging idea of LBSs is rapidly finding its path throughout our modern life, aiming to improve the quality of life by connecting many smart devices, technologies, and applications. Although 5G-Enabled LBSs are providing enhanced functionalities and convenience of ubiquitous computing, they open up new vulnerabilities that can be exploited to target the violation of the security and privacy of users. This vision has been supported and heavily invested by governments, interest groups, companies, and research institutions.

In this chapter, we have provided a comprehensive overview of the privacy-preserving techniques in 5G-enabled LBSs. We categorized and gave an inside-out look into existing techniques. By reviewing several typical works in each category, we sum up their basic principles and recent advances. We also highlight the use of privacy-preserving techniques in 5G-Enabled LBSs for enabling new research opportunities. Through the timely and comprehensive overview of the recent works, this survey may further encourage new research efforts into this promising field, eventually embracing the exciting vision of forming a society of "Internet of Things" in the near future.

References

1. Constandinos X. Mavromoustakis, George Mastorakis, and Jordi Mongay Batalla. *Internet of Things (IoT) in 5G Mobile Technologies*, Vol. 8. Cham, Switzerland: Springer, 2016.
2. Maria Rita Palattella, Mischa Dohler, Alfredo Grieco, Gianluca Rizzo, Johan Torsner, Thomas Engel, and Latif Ladid. Internet of things in the 5G era: Enablers, architecture, and business models. *IEEE Journal on Selected Areas in Communications*, 34(3): 510–527, 2016.
3. Georg Gartner and Haosheng Huang. *Progress in Location-Based Services 2014*. Cham, Switzerland: Springer, 2015.
4. Gerrit Heinemann and Christian Gaiser. *Social-Local-Mobile: The Future of Location-Based Services*. Heidelberg, Germany: Springer, 2014.
5. Haojun Huang, Hao Yin, Geyong Min, Hongbo Jiang, Junbao Zhang, and Yulei Wu. Data-driven information plane in software-defined networking. *IEEE Communications Magazine*, 55(6): 218–224, 2017.
6. Chen Wang, Hongzhi Lin, Rui Zhang, and Hongbo Jiang. SEND: A situation-aware emergency navigation algorithm with sensor networks. *IEEE Transactions on Mobile Computing*, 16(4): 1149–1162, 2017.
7. Chen Wang, Hongzhi Lin, and Hongbo Jiang. CANS: Towards congestion-adaptive and small stretch emergency navigation with wireless sensor networks. *IEEE Transactions on Mobile Computing*, 15(5): 1077–1089, 2016.

8. Hongbo Jiang, Ping Zhao, Chen Wang, and John C. S. Lui. Roblop: Towards robust privacy preserving against location dependent attacks in continuous LBS queries. *IEEE/ACM Transactions on Networking*, 26(2): 1018–1032, 2018.
9. Ping Zhao, Hongbo Jiang, Chen Wang, and Haojun Huang. Non-asymptotic bound on the performance of k-Anonymity against inference attacks. In *Proceedings of 20th IEEE HPCC*, Exeter, UK, 2018.
10. Ping Zhao, Hongbo Jiang, Chen Wang, Haojun Huang, Gaoyang Liu, and Yang Yang. On the performance of *k*-anonymity against inference attacks with background information. *IEEE Internet of Tings Journal*, 6(1): 808–819, 2019.
11. Ping Zhao, Jie Li, Fanzi Zeng, Fu Xiao, Chen Wang, and Hongbo Jiang. ILLIA: Enabling k-anonymity-based privacy preserving against location injection attacks in continuous LBS queries. *IEEE Internet of Things Journal*, 5(2): 1033–1042, 2018.
12. Michael Backes, Manuel Barbosa, Dario Fiore, and Raphael M Reischuk. Adsnark: Nearly practical and privacy-preserving proofs on authenticated data. In *Proceedings of IEEE S&P*, 2015.
13. Kassem Fawaz, Huan Feng, and Kang G. Shin. Anatomization and protection of mobile apps location privacy threats. In *Proceedings of USENIX Security Symposium*, pp. 753–768, 2015.
14. Susan Landau. Control use of data to protect privacy. *Science*, 347(6221): 504–506, 2015.
15. Vincent Primault, Sonia Ben Mokhtar, and Lionel Brunie. Privacy-preserving publication of mobility data with high utility. In *Proceedings of IEEE ICDCS*, 2015.
16. Dark net LinkedIn sale looks like the real deal, Available at http://www.theregister.co.uk/2016/05/18/linkedin/.
17. Recently confirmed Myspace hack could be the largest yet, Available at https://techcrunch.com/2016/05/31/recently-confirmed-myspace-hack-could-be-the-largest-yet/.
18. Baik Hoh, Marco Gruteser, Hui Xiong, and Ansaf Alrabady. Enhancing security and privacy in traffic-monitoring systems. *IEEE Pervasive Computing*, 5(4): 38–46, 2006.
19. John Krumm. Inference attacks on location tracks. In *International Conference on Pervasive Computing*, pp. 127–143, 2007.
20. Yutaka Matsuo, Naoaki Okazaki, Kiyoshi Izumi, Yoshiyuki Nakamura, Takuichi Nishimura, K^oiti Hasida, and Hideyuki Nakashima. Inferring long-term user properties based on users' location history. In *International Joint Conference on Artificial Intelligence*, pp. 2159–2165, 2007.
21. Marco Gruteser and Baik Hoh. On the anonymity of periodic location samples. In *International Conference on Security in Pervasive Computing*, Vol. 3450, pp. 179–192, 2005.
22. Baik Hoh and Marco Gruteser. Protecting location privacy through path confusion. In *International Conference on Security and Privacy for Emerging Areas in Communications Networks*, pp. 194–205, 2005.
23. Donald Reid. An algorithm for tracking multiple targets. *IEEE Transactions on Automatic Control*, 24(6): 843–854, 1979.
24. Location-based applications popular, despite privacy concerns: ISACA, Available at http://www.eweek.com/c/a/Mobile-and-Wireless/LocationBased-Applications-Popular-Despite-Privacy-Concerns-ISACA-181357.
25. Jun Tang, Yong Cui, Qi Li, Kui Ren, Jiangchuan Liu, and Rajkumar Buyya. Ensuring security and privacy preservation for cloud data services. *ACM Computing Surveys*, 49(1): 13, 2016.

26. Yuan Zhang, Qingjun Chen, and Sheng Zhong. Efficient and privacy-preserving min and *k*-th min computations in mobile sensing systems. *IEEE Transactions on Dependable and Secure Computing*, 14: 9–21, 2017.
27. Reza Shokri, George Theodorakopoulos, Carmela Troncoso, Jean-Pierre Hubaux, and Jean-Yves Le Boudec. Protecting location privacy: Optimal strategy against localization attacks. In *Proceedings of the 2012 ACM conference on Computer and Communications Security*, pp. 617–627. ACM, 2012.
28. Yuchen Yang, Longfei Wu, Guisheng Yin, Lijie Li, and Hongbin Zhao. A Survey on Security and Privacy Issues in Internet-of-Things. *IEEE Internet of Things Journal*, 4: 1250–1258, 2017.
29. Alastair R. Beresford, Andrew Rice, Nicholas Skehin, and Ripduman Sohan. MockDroid: Trading privacy for application functionality on smartphones. In *Proceedings of the 12th Workshop on Mobile Computing Systems and Applications*, pp. 49–54, 2011.
30. Einar Snekkenes. Concepts for personal location privacy policies. In *Proceedings of ACM Conference on Electronic Commerce*, pp. 48–57, 2001.
31. Hua Sun and Syed A. Jafar. The capacity of private information retrieval. *IEEE Transactions on Information Theory*, 63: 4075–4088, 2017.
32. Hua Sun and Syed A. Jafar. Private information retrieval from MDS coded data with colluding servers: Settling a conjecture by Freij-Hollanti et al. *arXiv preprint arXiv:1701.07807*, 2017.
33. Chengyang Zhang and Yan Huang. Cloaking locations for anonymous location based services: A hybrid approach. *GeoInformatica*, 13(2): 159–182, 2009.
34. Mohamed F. Mokbel, Chi-Yin Chow, and Walid G. Aref. The new Casper: Query processing for location services without compromising privacy. In *Proceedings of the 32nd International Conference on Very Large Data Bases*, pp. 763–774. VLDB Endowment, 2006.
35. Reza Shokri, Carmela Troncoso, Claudia Diaz, Julien Freudiger, and Jean-Pierre Hubaux. Unraveling an old cloak: *k*-anonymity for location privacy. In *Proceedings of the 9th Annual ACM Workshop on Privacy in the Electronic Society*, pp. 115–118. ACM, 2010.
36. Kar Way Tan, Yimin Lin, and Kyriakos Mouratidis. Spatial cloaking revisited: Distinguishing information leakage from anonymity. In *International Symposium on Spatial and Temporal Databases*, pp. 117–134. Springer, 2009.
37. Mingqiang Xue, Panos Kalnis, and Hung Keng Pung. Location diversity: Enhanced privacy protection in location based services. In *International Symposium on Location-and Context-Awareness*, pp. 70–87. Springer, 2009.
38. IETF. Geographic location/Privacy working group, Available at https://datatracker.ietf.org/wg/geopriv/charter/. Accessed on April 2019.
39. W3C. Platform for privacy preferences (p3p) project, Available at https://www.w3.org/P3P/. Accessed on April 2019.
40. Ali Khoshgozaran, Houtan Shirani-Mehr, and Cyrus Shahabi. Blind evaluation of location based queries using space transformation to preserve location privacy. *GeoInformatica*, 17(4): 599–634, 2013.
41. Feng Tian, Xiao-Lin Gui, Xue-Jun Zhang, Jian-Wei Yang, Pan Yang, and Si Yu. Privacy-preserving approach for outsourced spatial data based on poi distribution. *Chinese Journal of Computers*, 37(1): 123–138, 2014.
42. Leping Huang, Hiroshi Yamane, Kanta Matsuura, and Kaoru Sezaki. Towards modeling wireless location privacy. In *International Workshop on Privacy Enhancing Technologies*, pp. 59–77. Springer, 2005.

43. Florian Schaub, Zhendong Ma, and Frank Kargl. Privacy requirements in vehicular communication systems. In *Computational Science and Engineering, 2009. CSE'09. International Conference on*, Vol. 3, pp. 139–145. IEEE, 2009.
44. Levente Butty'an, Tam'as Holczer, and Istv'an Vajda. On the effectiveness of changing pseudonyms to provide location privacy in vanets. In *European Workshop on Security in Ad-hoc and Sensor Networks*, pp. 129–141. Berlin, Germany: Springer, 2007.
45. Marco Gruteser and Dirk Grunwald. Anonymous usage of location-based services through spatial and temporal cloaking. In *Proceedings of the 1st International Conference on Mobile Systems, Applications and Services*, pp. 31–42, 2003.
46. Jason I. Hong and James A. Landay. An architecture for privacy-sensitive ubiquitous computing. In *Proceedings of the 2nd International Conference on Mobile Systems, Applications, and Services*, pp. 177–189. ACM, 2004.
47. Hidetoshi Kido, Yutaka Yanagisawa, and Tetsuji Satoh. Protection of location privacy using dummies for location-based services. In *Data Engineering Workshops, 2005. 21st International Conference on*, pp. 1248–1248. IEEE, 2005.
48. Xinxin Liu, Han Zhao, Miao Pan, Hao Yue, Xiaolin Li, and Yuguang Fang. Traffic-aware multiple mix zone placement for protecting location privacy. In *INFOCOM, 2012 Proceedings IEEE*, pp. 972–980. IEEE, 2012.
49. Xiao Pan, Jianliang Xu, and Xiaofeng Meng. Protecting location privacy against location-dependent attacks in mobile services. *IEEE Transactions on Knowledge and Data Engineering*, 24(8): 1506–1519, 2012.
50. Miguel E. Andres, Nicolás E. Bordenabe, Konstantinos Chatzikokolakis and Catuscia Palamidessi. Geo-indistinguishability: Differential privacy for location-based systems. In *Proceedings of the 2013 ACM SIGSAC Conference on Computer and Communications Security* (*CCS'13*), pp. 901–914, 2012.
51. Shahriyar Amini, Janne Lindqvist, Jason Hong, Jialiu Lin, Eran Toch, and Norman Sadeh. Cache: Caching location-enhanced content to improve user privacy. In *International Conference on Mobile Systems, Applications, and Services*, pp. 197–210, 2011.
52. Reza Shokri, Panos Papadimitratos, George Theodorakopoulos, and Jean-Pierre Hubaux. Collaborative location privacy. In *Mobile Adhoc and Sensor Systems* (*MASS*), *2011 IEEE 8th International Conference on*, pp. 500–509. IEEE, 2011.
53. Christian Hauser and Matthias Kabatnik. Towards privacy support in a global location service. In *Proceedings of the IFIP Workshop on IP and ATM Traffic Management*, 2001.
54. Marco Gruteser and Xuan Liu. Protecting privacy, in continuous location-tracking applications. *IEEE Security & Privacy*, 2(2): 28–34, 2004.
55. William Enck, Peter Gilbert, Seungyeop Han, Vasant Tendulkar, Byung-Gon Chun, Landon P. Cox, Jaeyeon Jung, Patrick McDaniel, and Anmol N. Sheth. TaintDroid: An information-flow tracking system for realtime privacy monitoring on smartphones. *ACM Transactions on Computer Systems*, 32(2):5: 1–5:29, 2014.
56. The heartbleed bug. Available at http://heartbleed.com. Accessed on April 2019.
57. Benny Chor, Oded Goldreich, Eyal Kushilevitz, and Madhu Sudan. Private information retrieval. J ACM. In *Proceedings of the Symposium on Foundations of Computer Science*, pp. 41–50, 1995.
58. Eyal Kushilevitz and Rafail Ostrovsky. Replication is not needed: Single database, computationally-private information retrieval. In *Proceedings of the Symposium on Foundations of Computer Science*, pp. 364–373, 1997.

59. Alastair R. Beresford and Frank Stajano. Location privacy in pervasive computing. *IEEE Pervasive Computing*, 2(1): 46–55, 2003.
60. Alastair R. Beresford and Frank Stajano. Mix zones: User privacy in location-aware services. In *Pervasive Computing and Communications Workshops, 2004. Proceedings of the Second IEEE Annual Conference on*, pp. 127–131. IEEE, 2004.
61. Mingyan Li, Krishna Sampigethaya, Leping Huang, and Radha Poovendran. Swing & swap: User-centric approaches towards maximizing location privacy. In *Proceedings of the 5th ACM Workshop on Privacy in Electronic Society*, pp. 19–28. ACM, 2006.
62. Matthias Gerlach. Assessing and improving privacy in VANETs. In *Embedded Security in Cars (ESCAR)*, 2006.
63. Julien Freudiger, Mohammad Hossein Manshaei, Jean-Pierre Hubaux, and David C. Parkes. On non-cooperative location privacy: A game-theoretic analysis. In *Proceedings of the 16th ACM Conference on Computer and Communications Security*, pp. 324–337. ACM, 2009.
64. Yipin Sun, Xiangyu Su, Baokang Zhao, and Jinshu Su. Mix-zones deployment for location privacy preservation in vehicular communications. In *Computer and Information Technology (CIT), 2010 IEEE 10th International Conference on*, pp. 2825–2830. IEEE, 2010.
65. Antonio M. Carianha, Luciano Porto Barreto, and George Lima. Improving location privacy in mix-zones for vanets. In *Performance Computing and Communications Conference (IPCCC), 2011 IEEE 30th International*, pp. 1–6. IEEE, 2011.
66. Balaji Palanisamy, Ling Liu, Kisung Lee, Aameek Singh, and Yuzhe Tang. Location privacy with road network mix-zones. In *Mobile Ad-hoc and Sensor Networks (MSN), 2012 Eighth International Conference on*, pp. 124–131. IEEE, 2012.
67. Julien Freudiger, Maxim Raya, M′ark F′elegyh′azi, Panos Papadimitratos, and Jean-Pierre Hubaux. Mix-zones for location privacy in vehicular networks. In *ACM Workshop on Wireless Networking for Intelligent Transportation Systems (WiN-ITS)*, number LCA-CONF 2007-016, 2007.
68. Julien Freudiger, Reza Shokri, and Jean-Pierre Hubaux. On the optimal placement of mix zones. In *International Symposium on Privacy Enhancing Technologies Symposium*, pp. 216–234. Springer, 2009.
69. Balaji Palanisamy and Ling Liu. Mobimix: Protecting location privacy with mix-zones over road networks. In *Data Engineering (ICDE), 2011 IEEE 27th International Conference on*, pp. 494–505. IEEE, 2011.
70. Balaji Palanisamy and Ling Liu. Attack-resilient mix-zones over road networks: Architecture and algorithms. *IEEE Transactions on Mobile Computing*, 14(3): 495–508, 2015.
71. Balaji Palanisamy, Ling Liu, Kisung Lee, Shicong Meng, Yuzhe Tang, and Yang Zhou. Anonymizing continuous queries with delay-tolerant mix-zones over road networks. *Distributed and Parallel Databases*, 32(1): 91–118, 2014.
72. Xiaoyan Zhu, Haotian Chi, Shunrong Jiang, Xiaosan Lei, and Hui Li. Using dynamic pseudo-ids to protect privacy in location-based services. In *Communications (ICC), 2014 IEEE International Conference on*, pp. 2307–2312. IEEE, 2014.
73. Zhikai Xu, Hongli Zhang, and Xiangzhan Yu. Multiple mix-zones deployment for continuous location privacy protection. In *Trustcom/BigDataSE/ISPA, 2016 IEEE*, pp. 760–766. IEEE, 2016.

74. Imran Memon, Qasim Ali, Asma Zubedi, and Farman Ali Mangi. DPMM: Dynamic pseudonym-based multiple mix-zones generation for mobile traveler. In *Multimedia Tools and Applications*, pp. 1–30, 2016.
75. Pierangela Samarati and Latanya Sweeney. Protecting privacy when disclosing information: *k*-anonymity and its enforcement through generalization and suppression. Technical Report SRI-CSL-98-04, 1998.
76. Panos Kalnis, Gabriel Ghinita, Kyriakos Mouratidis, and Dimitris Papadias. Preventing location-based identity inference in anonymous spatial queries. *IEEE Transactions on Knowledge and Data Engineering*, 19(12): 1719–1733, 2007.
77. Toby Xu and Ying Cai. Exploring historical location data for anonymity preservation in location-based services. In *INFOCOM 2008. The 27th Conference on Computer Communications. IEEE*, pp. 547–555. IEEE, 2008.
78. Bugra Gedik and Ling Liu. Location privacy in mobile systems: A personalized anonymization model. In *Distributed Computing Systems, 2005. ICDCS 2005. Proceedings. 25th IEEE International Conference on*, pp. 620–629. IEEE, 2005.
79. Chi Wing Wong, Jiuyong Li, Wai Chee Fu, and Ke Wang. (α, k)-anonymity: An enhanced *k*-anonymity model for privacy preserving data publishing. In *ACM SIGKDD International Conference on Knowledge Discovery and Data Mining*, pp. 754–759, 2006.
80. Ashwin Machanavajjhala, Johannes Gehrke, Daniel Kifer, and Muthuramakrishnan Venkitasubramaniam. *L*-diversity: Privacy beyond *k*-anonymity. In *International Conference on Data Engineering*, p. 3, 2006.
81. Ninghui Li, Tiancheng Li, and Suresh Venkatasubramanian. *t*-closeness: Privacy beyond *k*-anonymity and *l*-diversity. In *IEEE International Conference on Data Engineering*, pp. 106–115, 2007.
82. Zhen Xiao, Jianliang Xu, and Xiaofeng Meng. p-sensitivity: A semantic privacy-protection model for location-based services. In *International Conference on Mobile Data Management Workshops*, pp. 47–54, 2008.
83. Xiaokui Xiao and Yufei Tao. M-invariance: Towards privacy preserving re-publication of dynamic datasets. In *ACM SIGMOD International Conference on Management of Data*, Beijing, China, pp. 689–700, June 2007.
84. Hidetoshi Kido, Yutaka Yanagisawa, and Tetsuji Satoh. An anonymous communication technique using dummies for location-based services. In *Pervasive Services, 2005. ICPS'05. Proceedings. International Conference on*, pp. 88–97. IEEE, 2005.
85. Hua Lu, Christian S. Jensen, and Man Lung Yiu. Pad: Privacy-area aware, dummy-based location privacy in mobile services. In *Proceedings of the Seventh ACM International Workshop on Data Engineering for Wireless and Mobile Access*, pp. 16–23. ACM, 2008.
86. Aris Gkoulalas-Divanis and Vassilios S. Verykios. A privacy-aware trajectory tracking query engine. *ACM SIGKDD Explorations Newsletter*, 10(1): 40–49, 2008.
87. Tun-Hao You, Wen-Chih Peng, and Wang-Chien Lee. Protecting moving trajectories with dummies. In *Mobile Data Management, 2007 International Conference on*, pp. 278–282. IEEE, 2007.
88. Ryo Kato, Mayu Iwata, Takahiro Hara, Akiyoshi Suzuki, Xing Xie, Yuki Arase, and Shojiro Nishio. A dummy-based anonymization method based on user trajectory with pauses. In *Proceedings of the 20th International Conference on Advances in Geographic Information Systems*, pp. 249–258. ACM, 2012.
89. Aniket Pingley, Nan Zhang, Xinwen Fu, Hyeong-Ah Choi, Suresh Subramaniam, and Wei Zhao. Protection of query privacy for continuous location based services. In *INFOCOM, 2011 Proceedings IEEE*, pp. 1710–1718. IEEE, 2011.

90. Akiyoshi Suzuki, Mayu Iwata, Yuki Arase, Takahiro Hara, Xing Xie, and Shojiro Nishio. A user location anonymization method for location based services in a real environment. In *Proceedings of the 18th SIGSPATIAL International Conference on Advances in Geographic Information Systems*, pp. 398–401. ACM, 2010.
91. Sergio Ilarri, Eduardo Mena, and Arantza Illarramendi. Location-dependent query processing: Where we are and where we are heading. *ACM Computing Surveys (CSUR)*, 42(3): 1–12, 2010.
92. Ben Niu, Qinghua Li, Xiaoyan Zhu, Guohong Cao, and Hui Li. Achieving *k*-anonymity in privacy-aware location-based services. In *INFOCOM, 2014 Proceedings IEEE*, pp. 754–762. IEEE, 2014.
93. Ben Niu, Zhengyan Zhang, Xiaoqing Li, and Hui Li. Privacy-area aware dummy generation algorithms for location-based services. In *Communications (ICC), 2014 IEEE International Conference on*, pp. 957–962. IEEE, 2014.
94. Vincent Bindschaedler and Reza Shokri. Synthesizing plausible privacy-preserving location traces. In *IEEE Symposium on Security and Privacy*, pp. 546–563, 2016.
95. Rakesh Agrawal and Ramakrishnan Srikant. Privacy-preserving data mining. In *ACM SIGMOD International Conference on Management of Data*, pp. 439–450, 2000.
96. Pedro Wightman, Winston Coronell, Daladier Jabba, Miguel Jimeno, and Miguel Labrador. Evaluation of location obfuscation techniques for privacy in location based information systems. In *Communications*, pp. 1–6, 2011.
97. Lung Yiu Man, Christian S. Jensen, Xuegang Huang, and Hua Lu. Spacetwist: Managing the trade-offs among location privacy, query performance, and query accuracy in mobile services. In *IEEE International Conference on Data Engineering*, pp. 366–375, 2008.
98. Cynthia Dwork. Differential privacy. In *Proceedings of 33rd International Colloquium on Automata, Languages and Programming*, pp. 1–12, 2006.
99. Ashwin Machanavajjhala, Daniel Kifer, John Abowd, Johannes Gehrke, and Lars Vilhuber. Privacy: Theory meets practice on the map. In *IEEE International Conference on Data Engineering*, pp. 277–286, 2008.
100. Shen Shyang Ho and Shuhua Ruan. Differential privacy for location pattern mining. In *ACM SIGSPATIAL International Workshop on Security and Privacy in GIS and LBS*, pp. 17–24, 2011.
101. Reza Shokri, George Theodorakopoulos, Panos Papadimitratos, Ehsan Kazemi, and Jean Pierre Hubaux. Hiding in the mobile crowd: Location privacy through collaboration. *IEEE Transactions on Dependable and Secure Computing*, 11(3): 266–279, 2014.
102. Xiaoyan Zhu, Haotian Chi, Ben Niu, Weidong Zhang, Zan Li, and Hui Li. Mobicache: When k-anonymity meets cache. In *Global Communications Conference*, pp. 820–825, 2014.
103. Ben Niu, Qinghua Li, Xiaoyan Zhu, and Guohong Cao. Enhancing privacy through caching in location-based services. In *IEEE Conference on Computer Communications*, pp. 1017–1025, 2015.
104. Mudhakar Srivatsa and Mike Hicks. Deanonymizing mobility traces: Using social network as a side-channel. In *Proceedings of the 2012 ACM Conference on Computer and Communications Security*, pp. 628–637. ACM, 2012.
105. Michael Naehrig, Kristin Lauter, and Vinod Vaikuntanathan. Can homomorphic encryption be practical? In *Proceedings of the 3rd ACM Workshop on Cloud Computing Security Workshop*, pp. 113–124. ACM, 2011.

Chapter 14

Blockchain Technology for the 5G-Enabled Internet of Things Systems: Principle, Applications and Challenges

Tianqi Yu, Xianbin Wang, and Yongxu Zhu

Contents

14.1 Blockchain Technology

Compared with many other technologies, the history of blockchain is relatively short. The concept of blockchain was first proposed by Satoshi Nakamoto in 2008 in his paper titled "Bitcoin: A Peer-to-Peer Electronic Cash System" [1]. This paper conceived a purely P2P version of cryptocurrency, namely, the renowned term "Bitcoin," which allowed online payments to be sent directly from one participant to another without going through a financial institution. Here, a participant was termed as a node in the P2P Bitcoin network. Blockchain was proposed as the public ledger of Bitcoin, which recorded all the transactions executed in the Bitcoin network. As the public ledger, blockchain was autonomously shared among all the nodes in the network for decentralized consensus. The transactions were thus verified and authorized by the decentralized consensus mechanism in the blockchain technology, without the necessity of any central authority or third party for verification.

Despite its short history of around 10 years, blockchain technology has been well developed and widely applied to many other areas well beyond Bitcoin. According to Statista [2], the investments on blockchain startups have grown from $210 million to $548 millions from 2016 to 2018, expecting to reach $2.3 billion in 2021. Besides its function as a public ledger for other cryptocurrencies similar to Bitcoin, blockchain has also been utilized as a secure and decentralized database to overcome the challenges in many emerging non-financial applications, such as decentralized proof of existence of documents in Proof-of-Existence, distributed data storage in Storj, and anti-counterfeit product verification in BlockVerify [3]. The decentralized, secure and autonomous characteristics of blockchain technology make it a promising solution in distributed ICT systems as well, particularly, the IoT systems. IBM has proposed an autonomous decentralized P2P telemetry (ADEPT) platform based on the blockchain technology, which aims at providing system architecture with greater scalability and security for IoT systems [4].

By analyzing its history and related applications, it can be inferred that the essence of blockchain is a decentralized database duplicated and shared among all distributed nodes within a network. To elaborate its principle, the general structure of blockchain is sketched in Figure 14.1. It can be seen that each of chained blocks comprises a block header and a set of transactions, where the block header

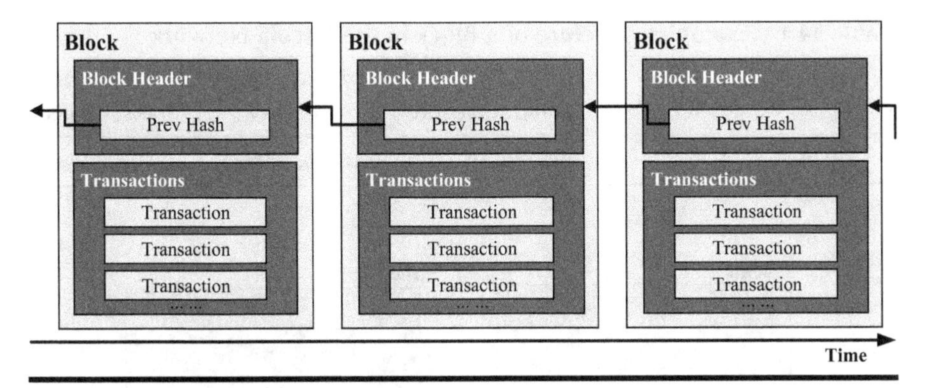

Figure 14.1 Structure of blockchain.

and the transaction are both predefined data structures. The block header contains the descriptive information of the block. Transactions are used to record the data exchanges executed within the network. Figure 14.1 also illustrates that the chain of blocks is established by referring to the hash value of previous block in the current block while blocks are managed in chronological order. When a block has been added to the chain for a period of time, it would be impractical to modify it because all the blocks chained after it would have to be regenerated and the adversaries can hardly afford the overall expense. The transactions recorded on the blockchain are thus protected from tampering.

More details of blockchain technology are further explained in this section from the following aspects. The structure of block is first stated in Section 14.1.1, where the overall block structure, block header and transaction are presented in detail. Decentralized consensus and mining are further stated in Section 14.1.2, where the procedure of generating and adding a new block to the blockchain is analyzed.

14.1.1 Block Structure

14.1.1.1 Overview of Block Structure

Block is the fundamental item of a blockchain, which generally comprises a block header and a set of transactions, as shown in Figure 14.1. Thus a block is normally defined as a container type of data structure in practical applications. The specific structure of block needs to be seriously designed according to the requirements of certain systems. Designs of the block structure in the Bitcoin network and an IoT-enabled smart home system are provided here as examples.

The block structure in Bitcoin network is given in Table 14.1 [5]. It can be seen that the first field is filled out with the block size followed by the block header with a fixed size of 80 bytes. The transaction counter and transactions are the number

Table 14.1 Example: Structure of a Block in the Bitcoin Network

Field	*Block Size*	*Block Header*	*Transaction Counter*	*Transactions*
Size (bytes)	4	80	1–9 (variable)	Variable

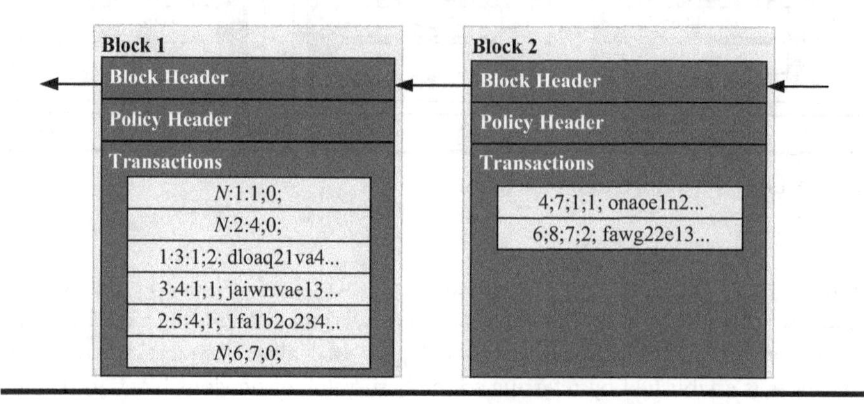

Figure 14.2 Example: structure of blockchain in IoT-enabled smart home system.

and records of transactions contained in the block respectively. The size of transaction is variable, with an average value of 400 bytes, which infers that transactions occupy most of the memory in a block.

In the IoT-enabled smart home system, blockchain technology is utilized to enhance the secure access control over IoT devices and IoT data [6]. As shown in Figure 14.2, a block comprises a block header, a policy header and a set of transactions. The block header containing the hash value of previous block is used to maintain the blockchain. The policy header is particularly designed to control the IoT devices. In such IoT-enabled smart home systems, the transactions are defined as the communications between IoT devices.

14.1.1.2 Block Header

The block header contains the descriptive information of the block, which consists of the hash value of the previous block and some self-defined fields.

The hash value of the previous block as mentioned is utilized to establish the chain of blocks. More specifically, each block is identified by its block height and hash value, as shown in Figure 14.3. Every time a new block is generated and added to the chain, the block height is automatically incremented by one. The chain of blocks is managed in chronological order. Particularly, the first block is termed the genesis block, and the height of genesis block is 0. In terms of the hash value, it is generated by hashing either the block header or the whole block, which is

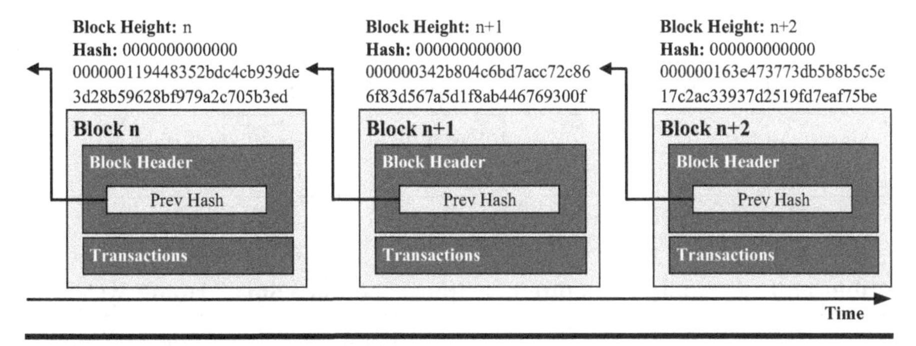

Figure 14.3 Identifiers of a block.

dependent on the choice of specific design. For example, the 256-bit hash value in Figure 14.3 is generated by hashing the block header with SHA256, namely, *Hash Value* = *SHA256* (*Block Header*). SHA256 is the secrete hash algorithm [7].

The self-defined fields are optional. If necessary, the fields are defined according to the requirements of certain applications. In the IoT-enabled smart home system, the block header contains the hash value of the previous block only [6], whereas in the Bitcoin network, besides the hash value of previous block, additional fields are defined in the block header and are utilized to identify the timestamp and contributions of generating the block [5].

14.1.1.3 Transaction

The essence of a transaction is transferring the ownership of a certain amount of property between participants. The property can be bitcoins in the Bitcoin network, IoT data in IoT systems, or any other actual or virtual commodities. In blockchain technology, a transaction is defined as a data structure with a stereotype. Identities of the nodes involved in the transaction and the trading between the nodes are clarified in the transaction. A transaction can be created by anyone, not necessarily the nodes involved in the transaction. The created transaction is then openly broadcast throughout the network so that every node in the network can witness the transaction. Only the nodes involved in the transaction can transfer the property by providing their private keys because the ownership of property is locked by the digital signatures generated from the private keys.

The structure of transaction defined in the Bitcoin network is shown in Table 14.2 as an example [5]. The field of version is used to track the upgrade of software. The input counter and output counter illustrate the number of inputs and outputs included in the transaction respectively. The fields of input and output identify the nodes involved the transaction with their Bitcoin addresses and also clarify the amount of bitcoins transferred between the nodes. The lock time refers to the earliest time that a transaction can be added to the blockchain.

Table 14.2 Example: Structure of a Transaction in the Bitcoin Network

Field	*Version*	*Input Counter*	*Input*	*Output Counter*	*Output*	*Lock Time*
Size (bytes)	4	1–9 (variable)	Variable	1–9 (variable)	Variable	4

Table 14.3 Example: Structure of a Transaction in a Smart Home System

Previous Transaction	*Transaction Number*	*Device ID*	*Transaction Type*	*Corresponding Multisig Transaction*
N = genesis transaction			Genesis: 0; Access: 1; Store: 2; Monitor: 3	If any (for keeping signature of requester)

Another example is the structure of transaction as defined in the IoT-enabled smart home system, which is shown in Table 14.3 [6]. Transactions are defined as the communications between IoT devices. Four types of transactions are considered: genesis, access, store and monitor. A genesis transaction is generated when a new device joins the system. Store transactions are generated by IoT devices to store data. An access transaction is generated when the data storage is accessed. Monitor transactions are periodically generated to monitor the information of devices. In addition, a transaction number is used to identify the current transaction, and the field of previous transaction refers to the transaction number of the previous transaction generated by the same device. In particular, for a genesis transaction, the field of previous transaction is filled out with "*N*." Device ID is the system identification of the device generating the current transaction. The last field is utilized to collect the signature of the requester if the transaction is generated outside of the local blockchain network.

14.1.2 Decentralized Consensus and Mining

The general structure of block was presented in Section 14.1.1. This subsection continues with how to generate and add a new block to the blockchain and reach the decentralized consensus.

Decentralized consensus is a mechanism that determines the conditions needed to be reached in order to make an agreement among all the nodes within the network on the validations of the blocks to be added to the blockchain. Thus, decentralized consensus is the exact mechanism in the blockchain technology that validates and secures the transactions and blocks in substitution of the traditional central authority and third-party verification. In blockchain technology, the procedure of generating and adding a new block to the blockchain is termed as "mining." The nodes that spare their efforts on block mining are termed as "miners." In other words, mining is the procedure used to meet the predefined conditions and finally reach the decentralized consensus.

Several decentralized consensus algorithms have been proposed already, such as proof-of-work (PoW), proof-of-stake (PoS), proof-of-activity (PoA), and practical Byzantine fault tolerance (PBFT) [8]. Because the conditions of reaching consensus defined in these algorithms are not the same, the corresponding ways of mining are also different.

The PoW algorithm and the way of mining utilized in the Bitcoin network are provided here as an example [5]. In PoW algorithm, a cryptographic puzzle is defined as finding a certain integer (termed as *nonce*) that can make the hash value of the block header less than or equal to a predefined threshold (termed the *difficulty target*), namely, *Hash* (*nonce+other contents in block header*) ≤ *difficulty target.* During the procedure of mining, the miners have to spend their computational power on repeating the hashing calculations with different values of *nonce* until the result meets the *different target.* The *nonce* and *different target* are included in the block header of the newly mined block as proof of work, so that other nodes can verify the work of the miner. The specific procedure of mining is explained in the following paragraph.

As introduced in Section 14.1.1.3, transactions are openly broadcast for tamper resistance. A miner then adds the transparent transactions to its transaction pool. By the time of previous block generated, a new round of mining competition among the miners starts. The miner encapsulates the transactions collected in its transaction pool into a new block and spares its computational power on solving the cryptographic puzzle. The miner who solves the puzzle with the shortest time wins the competition, and then broadcasts the newly mined block to the whole network. When a peer node in the network receives the new block, verification of the block is executed locally; that is, the work of the miner identified in the block is verified. If the block passes the verification, the node further propagates the block to other peers in the network. In the meantime, the node adds the verified block to its own copy of blockchain, where the longest chain accumulated with the largest amount of computational power is selected. If the node is also a competing miner, it stops the mining efforts on the same block height and begins the next round of mining regarding the received block as the previous block.

When a block has been added to the chain for a period of time, it would be impractical to modify it because all the blocks chained after it would have to be regenerated. Computational power spared on mining each block is rather expensive, the adversaries can hardly afford the overall expenses. Therefore, the transactions recorded on the blockchain can be protected from being altered or removed by unauthorized parties.

14.2 Applications of Blockchain Technology in Internet of Things Systems

The principle of blockchain technology was presented in Section 14.1. It has also been mentioned that blockchain technology has already been developed and applied in many other applications beyond Bitcoin, particularly in the IoT systems. The leading technology companies, for example, IBM, have already begun to study the utilization of blockchain technology on the future IoT.

In fact, in addition to the efforts from industry, there have been several technical papers published from academic institutions on applying blockchain technology into IoT systems. A systematic literature review has been conducted based on the technical papers [9] and infers that most of the existing works are focused on the blockchain technology–enabled new mechanisms in IoT systems, including data storage management, trade of goods and data, rating systems and identity management. Tiago and Paula have done a further review on the challenges of integrating blockchain technology into IoT systems, such as privacy, security, energy efficiency and scalability [8]. Technical surveys from the specific aspects of decentralized consensus [10] and smart contracts [11] have been carried out as well.

Therefore, a comprehensive summary on the existing applications of blockchain technology in IoT systems is conducted in this section and both industrial startups and research proposals from academic institutions are covered. According to the different technical functionalities of blockchain technology performed in the IoT systems, the applications are classified into the following categories: smart property trading, intelligent manufacturing, distributed data storage, and privacy and security management. The statement of each category comprises both the theoretical principle and the application instances.

14.2.1 Smart Contract-Based Property Trading

With the emergence of blockchain technology, the smart contract has been further developed, and it makes the blockchain-based smart contract a suitable solution to the decentralized smart property trading in IoT systems. The concept of the smart contract was initially defined as a set of promises recorded in digital form; that is, converting paper contracts into code and embedding them into hardware so that the transactions could be secured without the necessity of a trusted third party such as a lawyer, government and so on. [12]. By exploitation of the blockchain as public ledger, the concept of the smart contract is realized. More specifically, a potential contract jointly created by the traders is programmed as a piece of code, and then the code is encrypted and recorded on the public ledger. The public ledger is shared among all the participants in the network, ensuring that the contract is transparent and tamper resistant. The contract is automatically executed when all the predefined conditions are met. Only the parties involved in the transaction can decrypt the contract so that the trading property can be protected. The comparison between traditional and smart contracts are listed in Table 14.4 [13]. It can be seen that the smart contract is executed in much shorter time, with less cost and without the necessity of a trusted third party.

A general smart contract–based property trading model for IoT systems has been proposed [14] and is termed the e-commerce model. The IoT e-commerce model is demonstrated in Figure 14.4, which shows that the seller and customer execute transactions as trading entities through the smart contract without relying on any third-party authority. The trading commodities in IoT systems can be any registered properties, including IoT data and services. In terms of the currency involved in the trading, it can be Bitcoin or some other self-defined cryptocurrencies, for example, IoTcoin.

To better explain the function of the smart contract, two application examples are further described in the following paragraphs: Slock.it and Filament.

Table 14.4 Comparison between Traditional and Smart Contracts

Traditional Contract	*Smart Contract*
1–3 days	Minutes
Manual remittance	Automatic remittance
Expensive	Fraction of the cost
Physical presence (wet signature)	Virtual presence (digital signature)
Escrow necessary	Escrow may not be necessary
Layers necessary	Layers may not be necessary

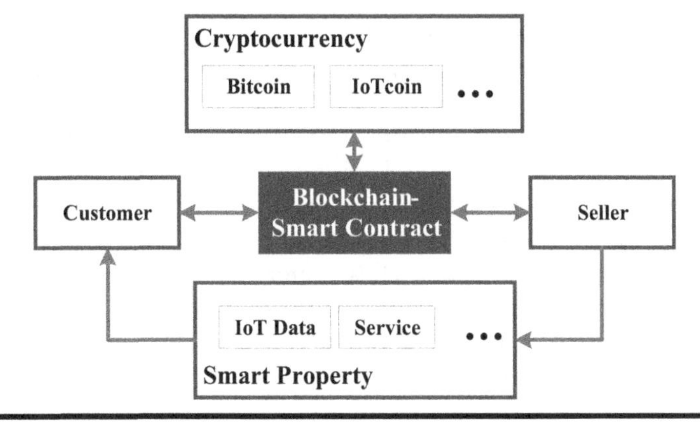

Figure 14.4 Smart contract-based IoT e-commerce model.

Slock.it is a startup working on the infrastructure of the shared economy [15]. One existing product of Slock.it is the IoT-enabled smart lock, which has already been applied to Airbnb households. By utilizing the smart contract, Airbnb renters can obtain electronic access to a house from the householder on a smartphone by paying equivalent cryptocurrency without the necessity of meeting in person. Slock.it is supported by Ethereum, a public blockchain-based distributed computing platform set up by Vatalik Buterin in 2013 [16]. Several blockchain-based startups run their products on the basis of Ethereum.

Filament is a company nominated by "Building the future of IoT" [17]. In Filament, sensor data are treated as smart property and traded for profit using smart contracts. Different from most of the blockchain-enabled software service providers, Filament designs its own hardware product. The product comprises multiple sensors and a USB communication interface, which is installed on the target facilities for machine health monitoring. The sensor data can be accessed by the product owners directly from the encrypted hardware without using the third-party cloud platform.

14.2.2 Intelligent Manufacturing

IoT technology has already been widely applied in the manufacturing industry for logistics tracking and quality assurance. By integrating blockchain technology into IoT systems, not only the procedure of logistics, but also the whole lifecycle and detailed information of a product can be tracked. The blockchain and IoT jointly enabled systems may be able to facilitate the manufacturing industry in the following aspects:

- *On-Demand Manufacturing*: In the new blockchain-IoT–enabled manufacturing industry system, all the manufacturers and customers involved can be treated as peer participants and keep updating their local copies of blockchain. Products can be produced automatically according to the needs of customers written into the blockchain [18].
- *Asset Tracking*: By using blockchain as a public record, all the information of a product in its lifecycle can be tracked. More specifically, materials, manufacturing date and factory information of a product are registered into the blockchain once it is produced. Logistics and retailer information is recorded on the blockchain as well. Thus it would be much easier for customers to track a product and protect their profits.
- *Product Certification*: Because all the information of a product is traceable, certifications for products are more trustworthy, which can protect the customers from the counterfeits jointly conducted by manufacturers and third-party certification institutions.

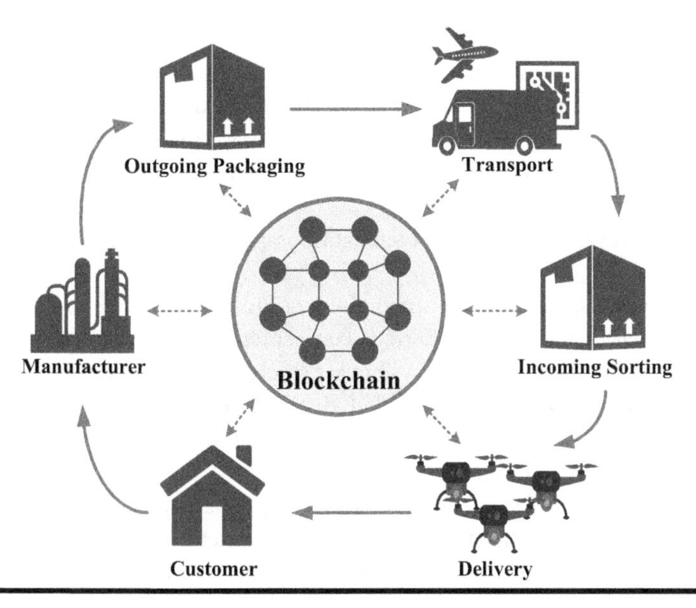

Figure 14.5 Blockchain and IoT technology–enabled asset tracking.

Figure 14.5 shows how blockchain and IoT technology can enable asset tracking [11]. A product is registered on the blockchain with all its manufacturing information when it is produced. The trading of this product is processed based on a smart contract. Once the smart contract is executed, the product is transported and delivered from the manufacturer to the customer. The information of the whole procedure of logistics is also recorded on the blockchain. Therefore, all the information of such a product is trackable on the blockchain.

Everledger is a company that provides diamond certification by creating a permanent ledger of the production and transaction history of each diamond using blockchain [19]. The unique identifications of the diamond such as weight, size and color are hashed and registered on the permanent ledger. The verification of diamonds can be easily done by any parties, including owners, financial institutions and law enforcement agencies because Everledger provides an application programming interface (API) for looking at diamonds and creating formal reports on diamonds.

14.2.3 Distributed Data Storage

In the current architectures of IoT systems, the cloud platform generally serves as the centralized data storage and processing center. Because most IoT users cannot afford the cost of setting up their own cloud server, data collected from the IoT devices are normally uploaded to a third-party cloud server. Failure or compromise of the centralized cloud platform can expose the whole IoT system to various kinds of attacks.

With the development of blockchain technology, the blockchain-enabled distributed data storage is becoming a promising substitution of the centralized cloud data center so as to overcome the problems met in the cloud-orchestrated IoT systems [20] (Figure 14.6).

The major advantages of distributed data storage are summarized as follows [21]:

- *Decentralization and Robustness*: In a cloud-orchestrated IoT system, the cloud platform is the centralized data center. If the data center is compromised, all the data of the IoT system are susceptible to various kinds of attacks. By using the blockchain-enabled distributed data storage, the data are stored on individual, distributed nodes, making the whole network more robust to central failure.
- *Privacy and Security*: All the data are encrypted and stored in the distributed storage spaces. No third party, such as the owner or manager in the centralized cloud platform, can access the distributed stored data. Hence, the privacy and security of data are enhanced.
- *Price*: Compared with Amazon's centralized cloud server that costs $25 per TB per month, blockchain storage costs only $2 per TB per month.

Storj is a startup working on the development of blockchain-based distributed data storage [22]. In the Storj network, each user shares its own spare disk space. All the distributed disk spaces are jointly utilized as the decentralized cloud platform. Data uploaded by users are sliced, encrypted and stored into the distributed spaces. The data are then secured and managed by the cryptographic keys that are privately owned by the users. In order to ensure the data download speed, data are generally stored in neighborhood spots.

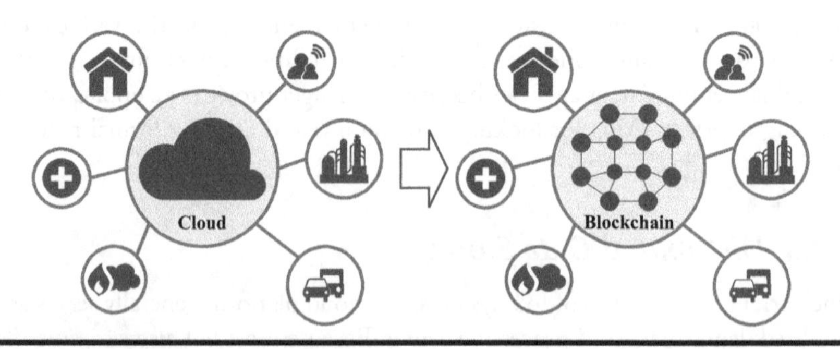

Figure 14.6 Migration from centralized cloud storage to blockchain-based distributed storage.

Filecoin is a company working in the same area as Storj [23]. The decentralized storage network of Filecoin was developed based on the combination of blockchain technology and interplanetary file system (IPFS), a distributed web-based file sharing platform [24]. Both Filecoin and Storj build their decentralized cloud platform on distributed disk spaces that are shared among users. Because of this, central failure, as has occurred on the traditional centralized cloud platform, can be prevented.

14.2.4 Privacy and Security Management

Privacy and security are critical issues in IoT systems at all times because IoT devices not only register a user's personal data, but they also monitor the user's activities. In the current IoT system, the privacy and security are normally managed by a centralized authority or third-party verification, which exposes the whole system to the potential risk of central failure. The decentralized and secure characteristics of blockchain technology can facilitate the privacy and security management in IoT systems from the following aspects.

- *Data Integrity*: In the study of security, data integrity means that the data collected from IoT devices should be protected from alteration or removal by unauthorized parties. By taking advantage of the tamper-proof feature of blockchain technology, IoT data integrity can be guaranteed. Guardtime is a company working on keyless signature infrastructure (KSI) using blockchain technology for its customers who need data integrity and security, termed as "Guardtime KSI Blockchain," where all data requests are hashed and recorded into a timestamped Merkle tree [25]. A Merkle tree is a binary hash tree for storing and searching data elements. By exploitation of timestamps, the Merkle tree is managed in chronological order and data can be further secured from alteration. The KSI blockchain has already been developed to overcome the loss of data control and protect the integrity of digital data in IoT systems [26].
- *Identity Management*: In the current IoT systems, identification of an IoT device is normally managed by the centralized authentication. The potential risk is that if the authority is compromised, the legal IoT devices would possibly be blocked from the system while adversaries would be able to obtain access. By utilizing blockchain technology, a user's identity can be easily established through a unique cryptographic address, and the identity is then managed by tailored transactions, identity verification and identity revocation without the necessity of third-party certification authority. Blockchain-based identity management has been applied in [27], where Pretty Good Privacy (PGP) is built on a Distributed Web of Trust, in which a user's

trustworthiness is established by others who could vouch through a digital signature for that user's identity and no central certification or third-party authority is involved.

- *Key Management*: Public and private keys are critical for an individual user in the IoT system because personal identity, data and all belongings are encrypted and managed by the key pairs. By utilizing the inherent cryptographic mechanism, blockchain technology can provide a more efficient solution to dynamic key management. Take the blockchain-based dynamic key management in intelligent transportation systems (ITS) as an example [28]. Traditionally, the key management in ITS relies on a third-party certificate authority. By exploiting the blockchain-based decentralized authority, the risk of central failure on the third-party authority is reduced and the time required for the key transfer procedure can be dramatically decreased, especially when the participants of key transfer are from different security domains. Another example is an IoT device management scheme that was developed together with the Ethereum-based smart home system [29]. In the scheme, the key is managed by the Rivest-Shamir-Adleman (RSA) public key cryptosystems, where public keys are stored on the blockchain and private keys are saved on the individual IoT devices. Data exchanges within the system are secured by certain keys.

In addition to smart property trading, intelligent manufacturing, distributed storage and privacy and security management summarized in this section, there are still many other applications of blockchain technology in IoT systems that have already been studied, such as IoT agricultural system [30], smart wearables [31], smart living system [32] and mobile crowdsensing [33].

14.3 Challenges and Research Directions

Blockchain technology has been widely applied in IoT systems as stated in Section 14.2, but there are still several aspects that can be improved upon to enhance the performance of blockchain-based IoT systems. In this section, the challenges and research directions are analyzed.

14.3.1 Heterogeneous Internet of Things Devices

With the development of IoT technology, the number and the diversity of devices deployed in the IoT systems is increasing dramatically. The capabilities of heterogeneous IoT devices are quite different in the following aspects: computational power, communicating capability, memory size, power supply, mobility and so on. In particular, some IoT devices with limited capabilities can barely afford the full record of blockchain. Thus, heterogeneity has to be considered when integrating blockchain technology into IoT systems.

The IBM ADEPT platform has provided a potential solution to the issue of heterogeneous IoT devices in the blockchain-based IoT systems [4]. In the IBM ADEPT platform, devices are classified into three categories—light peer, standard peer and peer exchange—according to their different levels of computational power and memory size. More specifically, the light peer category, with the weakest capability, does not hold any portion of blockchain and only maintains its blockchain identity. The relevant transactions are obtained by communicating with the more powerful peers, either a standard peer or a peer exchange. A standard peer can retain part of blockchain with its own recent transactions and some transactions for its light peers. A peer exchange, as the most powerful type of node, can synchronize the full record of blockchain in support of its weak peers. However, the IBM ADEPT platform is currently under development and still faces several challenges, including issues of scalability and anonymity are analyzed in the following subsections.

14.3.2 Blockchain Size and Scalability

Because blocks containing the transactions are periodically generated and added to the chain, the size of blockchain keeps enlarging, consequently making the cost of time and memory on the initial download and synchronization of blockchain overlarge. The overlarge blockchain size has already resulted in issues of scalability in the blockchain-based networks. Simultaneously, as analyzed in the issue of heterogeneous IoT devices, some of the devices can hardly afford the full record of blockchain. Hence, scaling the blockchain has to be taken into consideration during the design of blockchain-based IoT systems.

The mini-blockchain is among the most promising solutions to the issue of scalability [34]. In mini-blockchain, the balance of each account is recorded and managed by an "account tree," which is a Merkle binary hash tree. Old transactions no longer need to be recorded. Only the account tree and the newest portion of blockchain comprising recent transactions are kept synchronized among the network. This mechanism makes the mini-blockchain more scalable than the original blockchain.

The mini-blockchain or some other lightweight versions of blockchain seem to be better options for IoT systems than the original blockchain. Thus, scaling the blockchain to make it more suitable for IoT systems needs to be further studied.

14.3.3 Privacy and Security

It has been stated that blockchain technology can facilitate IoT systems with privacy and security management (Section 14.2.4). However, blockchain technology itself still faces some potential risks on privacy and security. Because IoT data are highly related to the privacy of users, any privacy issue of blockchain technology can finally expose the information of IoT users to adversaries. In addition, the decentralized consensus mechanism in blockchain technology makes it able to be widely applied in the scenarios without third-party authority. However, the decentralized consensus mechanism makes the blockchain technology vulnerable to the consensus attacks, such as 51% attack. These potential security issues are brought into the blockchain-based IoT systems along with the multiple advantages. Therefore, the following privacy and security issues need to be well evaluated before integrating blockchain technology into IoT systems.

- *Anonymity*: In the blockchain-based networks, transactions are openly broadcast and the records of blockchain are shared among all the nodes. Although this approach can protect the transactions from being tampered with, it may also lead to the issue of anonymity. Transparent transactions can be visualized by all the nodes in the blockchain-based network, providing an opportunity to malicious attackers. The adversaries can record transactions and infer the actual identities behind the nodes by pattern recognition. Anonymity protection is critical in IoT systems because IoT data are highly related to the personal identities and daily activities of users [35]. Therefore, anonymity enhancement needs to be studied for the blockchain-based IoT systems. [36] proposed a node cooperation verification approach to achieve privacy protection for the blockchain-based crowdsensing applications. In addition, some solutions to the issue of anonymity in the Bitcoin network may be developed and migrated to the blockchain-based IoT systems as well. In Satoshi's original Bitcoin white paper [1], he suggested an approach that would generate a new address for each individual transaction, which could possibly relieve the risk. However, because the Bitcoin address generated by a user's private key is pseudo-anonymous, it is still possible for the adversary to infer the actual identity by longer observation. Ring signature–based CryptoNote [37] is also a promising approach to enhancing anonymity, here a transaction is signed by the payer with a ring signature generated together with its group so that the payer is difficult to identify.
- *Authentication*: For each participant in blockchain-based networks, a private key is among the most important information because the private key is the main element used to do authentication. However, in the short history of blockchain technology, a Bitcoin wallet company, Mt. Gox, was severely attacked and all the private keys of its customers were stolen [38]. This well-known attack reminds all engineers and researchers that the elliptic curve cryptography used in the current blockchain technology is not strong enough to protect private keys. Therefore, when the blockchain technology is integrated into IoT systems, how to protect the private keys, or how to propose another way to authenticate the IoT devices, must be considered. Rotating asymmetric keys [39] and automatic

authentication [40] are among the promising solutions, where novel identity management and authentication methods are proposed.

- *51% Attack*: The security of blockchain technology is built on the decentralized consensus mechanism, which leads to a potential risk that if an attacker owns the major portion of the computational power of the network, it may be able to manipulate the blockchain [41]. It is termed the 51% attack because when the adversary controls over 51% of the computational power, it almost guarantees the success of a consensus attack. The 51% attack has been a critical research direction since the emergence of the Bitcoin network. There are still no effective solutions to the issue itself. However, in terms of the Bitcoin network, with the dramatic increase in the difficulty of mining, it is hard for a single or a few adversaries to occupy over 51% of the computational power and initiate a consensus attack, whereas for small-scale blockchain-based IoT systems, effort is still needed on the study of protecting the systems from the 51% attack.
- *Malleability Attack*: Data integrity is among the major concerns of security, which ensures that the data records are not altered or tampered by unauthorized parties. A malleability attack brings huge a challenge to the data integrity in the blockchain-based networks [42]. In a malleability attack, the adversary cheats the transaction issuer by intercepting, modifying and rebroadcasting a transaction, which can finally lead to the termination of all new transactions. In order to deal with the malleability attack, a deposit protocol with a timed commitment scheme has been proposed for the Bitcoin network; this would enable a malleability-resilient refund transaction [43]. However, migration of the proposed deposit protocol into the blockchain-based IoT systems needs to be further studied.

14.3.4 Network Performance

Integration of blockchain technology can tackle several issues in the IoT systems indeed, but the influences on the network performance induced by the integration have to be considered as well. The issues of throughput, latency and energy efficiency are analyzed in this subsection.

- *Throughput*: Considering the large scale of IoT systems, the amount of transactions that need to be processed per second is huge. However, this can be a critical challenge with the current blockchain technology. In the current blockchain-based networks, PoW is normally adopted as the consensus algorithm, only a maximum of seven transactions can be processed per second, which can hardly meet the requirements of throughput in the large-scale IoT systems [8].
- *Latency*: In certain application scenarios of IoT systems, real-time processing is a necessity. For example, in the smart healthcare system, real-time detection failure on emergent events can lead to severe results, especially

for elders living alone. In the current blockchain-based networks, each new block is mined with an expectation of 10 min and a transaction cannot be promised to be added to the next block. That imposes a long latency on network processing and makes it unsuitable for real-time IoT systems [44].

- *Energy Efficiency*: Due to the requirements of low power and low cost, IoT devices are normally built with limited resources, such as low computational capability and constrained power supply. However, the blockchain technology is quite energy hungry, particularly the procedures of mining and communicating. In the mining procedure, all the miners compete on solving the cryptographic puzzle, which consumes a large amount of computational power and electrical energy. In terms of communication, blockchain-based networks work in the P2P mode, whereas the IoT devices have to keep awake and spend extra energy on listening to the communication channels [45]. Therefore, energy efficiency is also a critical challenge for the blockchain-based IoT systems.

One potential way to improve the performance of blockchain-based IoT systems is to adopt a consensus algorithm with lower cost and higher efficiency because most of the existing shortcomings are caused by the high complexity and high cost of the PoW algorithm. For example, a lightweight scalable blockchain (LSB) has been proposed for an IoT-enabled smart home system in [46] where novel algorithms for lightweight consensus and distributed throughput management have been particularly developed. Simulations have been conducted using Cooja and NS3, and the results illustrated that the newly proposed LSB has indeed reduced the packet overhead in the smart home system.

In the future, lightweight consensus algorithms and even lightweight versions of blockchain need to be further developed to meet the requirements of more general IoT systems.

14.4 Summary

In this chapter, blockchain technology for IoT systems were analyzed from the aspects of principle, existing applications and potential research directions. The principle of blockchain technology was first presented with the block structure, block mining and decentralized consensus mechanism. Existing applications of blockchain technology on strengthening the IoT systems were then summarized, where smart property trading, intelligent manufacturing, distributed data storage and privacy and security management were stated in detail with both theoretical principle and application instances. Last but not least, the main challenges and research potentials of integrating blockchain technology into IoT systems were well analyzed in the following aspects: heterogeneous IoT devices, blockchain size and scalability, privacy and security, and the network performance, including throughput, latency and energy efficiency.

References

1. S. Nakamoto, "Bitcoin: A Peer-to-Peer Electronic Cash System." Available: https://bitcoin.org/bitcoin.pdf, 2008.
2. Markets and Markets, Statista Estimates, "Market for Blockchain Technology Worldwide." Available: https://www.statista.com/statistics/647231/worldwide-blockchaintechnology-market-size/, 2018.
3. M. Crosby, P. Pattanayak, S. Verma, and V. Kalyanaraman, "Blockchain Technology: Beyond Bitcoin," *Appl. Innovation*, vol. 2, pp. 6–10, 2016.
4. S. Panikkar, S. Nair, P. Brody, and V. Pureswaran, *ADEPT: An IoT Practitioner Perspective*. IBM, 2015.
5. A. M. Antonopoulos, *Mastering Bitcoin: Unlocking Digital Cryptocurrencies*. Sebastopol, CA: O'Reilly Media, 2014.
6. A. Dorri, S. S. Kanhere, R. Jurdak, and P. Gauravaram, "Blockchain for IoT Security and Privacy: The Case Study of a Smart Home," in *Proceedings of the IEEE International Conference on Pervasive Computing and Communication Workshops (PerCom Workshops)*. IEEE, 2017, pp. 618–623.
7. C. Dobraunig, M. Eichlseder, and F. Mendel, "Analysis of SHA-512/224 and SHA-512/256," in *International Conference on the Theory and Application of Cryptology and Information Security*. Springer, 2014, pp. 612–630.
8. T. M. Fernandez-Carames and P. Fraga-Lamas, "A Review on the Use of Blockchain for the Internet of Things," *IEEE Access*, vol. 6, pp. 32979–33001, 2018.
9. M. Conoscenti, A. Vetro, and J. C. De Martin, "Blockchain for the Internet of Things: A Systematic Literature Review," in *Proceedings of the IEEE/ACS 13th International Conference on Computer System and Applications (AICCSA)*. IEEE, 2016, pp. 1–6.
10. K. Yeow, A. Gani, R. W. Ahmad, J. J. Rodrigues, and K. Ko, "Decentralized Consensus for Edge-Centric Internet of Things: A Review, Taxonomy, and Research Issues," *IEEE Access*, vol. 6, pp. 1513–1524, 2018.
11. K. Christidis and M. Devetsikiotis, "Blockchains and Smart Contracts for the Internet of Things," *IEEE Access*, vol. 4, pp. 2292–2303, 2016.
12. N. Szabo, "Smart Contracts: Building Blocks for Digital Markets," *J. Transhumanist Thought*, vol. 16, 1996.
13. A. Morrison, "How Smart Contracts Automate Digital Business." Available: http://usblogs.pwc.com/emerging-technology/howsmart-contracts-automate-digital-business/, 2016.
14. Y. Zhang and J. Wen, "The IoT Electric Business Model: Using Blockchain Technology for the Internet of Things," *Peer-to-Peer Netw. Appl.*, vol. 10, no. 4, pp. 983–994, 2017.
15. Slock.it Official Website. Available: https://slock.it/.
16. Ethereum Official Website. Available: https://www.ethereum.org.
17. Filament Official Website. Available: https://filament.com/.

18. A. Bahga and V. K. Madisetti, "Blockchain Platform for Industrial Internet of Things," *J. Softw. Eng. Appl.*, vol. 9, no. 10, p. 533, 2016.
19. Everledger Official Website. Available: https://diamonds.everledger.io/.
20. B. Butler, "How IBM Wants to Bring Blockchain From Bitcon to Your Data Center." Available: http://www.networkworld.com/article/3182806/cloud-computing/how-ibm-wants-tobring-blockchain-from-bitcoin-to-your-data-center.html, 2017.
21. Z. Herbert, "Why Blockchains are the Future of Cloud Storage." Available: https://blog.sia.tech/why-blockchains-are-the-future-ofcloud-storage-91f0b48cfce9, 2017.
22. Storj Official Website. Available: https://storj.io/.
23. Filecoin Official Website. Available: https://filecoin.io/.
24. IPFS Official Website. Available: https://ipfs.io/.
25. Guardtime Official Website. Available: https://guardtime.com/.
26. H. Yin, D. Guo, K. Wang, Z. Jiang, Y. Lyu, and J. Xing, "Hyperconnected Network: A Decentralized Trusted Computing and Networking Paradigm," *IEEE Netw.*, vol. 32, no. 1, pp. 112–117, 2018.
27. D. Wilson and G. Ateniese, "From Pretty Good to Great: Enhancing PGP Using Bitcoin and the Blockchain," in *International Conference on Network and System Security*. Springer, 2015, pp. 368–375.
28. A. Lei, H. Cruickshank, Y. Cao, P. Asuquo, C. P. A. Ogah, and Z. Sun, "Blockchain-Based Dynamic Key Management for Heterogeneous Intelligent Transportation Systems," *IEEE Internet Things J.*, vol. 4, no. 6, pp. 1832–1843, 2017.
29. S. Huh, S. Cho, and S. Kim, "Managing IoT Devices Using Blockchain Platform," in *Proceedings of the IEEE 19th International Conference on Advanced Communications Technology (ICACT)*. IEEE, 2017, pp. 464–467.
30. F. Tian, "An Agri-Food Supply Chain Traceability System for China based on RFID & Blockchain Technology," in *Proceedings of the IEEE 13th International Conference on Service Systems and Service Management (ICSSSM)*. IEEE, 2016, pp. 1–6.
31. M. Siddiqi, S. T. All, and V. Sivaraman, "Secure Lightweight Context-Driven Data Logging for Bodyworn Sensing Devices," in *Proceedings of the IEEE 5th International Symposium on Digital Forensic and Security (ISDFS)*. IEEE, 2017, pp. 1–6.
32. D. Han, H. Kim, and J. Jang, "Blockchain Based Smart Door Lock System," in *Proceedings of the IEEE International Conference on Information and Communication Technology Convergence (ICTC)*. IEEE, 2017, pp. 1165–1167.
33. C. Tanas, S. Delgado-Segura, and J. Herrera-Joancomarti, "An Integrated Reward and Reputation Mechanism for MCS Preserving Users Privacy," in *Data Privacy Management and Security Assurance*. Springer, 2015, pp. 83–99.
34. J. Bruce, "The Mini-Blockchain Scheme." Available: http://cryptonite.info/files/mbc-scheme-rev3.pdf, 2014.
35. Y. Yang, L. Wu, G. Yin, L. Li, and H. Zhao, "A Survey on Security and Privacy Issues in Internet-of-Things," *IEEE Internet Things J.*, vol. 4, no. 5, pp. 1250–1258, 2017.
36. J. Wang, M. Li, Y. He, H. Li, K. Xiao, and C. Wang, "A Blockchain Based Privacy-Preserving Incentive Mechanism in Crowdsensing Applications," *IEEE Access*, vol. 6, pp. 17545–17556, 2018.
37. CryptoNote Official Website. Available: https://cryptonote.org/.
38. R. McMillan, "The Inside Story of Mt. Gox, Bitcoin's $460 Million Disaster," *Wired*, vol. 3, March 2014.

39. D. W. Kravitz and J. Cooper, "Securing User Identity and Transactions Symbiotically: IoT Meets Blockchain," in *Global Internet of Things Summit (GIoTS)*. IEEE, 2017, pp. 1–6.
40. A. Dorri, S. S. Kanhere, and R. Jurdak, "Towards an Optimized Blockchain for IoT," in *Proceedings of the ACM 2nd International Conference on Internet-of-Things Design and Implementation*. ACM, 2017, pp. 173–178.
41. J. Yli-Huumo, D. Ko, S. Choi, S. Park, and K. Smolander, "Where is Current Research on Blockchain Technology? A Systematic Review," *PLOS one*, vol. 11, no. 10, p. e0163477, 2016.
42. C. Decker and R. Wattenhofer, "Bitcoin Transaction Malleability and Mt Gox," in *European Symposium on Research in Computer Security*. Springer, 2014, pp. 313–326.
43. M. Andrychowicz, S. Dziembowski, D. Malinowski, and L. Mazurek, "On the Malleability of Bitcoin Transactions," in *International Conference on Financial Cryptography Data Security*. Springer, 2015, pp. 1–18.
44. J. Bonneau, A. Miller, J. Clark, A. Narayanan, J. A. Kroll, and E. W. Felten, "SoK: Research Perspectives and Challenges for Bitcoin and Cryptocurrencies," in *Proceedings of the IEEE Symposium on Security Privacy (SP)*. IEEE, 2015, pp. 104–121.
45. Z. Zhou, M. Xie, T. Zhu, W. Xu, P. Yi, Z. Huang, Q. Zhang, and S. Xiao, "EEP2P: An Energy-Efficient and Economy-Efficient P2P Network Protocol," in *Proceedings of the IEEE International Green Computing Conference (IGCC)*. IEEE, 2014, pp. 1–6.
46. A. Dorri, S. S. Kanhere, R. Jurdak, and P. Gauravaram, "LSB: A Lightweight Scalable Blockchain for IoT Security and Privacy," CoRR, abs/1712.02969, 2017.

IV

EMERGING APPLICATIONS OF THE 5G-ENABLED INTERNET OF THINGS

Chapter 15

Searching for Internet-of-Things Resources: Requirements and Outlook

Yasmin Fathy

Contents

15.1 Introduction

The Internet of Things (IoT) has emerged as a paradigm that bridges the gap between physical and digital worlds by providing connections and communication between physical objects (things) in the real world to the digital world (Internet).

Leveraging Web standards and communication technologies that allow IoT resources to collect and exchange their data with other resources on the Web is often described under the umbrella term of the Web of Things (WoT). The core concept of the IoT/WoT is an evolution of using various networking and communication technologies and data analytics. For example, radio frequency identification (RFID) is utilized to capture object context (e.g., location), and sensor motes are attached to vehicles and trains to monitor and analyze transportation systems. Machine learning methods are applied to IoT data and resources to make predictions and provide recommendations and prescriptive actions and analytics.

The availability of IoT data opens up new business opportunities across multiple verticals in such areas as environment (e.g., smart metering and agriculture), industry (e.g., supply chain and intelligent transportation systems) and healthcare (e.g., activity tracking and healthcare monitoring). It is expected that 50–100 billion Internet-enabled devices will be connected to the Internet by 2020 [1]. As a result, a myriad of streaming, high-scale, heterogeneous and spatiotemporal data will be collected and published by billions of heterogeneous sensors and resources in a real-time manner. The dynamicity and ad hoc nature of underlying IoT resources make accessing and processing their data and services a challenging task [2]. Moreover, the high latency (i.e., the time the data takes to reach a destination and come back to its source) in the cellular networks limits the effectiveness of IoT solutions and applications. To unleash the potential of IoT, there is a need for fast, reliable, efficient and high-capacity networks.

The current cellular networks' capabilities limit the effectiveness of IoT solutions and applications in terms of capacity, latency, mobility, reliability and speed [3]. 5G networks are expected to revolutionize and lead to significant growth in the IoT across multiple verticals. The IoT leverages the exploitation of 5G network capabilities by allowing more devices and objects to connect to the network and communicate with each other, with low latency and high response time. The network is expected to meet different performance requirements for various IoT applications ranging from low latency and mobility on demand of tracking vehicles (e.g., traffic control applications) to reliable, secure and delay-sensitive applications (e.g., healthcare monitoring) [4]. This enables the development of smart and powerful IoT ecosystems.

The architecture of 5G networks has to address the limitations associated with the current cellular networks on the communication level; however, the deluge of data produced by a large number of resources make accessing and analyzing this data a challenging task. The current information access and retrieval solutions on the Web are far from ideal. Most of the existing solutions rely on predefined links between resources and are mainly tailored to process text-based data [5]. To this end, indexing, search, and discovery methods to address the inherent features and characteristics of IoT data in dynamic and ad hoc networks have a potential impact on the efficient access and use of available IoT data and resources [6].

This chapter is organized as follows: Section 15.2 states the special characteristics of IoT data. Section 15.3 discusses the key components and requirements for developing effective IoT applications. Searching for IoT resources and data and its key challenges are discussed in Section 15.4. Section 15.5 concludes and discusses areas for further research.

15.2 Special Characteristics of Internet-of-Things Data

IoT resources are continuously generated data as data streams. The massive amounts of data streams have particular characteristics that are different from conventional data streams [7]. The data are collected from heterogeneous resources with different formats and various quality and granularities. The data are not only large in volume, but they are also continuous and dynamic, with spatial and temporal dependency (i.e., spatiotemporal) [6,8]. IoT data streams are often collected and published with meta-data, and consequently, the streams have a wide variety of representations.

IoT data are a type of Big Data. There are five intrinsic characteristics of Big Data [9]: Volume, Variety, Velocity, Veracity and Value (the 5 Vs). The variety is increasing while technology is advancing, and the amount of data are growing while network-enabled devices are connecting to the Internet. IoT data not only have Big Data characteristics, but IoT data also have dynamicity, distribution and spatiotemporality characteristics [6]. Different IoT applications have different requirements and collect various types of data. For example, different services communicate directly with mobile devices to track their locations in smart connected vehicles and traffic monitoring applications [10]. In real-time railway applications, a Global Positioning System (GPS) unit is associated with each train to help users find departures and arrivals in a real-time. The characteristics of IoT data are summarized in Figure 15.1.

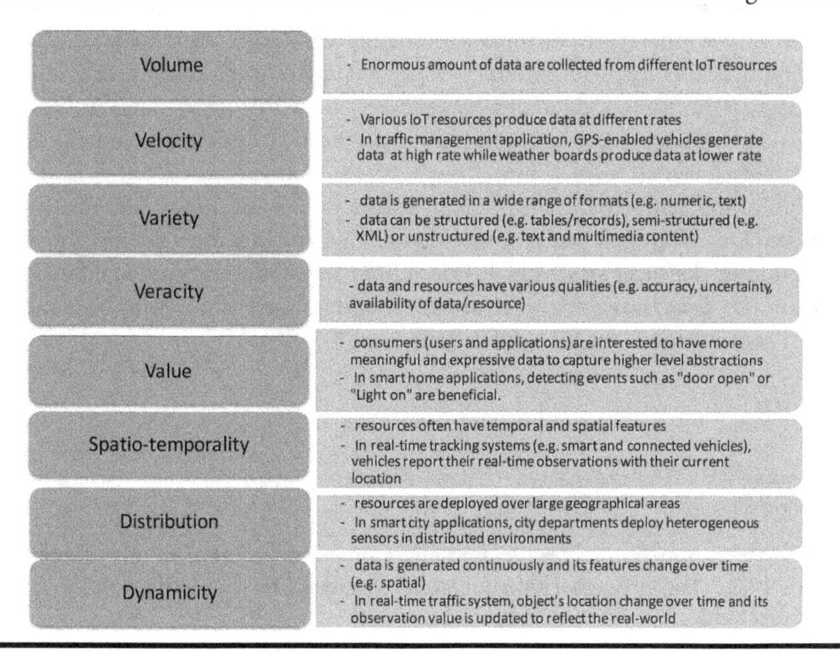

Figure 15.1 Characteristics of IoT data.

15.3 Key Design Requirements for Internet-of-Things Applications

The large scale of the network-enabled devices imposes challenges in collecting, aggregating, and processing data and services on the network. The process chain from collecting real-world observation data up to making the data accessible on the Web is discussed in [8]. However, indexing and ranking processes are not considered within the process. On the other hand, the IoT lifecycle—from data production to data querying and analysis—is presented in [11]. Indexing and ranking processes are not included in the cycle.

Similar to [6], the process chain from sensing and collecting IoT measurement and observation data up to making them discoverable for end users is shown in Figure 15.2. IoT heterogeneous devices have to connect and communicate autonomously with each other and with the Internet. Data can be aggregated and summarized from different data sources with various data types. Meaningful abstractions from raw sensor data allow inferring a higher-level description of the data and also enable representing the data in a machine-readable/human understandable format.

Data can be published with or without abstraction and aggregation. For example, different sources of data are aggregated in traffic monitoring applications (e.g., vehicle traffic flow, pedestrian activity). The data can then be abstracted to inform users about traffic status updates (e.g., low/high traffic, recommended routes). Getting the temperature of a particular room requires raw sensory temperature readings without aggregation or abstraction. Different IoT applications require different representation of data to users (human and applications). IoT data can be archived (i.e., historical data) and stored in information repositories for building predictive models to monitor/detect phenomena/events.

Publishing sensor data with semantic annotation and meta-data enrich data representation and make data more interpretable and interoperable [8]. Indexing and ranking the mechanisms of IoT resources are required to provide faster and efficient real-time data retrieval to answer user queries. The following lists the key design considerations for IoT applications.

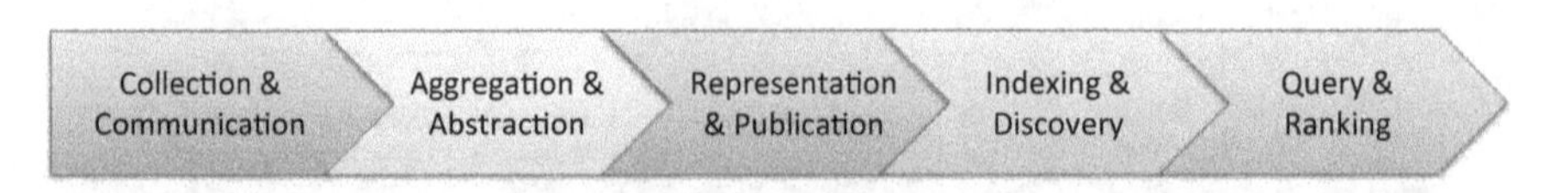

Figure 15.2 The process chain for collecting data to make them discoverable and searchable.

15.3.1 Collection and Communication

Sensory devices are key enablers to collect observation and measurement data about the real-world environment. The data collected by devices should be integrated with other resources on the Web; however, resources interact and communicate via different interfaces and protocols. Resources not only have different interfaces, but they also provide different types of data (e.g., numeric, text, media). The Sensor Web Enablement (SWE)[1] standard has been proposed to access and exchange data between heterogeneous resources in a standard way through a Web service interface.

IoT devices have limited power, memory and processing capabilities. Dealing with the data deluge generated by IoT resources at the source level is not a practical approach. The basic approach is to collect observation and measurement data from sensory devices at a base station to perform some processing on the data; however, continuous transmission of measurement and observation data incurs high communication costs for sensory devices. Some mechanisms and strategies have been proposed to reduce the communication between sensor nodes and a base station by predicting the recent sensor readings at both sensor and base station and sensory devices required to transmit their data if they deviate significantly from the predicted values [12,13]. Such approaches can effectively reduce the energy consumption for each sensory device and, consequently, prolong network lifetime, especially for battery-powered nodes.

The key requirements for sensing and communication include having a unified identifier for each resource, providing a way of interaction and communication between different IoT resources and identifying how the IoT resources should be discovered and accessed. Sensor devices can be detected automatically (i.e., active resource discovery) or devices can be registered manually by the device owner (i.e., passive resource discovery) [6,14].

15.3.2 Aggregation and Abstraction

IoT resources are distributed and provide data with different structure, qualities and granularities. The key challenge in data aggregation is that the data from multiple resources have different data models and formats. There are different approaches to aggregate data from various resources. For example, temperature measurements from multiple sensors in a monitored area are aggregated together by averaging measurement values. The aggregation can also be from different types of resources, such as traffic monitoring applications where data from various resources are aggregated to identify the traffic update status. There are also other in-network

[1] http://www.opengeospatial.org/projects/groups/sensorwebdwg

aggregation techniques where an aggregated node collects observations and measurements from different sensors and aggregates them into one data packet.

Abstractions are less granular data; an abstraction creates high-level description data from raw sensory data [15]. Abstractions can have two levels. Low-level abstraction is about inferring abstracted information from a single source of data. High-level abstraction is about deriving abstracted information from multiple sources. Meaningful abstraction allows deriving information from data such as events and patterns. We refer the interested reader to the detailed discussion in [6] on different aggregation and abstraction approaches.

The key requirements for data aggregation include aggregating and combining data from multiple resources (e.g., sensors, Web), where each resource might have different data models and structures and various qualities.

15.3.3 Representation and Publication

IoT data are collected from heterogeneous resources with different granularity, and various quality and data providers have different ways of published their data. The collected data are stored in a central repository or a distributed cloud [16]. The data can have different data attributes (e.g., numeric, categorical) and are stored in various formats (e.g., CSV, XML, JSON). The data can be enriched by meta-data to enable interoperability with other data resources [8]. The data are often accessed through Web service/application programming interface (API). The data should be represented and interpretable in a machine-understandable and human-readable format.

Several standards have been developed by different organizations such as the Internet Engineering Task Force (IETF) and the Internet Protocol for Smart Objects (IPSO) Alliance. The standards offer connectivity for IoT resources. Different organizations and standardization are discussed in [6]. Best practices for publishing and accessing spatial data on the Web are discussed in [17]. The way the data are published and represented has a direct impact on the crawling, discovery and access of the data. The key challenge in representing and publication is that we do not have a unified framework or standardization for publishing and representing the collected data.

15.3.4 Indexing and Discovery

In Web search engines, Web pages are indexed to allow fast search and data retrieval. Similarly, IoT data and resources are indexed for efficient access and discovery. In IoT, we can construct indexing based on the type of applications. Indexing can be built for IoT data or resources [6]. Data can be indexed based on their spatial and contextual features (e.g., location). For instance, the spatial property of the data can be presented in longitude, latitude and altitude coordinates. One way to index the data based on spatial feature is by space-filling curves (e.g., Z-order), where the coordinates are mapped into one dimension (i.e., a key) such as [2,7,18]. Such approaches allow discovery of the data based on a constructed key.

Indexing could be based on thematic data (e.g., specific term, field). For instance, [19] described indexes built on a selected set of XML fields of the meta-data of the connected resources to the network. Tree-based indexes have been proposed in literature; however, most of the existing approaches do not allow updates; for others, updating the constructed tree is computationally expensive [6]. Moreover, indexing can be constructed on resources such that the discovery is based on finding a resource that can answer a user query. For instance, a distributed in-network indexing approach has been proposed in [5]. The indexes are based at a gateway after clustering the resources based on their spatial properties. A tree structure is built based on the type of services the resources provide. However, the tree is based on a predefined type of services. Other approaches such as [20,21] are based on semantic indexing of the description of IoT resources.

Indexing IoT data and resources has a direct link with data discovery. Whereas indexing organizes IoT resources and data to allow efficient access and discovery, discovery utilizes the constructed indexes to respond to user queries. The dynamic nature of IoT data and resources requires having adaptive and efficient indexing approaches. The approaches should be updated efficiently taking into account that a new resource or data might be added to the constructed indexes.

15.3.5 Query and Ranking

Users expect to find meaningful information from raw sensory data. Therefore, the main objective of real-time processing and discovery the IoT data are to convert the contextual information of raw measurement and observation sensor data into useful and actionable knowledge to design and develop smart applications that can respond to end user queries. However, accessing a specific part of data to provide a real-time analysis with minimal network communications to adhere the communication, sensing, and power constraints of some sensory devices is a challenging task. Data query can be described by three parts: accessing the *right location* to get the *right data* at the *right time*. The accessing process is based on user queries (search). A query could be composed of type, location, and time attributes. In this case, it is called "Exact query." For instance, get the temperature value (type) in London (location) now (time). Other possible types of queries exist, including proximate, range, and composite queries [7]. Data discovery could be limited by the interval time (time between two consecutive data streams) in the processing of data for disaster monitoring.

Ranking is about prioritization the resources given a set of criteria such as data quality and resource availability. Different ranking mechanisms have been proposed based on multiple criteria such as quality of information (QoI), quality of service (QoS) and user feedback. In [22], data quality model is developed. The data quality model is based on a set of data quality attributes of the sensors such as accuracy, completeness and accuracy while responding to queries. However, the model

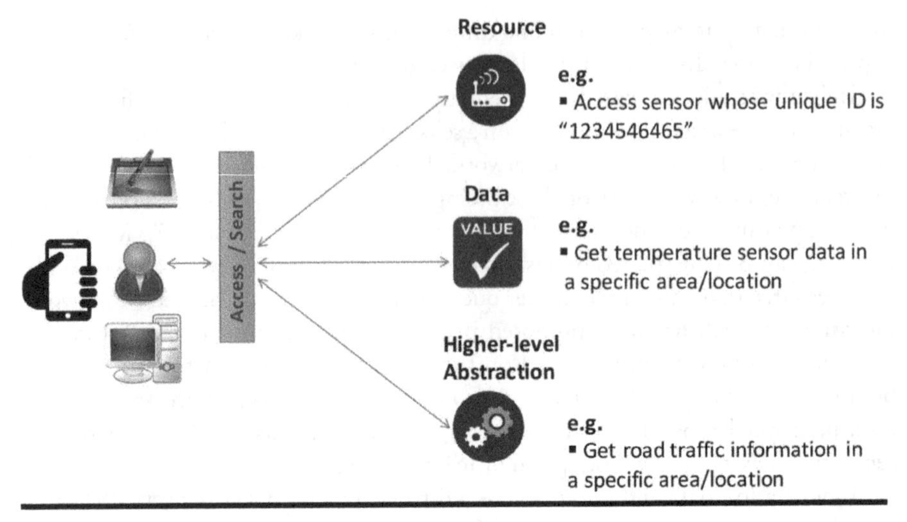

Figure 15.3 How users can interact with IoT applications. (From Fathy, Y. et al., *ACM Comput. Surv.*, 51, 29, 2018.)

is based on the sensors' precision as provided by their manufacturer. In such a case, a sensor provides a valid data even if it has failures.

Overall, in the IoT applications, it is essential to allow users to query (search) on a particular term and receive near (real-time) analysis response even as other data are arriving and devices are connecting. Query results should be based on some ranking criteria. This could be based on user requirements (e.g., fast response) or resource requirements (e.g., quality of services, resource availability). Most of the existing IoT systems and frameworks are complex or centralized, which make them a hindrance to scalability for large-scale and uncoordinated networks of devices and resources [14,23–25]. Figure 15.3 shows how users can interact with IoT applications. The diversity of how users can interact with IoT applications suffices for designing and developing a wide range of IoT applications and allows stakeholders to recognize the full potential of the IoT.

15.4 Searching for the Internet of Things

IoT data are often generated in ad hoc and dynamic environments. Various aspects have an impact when designing IoT discovery and search solutions. Reliability, uncertainty and quality, among others, are key concerns in designing IoT solutions [26]. Users should be able to specify different requirements of requested data concerning the accuracy, reliability, energy, and availability [27].

Providing data discovery in a real-time manner and gaining valuable insights and actionable knowledge is a challenging task. Search and query results should be ranked based on given requirements (e.g., resource quality, precision) [6]. Queries can be answered based on a similarity search in IoT resources [28]. A similarity search could be beneficial and provide a measurement of a neighborhood node if data from a specific sensor are unexpectedly not available. IoT solutions should have a balance between the expressiveness of data description and complexity of data model [8]. The application should also automatically abstract, aggregate, and integrate data from newly connected IoT devices and services into the network while continuously analyzing data and responding to user queries in a real-time manner. The solutions should also have highly optimized processing and discovery mechanisms that are tailored to eliminate response time and provide immediate results for user queries.

The scalability of data processing and searching is another important issue. Software-defined solutions have been proposed to address this issue [29]. The solutions enable distributing a processing task among multiple available resources. The spontaneous interaction between resources is another key issue [30]. IoT devices are often deployed in an ad hoc, dynamic environment. The devices have limited process and computation capabilities, and devices might spontaneously interact with each other. IoT solutions and networks should handle such interactions efficiently. On the other hand, mobile sensors produce different sensor readings over time due to changes in their locations. Conventional Web search engines are limited in addressing the dynamicity nature of IoT. They assume that Web page content is updated slowly, and consequently, Web indexes are updated and refreshed every couple of days [30]. IoT indexes should be adaptive and updated frequently. The update rate depends on how often the underlying data changes [6]. In addition, there is no standardized way of crawling IoT resources. Crawling is about how resources can be discovered and connected to the network, and their features can be integrated into indexes. Overall, IoT requires (auto) discovery and (near) real-time searching and processing mechanisms.

Some initial work has been done on developing IoT/WoT search engines. Dyser has been proposed [31]. Dyser assumes that each sensor has a meta-data description in the HTML page that can be crawled. Dyser employs a probabilistic model to predict the sensor that might be communicated with while answering a user query. The main shortcoming for the engine is that users must be aware of the state names for all objects to query them. Snoogle is another search engine for indexing and ranking entities [32]. Indexing entities are based on using IPs. Moreover, indexing relies on building inverted indexes for all connected entities, and the IPs are managed at a key index point (KeyIP). However, building indexes based on IPs that might change is not efficient. MAX is another human-centric search [33]. MAX assumes that each device has RFID tags and once a query is received, it then broadcasts to all physical devices to find a response from one of them which; this is computationally expensive. Moreover, the search engine does not support indexing or tanking mechanisms.

Wolfram-Alpha[2] and Thingful[3] are other examples of discovery knowledge and services engines. Thingful is considered a search engine for public IoT services. It allows users to search for a service that is provided by connected IoT devices of different categories such as environment (e.g., temperature, humidity, and pressure), energy (e.g., light, battery), and health (e.g., smart weight to report user's fitness) in different locations. Users can register and share observation and measurement data of their objects and devices. The major drawback of Thingful is the unavailability of real-time data. Search results include all data sources. However, these sources might not be available at search time. This does not guarantee real-time measurements, especially for environmental observations such as temperature and humidity that usually change over time. Other deficiencies are trust, security, and quality. Many resources provide the same type of measurement in the same location, but there is no guarantee to what extent these resources are trustworthy and with good quality and precision.

Wolfram-Alpha is a computational knowledge discovery engine. It is based on Mathematica, which is often known as Wolfram Language (http://www.wolfram.com/mathematica/). The engine allows a user to register their own Internet-connected things such as Twitter, email, and Raspberry Pi.[4] Each registered device/service has a unique "databin" and data are updated from each databin every 30 s.[5] The main shortcoming is that users can search and query resources if they know their databins. Moreover, Wolfram-Alpha has the Wolfram Data Framework (WDF) that summarizes and integrates data into a meaningful and expressive form.[6] However, information about its architecture and technical details is not available. We refer interested readers to a discussion on other IoT engines in [6,34].

Overall, the main problem with the current IoT/WoT search engines is that they do not provide efficient search and discovery for IoT data and resources. They allow querying data from resources; however, there is no deep analysis and mining of collected data to enable answer complex user queries. Scalability and/or distributed processing and analysis are still key issues [6].

15.5 Conclusion and Outlook

In ubiquitous computing, smart sensors are used to report, monitor, and track the physical environment. The IoT incorporates the concept from ubiquitous computing. However, there is still fuzziness on defining the real capabilities of IoT given

---

[2] http://www.wolframalpha.com/

[3] http://thingful.net/

[4] http://www.raspberrypi.org/

[5] http://blog.wolfram.com/2015/03/04/the-wolfram-data-drop-is-live/

[6] http://www.wolfram.com/data-framework/

the current status of real-time Big Data analytics research. Advances in Big Data analysis techniques is a key enabler for providing IoT environments and applications with (near) real-time analysis. On the other hand, the emergence of 5G technologies will unleash the potential of IoT applications by providing a new spectrum of effective communication models and usage and by allowing more devices and sensors to connect to the network. It is therefore foreseeable that the IoT will have high potential as well as an ecological impact on citizens in the physical environments with the advancement of 5G technologies and Big Data analytics in the near future.

Building IoT/WoT search engines is on-going research. In this chapter, we underline challenging in discovery and processing mechanisms. The future research directions of IoT will also depend on creating large-scale ecosystems of IoT systems that can work and collaborate with each other to share and exchange data and services. While scalability, (near) real-time analysis linked to quality and granularity of the data and access policies are critical components of designing future IoT systems, security, provisioning, reliability, and trust will also be crucial components of any design in future IoT data/service access and discovery systems. There has been recent work on adopting block-chain technologies into IoT. For accountability and audit-ability, data permissions based on smart contracts using block-chain can be achieved such that there is no need of central trusted authorities or intermediaries for providing reliable data integrity between device owners and data consumers and consequently reducing the cost of deployment. In critical infrastructure, block-chain solutions can provide a history of connected IoT resources for troubleshooting purposes.

References

1. Harald Sundmaeker, Patrick Guillemin, Peter Friess, and Sylvie Woelffl´e. *Vision and Challenges for Realising the Internet of Things*, volume 20. EUR-OP, 2010.
2. Yasmin Fathy, Payam Barnaghi, and Rahim Tafazolli. Distributed spatial indexing for the Internet of Things data management. In *Integrated Network and Service Management (IM), 2017 IFIP/IEEE Symposium on*, pp. 1246–1251. IEEE, 2017.
3. Godfrey Anuga Akpakwu, Bruno J. Silva, Gerhard P. Hancke, and Adnan M. Abu-Mahfouz. A survey on 5G networks for the Internet of Things: Communication technologies and challenges. *IEEE Access*, 6: 3619–3647, 2018.
4. Enida Cero, Jasmina Barakovi´c Husi´c, and Sabina Barakovi´c. IoTs tiny steps towards 5G: Telcos perspective. *Symmetry*, 9(10): 213, 2017.
5. Yasmin Fathy, Payam Barnaghi, Shirin Enshaeifar, and Rahim Tafazolli. A distributed in-network indexing mechanism for the internet of things. In *Internet of Things (WF-IoT), 2016 IEEE 3rd World Forum on*, pp. 585–590. IEEE, 2016.
6. Yasmin Fathy, Payam Barnaghi, and Rahim Tafazolli. Large-scale indexing, discovery, and ranking for the Internet of Things (IoT). *ACM Computing Surveys* (*CSUR*), 51(2): 29, 2018.

7. Payam Barnaghi, Wei Wang, Lijun Dong, and Chonggang Wang. A linked-data model for semantic sensor streams. In *Green Computing and Communications (GreenCom), 2013 IEEE and Internet of Things (iThings/CPSCom), IEEE International Conference on and IEEE Cyber, Physical and Social Computing*, pp. 468–475. IEEE, 2013.
8. Payam Barnaghi, Amit Sheth, and Cory Henson. From data to actionable knowledge: Big Data challenges in the web of things [Guest Editors' Introduction]. *IEEE Intelligent Systems*, 28(6): 6–11, 2013.
9. Yuri Demchenko, Paola Grosso, Cees De Laat, and Peter Membrey. Addressing Big Data issues in scientific data infrastructure. In *Collaboration Technologies and Systems (CTS), 2013 International Conference on*, pp. 48–55. IEEE, 2013.
10. Flavio Bonomi, Rodolfo Milito, Jiang Zhu, and Sateesh Addepalli. Fog computing and its role in the internet of things. In *Proceedings of the First Edition of the MCC Workshop on Mobile Cloud Computing*, pp. 13–16. ACM, 2012.
11. Mervat Abu-Elkheir, Mohammad Hayajneh, and Najah Abu Ali. Data management for the internet of things: Design primitives and solution. *Sensors*, 13(11): 15582–15612, 2013.
12. Yasmin Fathy, Payam Barnaghi, and Rahim Tafazolli. An adaptive method for data reduction in the internet of things. In *Proceedings of IEEE 4th World Forum on Internet of Things*. IEEE, 2018.
13. Silvia Santini and Kay Romer. An adaptive strategy for quality-based data reduction in wireless sensor networks. In *Proceedings of the 3rd International Conference on Networked Sensing Systems* (*INSS 2006*), pp. 29–36. TRF Mold, Chicago, IL, 2006.
14. Dominique Guinard, Vlad Trifa, Stamatis Karnouskos, Patrik Spiess, and Domnic Savio. Interacting with the SOA-based internet of things: Discovery, query, selection, and on-demand provisioning of web services. *Services Computing, IEEE Transactions on*, 3(3): 223–235, 2010.
15. Frieder Ganz, Daniel Puschmann, Payam Barnaghi, and Francois Carrez. A practical evaluation of information processing and abstraction techniques for the Internet of Things. *IEEE Internet of Things Journal*, 2(4): 340–354, 2015.
16. Vasileios Karagiannis, Periklis Chatzimisios, Francisco Vazquez-Gallego, and Jesus Alonso-Zarate. A survey on application layer protocols for the internet of things. *Transaction on IoT and Cloud Computing*, 3(1): 11–17, 2015.
17. Linda van den Brink, Payam Barnaghi, Jeremy Tandy, Ghislain Atemezing, Rob Atkinson, Byron Cochrane, Yasmin Fathy, Rau'l Garcia Castro, Armin Haller, Andreas Harth et al. Best practices for publishing, retrieving, and using spatial data on the web, *Semantic Web*, (Preprint), pp. 1-20, 2017.
18. Yuchao Zhou, Suparna De, Wei Wang, and Klaus Moessner. Enabling query of frequently updated data from mobile sensing sources. In *Computational Science and Engineering* (*CSE*), *2014 IEEE 17th International Conference on*, pp. 946–952. IEEE, 2014.
19. Willian T. Lunardi, Everton de Matos, Ramao Tiburski, Leonardo A. Amaral, Sabrina Marczak, and Fabiano Hessel. Context-based search engine for industrial IoT: Discovery, search, selection, and usage of devices. In *Emerging Technologies & Factory Automation* (*ETFA*), *2015 IEEE 20th Conference on*, pp. 1–8. IEEE, 2015.
20. Gilbert Cassar, Payam Barnaghi, and Klaus Moessner. Probabilistic methods for service clustering. In *Proceeding of the 4th International Workshop on Semantic Web Service Matchmaking and Resource Retrieval, Organised in Conjonction the ISWC*, 2010.

21. Wei Wang, Suparna De, Gilbert Cassar, and Klaus Moessner. An experimental study on geospatial indexing for sensor service discovery. *Expert Systems with Applications*, 42(7): 3528–3538, 2015.
22. Anja Klein and Wolfgang Lehner. Representing data quality in sensor data streaming environments. *Journal of Data and Information Quality (JDIQ)*, 1(2): 10, 2009.
23. Li Da Xu, Wu He, and Shancang Li. Internet of things in industries: A survey. *Industrial Informatics, IEEE Transactions on*, 10(4): 2233–2243, 2014.
24. Julien Mineraud, Oleksiy Mazhelis, Xiang Su, and Sasu Tarkoma. A gap analysis of Internet-of-Things platforms. arXiv preprint arXiv:1502.01181, 2015.
25. Daniele Miorandi, Sabrina Sicari, Francesco De Pellegrini, and Imrich Chlamtac. Internet of Things: Vision, applications and research challenges. *Ad Hoc Networks*, 10(7): 1497–1516, 2012.
26. Payam Barnaghi and Amit Sheth. On searching the internet of things: Requirements and challenges. *IEEE Intelligent Systems*, 31(6): 71–75, 2016.
27. Charith Perera, Arkady Zaslavsky, Peter Christen, Michael Compton, and Dimitrios Georgakopoulos. Context-aware sensor search, selection and ranking model for Internet of Things middleware. In *Mobile Data Management (MDM), 2013 IEEE 14th International Conference on*, volume 1, pp. 314–322. IEEE, 2013.
28. Cuong Truong, K. Romer, and Kai Chen. Fuzzy-based sensor search in the web of things. In *Internet of Things (IOT), 2012 3rd International Conference on the*, pp. 127–134. IEEE, 2012.
29. Soheil Hassas Yeganeh, Amin Tootoonchian, and Yashar Ganjali. On scalability of software-defined networking. *IEEE Communications Magazine*, 51(2): 136–141, 2013.
30. Daqiang Zhang, Laurence T. Yang, and Hongyu Huang. Searching in Internet of Things: Vision and challenges. In *Parallel and Distributed Processing with Applications (ISPA), 2011 IEEE 9th International Symposium on*, pp. 201–206. IEEE, 2011.
31. Benedikt Ostermaier, K. Romer, Friedemann Mattern, Michael Fahrmair, and Wolfgang Kellerer. A real-time search engine for the web of things. In *Internet of Things (IoT), 2010*, pp. 1–8. IEEE, 2010.
32. Haodong Wang, Chiu C. Tan, and Qun Li. Snoogle: A search engine for pervasive environments. *IEEE Transactions on Parallel and Distributed Systems*, 21(8): 1188–1202, 2010.
33. Kok-Kiong Yap, Vikram Srinivasan, and Mehul Motani. MAX: Human-centric search of the physical world. In *Proceedings of the 3rd International Conference on Embedded Networked Sensor Systems*, pp. 166–179. ACM, 2005.
34. Kay Romer, Benedikt Ostermaier, Friedemann Mattern, Michael Fahrmair, and Wolf- gang Kellerer. Real-time search for real-world entities: A survey. *Proceedings of the IEEE*, 98(11): 1887–1902, 2010.

Chapter 16

Applications of the Internet of Things and Fog Computing for Community Safety toward the 5G Era

Chenyang Wang, Dongyu Guo, Xu Tong, Ruibin Li, Xingli Gan, and Xiaofei Wang

Contents

16.1 Introduction

5G is the new generation of mobile communication technology with many unprecedented advantages, mainly including wide signal coverage, high network speed, high flux density, high mobility, and diversified application. [1]. Compared with 4G, 5G can achieve system capacity growth of 1,000 times and end-to-end latency reduction of 5 times, provide energy efficiency growth of at least 10 times, and area throughput growth of at least 25 times. It is more of a revolution that will change our way of life. 5G is thought of as the second Industrial Revolution. Very low latency, high throughput, reliability, security, and high mobility are characteristics of this upcoming technology. This revolution will enable the cellular IoT paradigm [2]. Main IoT requirements are to be met by this new mobile network. The IoT is expected to be integrated into the 5G mobile network.

Edge devices in IoT are often resource-constrained in energy storage and processing capabilities. The combination of cloud and fog computing provides some relief to these limitations, meeting requirements of 5G such as geo-distributed real-time processing and runtime adaptability [3].

Cloud computing has significantly enhanced the growth of the IoT by ensuring and supporting the QoS of IoT applications. However, cloud services are still far from IoT devices. Notably, the transmission of IoT data experiences network issues, such as high latency. In this case, cloud platforms cannot satisfy the IoT applications that require real-time response. Yet, the location of cloud services is one of the challenges encountered in the evolution of the IoT paradigm [4]. Recently, fog computing has been proposed to bring cloud services closer to the IoT end user, becoming a promising paradigm whose pitfalls and challenges are not yet well understood.

For fog computing, especially mobile fog computing, 5G technology has the potential to overcome the bottleneck of resource limitation and provide more and more resource-intensive services for mobile users [5,6]. It can also meet the demands for high-speed data applications, high-quality wireless communication, and low-latency services.

Performance of fog orchestration for the IoT faces several challenges within the context of 5G networks [7]. The high density of devices, together with the latency and reliability requirements of critical applications as well as node mobility, raise important issues concerning the monitoring of community/industrial safety, the utilization of home automation and the problems of device surveillance, and so on [8].

Compared with edge devices, fog nodes have more memory or storage ability for computing, making it possible to process a significant amount of data from edge devices. On the other hand, when the need is for a more complex and long-time computation, the computation work should be sent to the cloud by fog nodes through various available communications technologies, such as, 3G, 4G, and 5G cellular networks and wireless fidelity (Wi-Fi). Fog nodes are bridges between cloud and edge devices.

16.2 Community Safety by the Internet of Things and Fog Computing

16.2.1 Hierarchical Fog Computing Architecture for Community Safety

This topic mainly addresses the problem of access and integration of multi-source data information and achieving a "real-time in-network" function for a wide variety of community equipment facilities. The main principles include photoelectric conversion, image recognition, Option Character Recognition (OCR) technology, and so on. The data direct reading mode of traditional equipment facilities is converted into information data, which facilitates the transmission and processing of the IoT system [9]. The IoT networking transmission technologies of different equipment and facilities in the community in 5G are different, for example, narrow bandwidth Internet of Things (NB-IoT), enhanced Machine-Type Communication (eMTC), Long Range (LoRA), and so on. This makes it extremely difficult to achieve multimodal access and data fusion of data information in complex environments. With the development of society, new equipment and facilities are continuously integrated into the community environment and the existing methods cannot realize real-time acquisition, transmission and fusion processing of heterogeneous and converged IoT data under different environments, application scenarios and community user requirements [10,11].

In view of the above problems, the main purpose of this subtopic is to solve the access and fusion of multi-source data information. In order to achieve the acquisition and integration of multi-source information, we propose solutions for protocol and interface design, data acquisition and processing, data fusion, and self-organizing IoT systems in the transmission process of different equipment and facilities in the community, and the key technologies are shown. We optimize the networking of facilities at different levels in the community, designing scalable protocols and interface technologies for heterogeneous multi-source data collected by millimeter-wave radar electronic fences, power-aware anomaly detection devices, fire-fighting sensor detection devices, and other

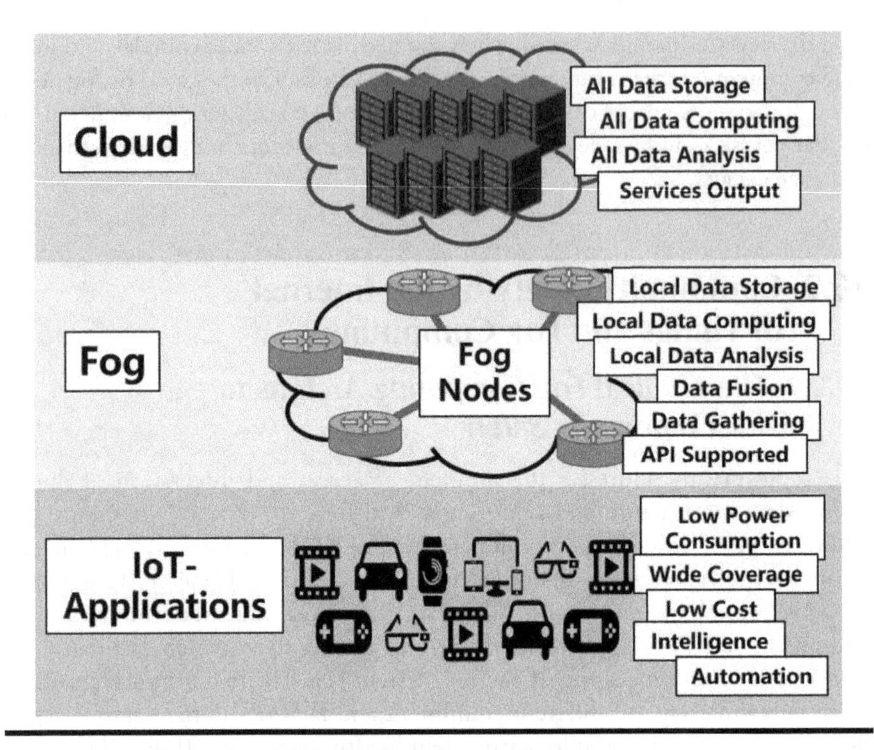

Figure 16.1 Illustration of hierarchical fog architecture.

user-aware devices. At the same time, these data are aggregated to the building fog gateway and processed locally according to the requirements. This layer has storage and computing capacity: It can realize access, real-time transmission and localization processing of multi-source information. We can convert the data mode of traditional equipment facilities into data mode based on machine-to-machine (M2M) technology, which facilitates further data processing of equipment-devices. Through IoT transmission technology such as NB-IoT, eMTC, and LoRA and the ad hoc network construction facility in 5G [12,13], we can achieve data fusion processing. Finally, after pre-processing in the fog computing layer, we will upload the data to the data center or cloud platform as required and proceed to the next step of data processing and analysis of the risk warning plan (Figure 16.1).

16.2.2 Fog Gateway

This topic mainly solves the problem of wireless transmission and network access in a complex social environment. Based on the technology of fog calculation, research, and development in intelligent gateways that can adapt to various scenarios, this gateway can unify the communication technology standards of various IoT devices in a multi-layer heterogeneous IoT environment and support multiple access methods. At the same time, because 5G networks mean ultra-fast data transfer speeds that can reach 10 Gbps [14], we will realize the unified conversion and packaging of various sensing device protocol standards based on the intelligent gateway. Based on the idea of "smart front-end" in fog computing, we deploy intelligent gateways between cloud servers and terminal devices. This method directly stores and calculates a large amount of information related to specific environments locally, thereby reducing computation and store overhead of the cloud server platform, improving system response speed and scenario quality of service and saving network bandwidth.

We implement a community IoT platform based on fog computing, which integrates and processes data locally to realize real-time monitoring of community facilities. Due to the complex nature of community equipment facilities, the diversity of data sources, and the complexity of network transmission, the development of the IoT platform is quite difficult. Therefore, this topic will make full use of the related IoT platform, use the idea of "data localization and intelligent front-end" in fog computing, and combine the characteristics of the community to optimize the database design and wireless network architecture. At the same time, we focus on the 5G standard NB-IoT as the IoT transport protocol standard, which standardizes and simplifies the network and access layers. The specific process is as shown in the Figure 16.2. The fog computing layer is set up between the application layer of the sensor deployment, such as NB-IoT and the cloud computing center; this layer has the function of connecting the upper and lower layers. The fog computing layer is deployed at the access layer of the application layer, and it can be directly connected to the application layer terminal through a fog node computing unit with storage and computing capabilities. In this way, on the one hand, the data collected by the sensor is aggregated and merged, and can cooperate with other fog node computing units; on the other hand, the layer plays a role of linking up and down and can cache data or application-layer data on the cloud computing server, where it can be calculated and migrated to localize computing tasks as much as possible.

At the same time, the network function virtualization (NFV) technology is combined with the network function virtualization infrastructure (NFVI) to integrate commercial off-the-shelf servers, switches, storage devices, and so on into the virtualization layer (such as Docker or vSwitch). Thus ensuring that virtual network functions such as data calculation and cache resource in the fog node computing unit are independent of the underlying physical platform (Figure 16.2).

As shown in Figure 16.2, the fog computing intelligent gateway is responsible for aggregating, merging, filtering, and pre-processing the multi-source

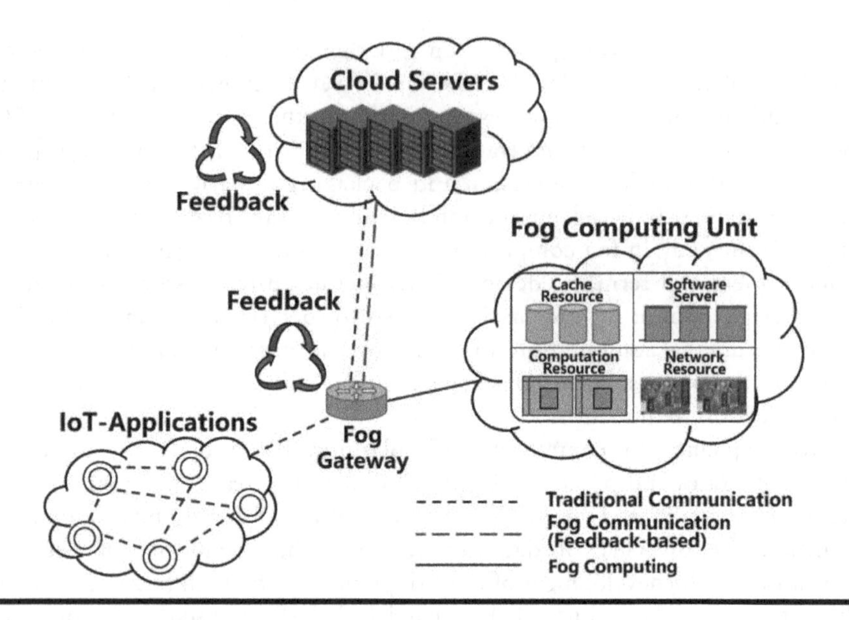

Figure 16.2 Feedback-based fog computing gateway.

heterogeneous IoT device sensing data, and sensing and controlling the energy consumption of the IoT sensor. The fog node computing unit (module) has the ability to calculate, store, and provide network services for multi-source IoT-aware data based on its powerful virtualization capabilities. Figure 16.3 shows the intelligent gateway based on fog calculation in hierarchical form. Among them, various IoT-aware devices, virtual nodes, and virtual-aware networks are integrated into the physical and the virtualization layers according to their respective needs; the monitoring layer provides monitoring functions for sensing device activity, energy consumption, datalog logs, and requirements; at the processing layer, we perform data collection, filtering, analysis, and the like; at the same time, according to different needs, some data will be temporarily stored in the temporary storage layer, so as to carry out the next step of data processing in a targeted manner. Finally, the data transmission layer transfers the preprocessed data to the cloud data platform.

The fog computing node based on community application needs to meet the following requirements: It must (1) meet the operational needs of fog calculation, (2) provide strong gateway processing, and (3) perform multi-peripheral protocol processing and interactive functions.

The device needs to undertake operations such as access, information aggregation, and information edge processing of multiple peripherals. In particular, the device should be able to support multiple wireless protocol access

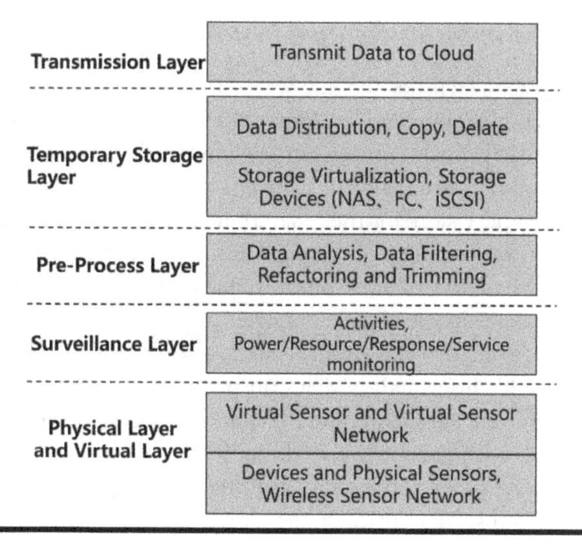

Figure 16.3 Demonstration of fog gateway level.

functions at the same time, have the Lora gate- way function, and implement device information access aggregation and relay distribution functions within the community.

The intelligent gateway receives a data packet and converts the format, and then fuses the data, and finally sends the data through other heterogeneous networks. Conversely, other heterogeneous networks can send control command information to the intelligent gateway and control the entire wireless sensor network. According to the general model of fog calculation, the local fog based on the intelligent gateway can be connected with the central fog deployed on the PC and the like, and the data can be uploaded to the central fog, which uploads the processed data to the cloud center. The application of the IoT obtains the required data through the interface provided by cloud computing or fog computing.

We developed an intelligent gateway based on fog calculation to realize hierarchical transmission, aggregation and processing of massive multi-source heterogeneous IoT data, and energy optimization control including (1) researching the fusion networking and aggregation hardware technologies of multiple IoT technologies and studying multi-task dynamic load balancing of fog computing, machine learning-specific acceleration processing and priority parallel processing technology on embedded systems, and (2) developing a fog gateway device with certain aggregation processing and task migration scheduling capabilities under a hierarchical organizational structure. In this way, tasks such as the transmission, aggregation, storage and processing analysis of the object data are dynamically balanced over the sensor front end, the fog meter gateway, and the cloud (Figure 16.3).

16.2.3 Models for Multi-feature Risk/Threats

Community equipment facilities include water, electricity, gas, network, building safety and other related supporting equipment or monitoring equipment required for community life [15]. Such equipment provides great convenience to community life, but it may also pose a major security risk to the community. It is necessary to strengthen the supervision of the operation of such equipment and facilities [16]. At present, the facilities of the community have the following characteristics: (1) There are many types of community facilities and equipment. Different equipment and installations are of different ages. It is very difficult to have uniform interfaces or access to equipment and facilities. (2) The supervision of community equipment and facilities mainly relies on physical meter reading, inspection. This method is inefficient and time intensive, and it is still in the after-treatment mode for faults or accidents, so it is very difficult to intervene in advance. (3) The distribution environment of community facilities is complex, including indoor, underground, tube wells, high-altitude and other regional distributions. It is difficult to achieve community-wide coverage by a single network transmission and communication method. (4) The equipment and facilities monitoring platform is not uniform. Even if remote meter reading or remote monitoring is adopted, it is only implemented in a relatively single facility or system. It is not possible to unify platform facilities in the community. (5) Facilities management is mainly based on data collection and display (upstream), it rarely has active management capability (downstream), and it does not allow for intelligent monitoring and management.

16.2.4 Big Data and Artificial Intelligence/Deep Learning for Fog Computing

In order to realize Big Data analysis on risk, intelligent risk prediction and early warning of community equipment and facilities, the system platform will need to integrate Big Data analysis and machine learning algorithms so as to import large-scale data collected by sample community equipment and facilities into the framework for training and testing and to ensure that the project is effective and feasible both in theory and in practice [17]. Based on high performance and the OpenStack-integrated cloud platform, we utilize Big Data analyzing platforms such as Kafka, HBase, Hadoop, and Spark to learn and train on big-scale data in the community IoT, video monitoring network and wireless data network; then we mine and analyze the security risk relationships among people, places, events, things and situations based on Big Data technology, and finally construct a public security steady-state model based on machine learning, Temporal-Difference Learning (TD-learning) and Deep Q-Network to meet the requirements of automatic and intelligent equipment and facilities operation condition monitoring and risk early warning [18] (Figure 16.4).

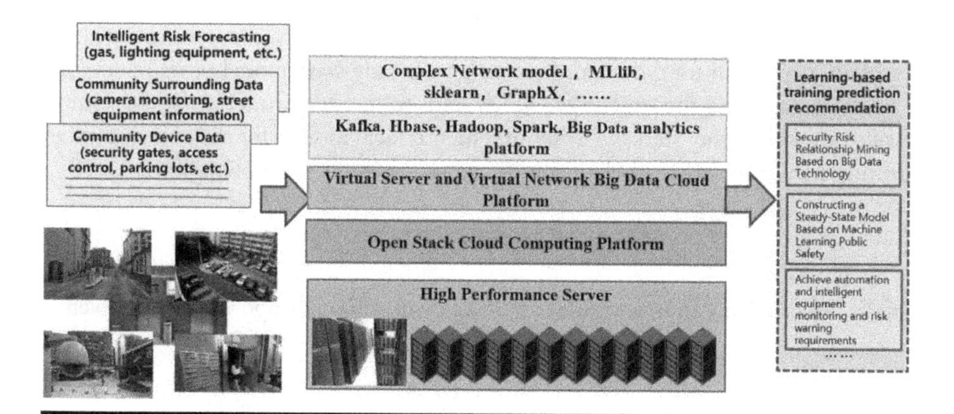

Figure 16.4 AI-based data processing system platform.

We plan to use TD reinforcement learning to analyze and model the risk level of the state of the community equipment operation. TD-Learning uses average revenue to output the utility value of the function after the operation state of the equipment changes. Therefore, we cannot only guarantee the fairness of the monitoring of equipment operation state but also realize the continuity of condition learning to a certain extent.

We use the Markov Decision Process (MDP), $\{S, A, \{P_{sa}\}, \gamma, R\}$ to model the community device running state. For a fixed strategy, the value function V_π satisfies Bellman Equations

$$V_\pi(s,a) = \mathrm{R}(s,a) + \gamma \sum_{s' \in S} P_{s\pi(s)}(s,a) \mathrm{V}_\pi(s',a') \tag{16.1}$$

In the same way, the state transition of TD-Learning can be formulated as follows:

$$V(S_t) \rightarrow V(S_t) + \alpha[\mathrm{R}_{t+1} + \gamma \mathrm{V}(\mathrm{S}_{t+1}) - \mathrm{V}(\mathrm{S}_t)], \tag{16.2}$$

where γ is the discount rate, α is the learning rate, R is the reward. Another important concept in TD-learning is TD error, δ_t, meaning the estimation error at this time, that is

$$\delta_t = R_{t+1} + \gamma V(S_{t+1}) - V(S_t), \tag{16.3}$$

where $\gamma V(S_{t+1}) - V(S_t)$ is the difference of the running state of devices at different times. Thus, we can finally obtain the update rule for the value function

$$V(S_t) \leftarrow V(S_t) + \alpha\left[R_{t+1} + \gamma \underbrace{V(S_{t+1}) - V(S_t)}_{\text{Difference}\,\textit{underdifferent}} \right] \tag{16.4}$$

Focusing on monitoring community equipment and facilities, this subproject aims to collect device running data and construct a steady-state model of its operation state. By considering the strategy π, which leads to the value function V_π under the device's worst running state, we will build hierarchical models for the overall device running state. Thus, the problem can be translated into the following form.

$$\pi^*(s) = \arg\min V^\pi(s) \tag{16.5}$$

Finally, the problem can be expressed as the following objective optimization problem.

$$\min V^\pi(s) \tag{16.6}$$

Because TD-Learning is the combination of Monte Carlo thought and dynamic planning, in the device running state optimization problem above, TD-Learning does not need to collect all the device running state parameters in a community to calculate the final profit. Focusing on different types of device, it is a bootstrapping method to get its average profit to update state by considering the input of the running state of the local equipment. Therefore, in risk-level modeling and optimization for the community device running state, we gradually adjust each optimization objective (i.e., the operation behavior leading to the worst state of equipment operation) so as to adjust the risk level model of different types of equipment operation states, and to a certain extent, realize the continuity and comprehensiveness of risk monitoring of community equipment operation state.

16.3 Progress and Results

The development of the IoT and the mobile Internet has greatly expanded the scale of the Internet and generated a large amount of information [19]. Especially in the oncoming 5G, the technology of ultra-dense network will elevate IoT data fusion and integrate the devices more efficiency [20]. A large amount of data is generated on various digital devices, such as our commonly used smart phones and smart bracelets. A large amount of data is collected and uploaded to the platform through our devices and sensors. By applying these dynamic-perceived environmental data, yes, services are becoming more intelligent, enabling the provision of user-centric, personalized intelligence services.

The "Electric Doctor" was developed by Tianjin University in conjunction with Tianjin Telecom, using NB-IoT, cloud computing and Big Data technologies to support remote monitoring of the status of various IoT devices. The platform displays statistical analysis results through the data platform to provide users with a multi-dimensional device interaction experience [21]. Through the Web platform, we can show the deployment of all the devices of the IoT device holding company in the Tianjin area we are monitoring, and we can remotely control or even repair the device directly through the Web platform. On our Android side, company ground staff with Android devices will perform the entry and maintenance of IoT devices, integrating online task management and offline personnel services.

We connect smart sockets and fire alarms in civilians by connecting our partner electrical sensors to our IoT platform. Compared with the traditional alarm, the alarm and mechanical means remind the user, but based on the NB-IoT, we realize that the platform and mobile APP monitoring sockets are leaking and healthy, and can directly monitor the home while going to work or on a business trip. In addition, we install fire alarms in schools, hospitals and communities to remotely monitor the safety status of various points in the community through smoke and temperature. In the event of fire or electric leakage, the platform automatically alarms to 119 and the fault location is accurately positioned by the platform and fed back to the fire department. As soon as a safety problem occurs, the fire department receives accurate alarm information, minimizing the loss of personal property and realizing fast, accurate and reliable intelligence.

The "Electric Doctor" platform is the core of "Internet+" based on the "Internet+" core thinking of China Telecom Tianjin Branch, using NB-IoT technology, mobile Internet technology and cloud computing technology to integrate personnel and task management of online and offline platforms. Service, a complete platform solution for the electrical safety IoT system of the partner Hongyuan Electric Co., Ltd. Based on the Telecom Tianyi Cloud, the platform uploads real-time status data of large-scale deployment of NB-IoT–based electrical equipment and facilities to China Telecom's IoT platform. After being processed by operators, it enters the company's designed IoT platform data collection point. More advanced load balancing technology, stream backup technology, dynamic message queuing technology, distributed data storage and analysis technology, dynamic two-dimensional code technology, RESTful open interface technology, and so on are used to build back-end systems, adaptive page interaction technology and other mobile terminals. Web front-end technology, data processing, analysis, visualization, and so on are based on mobile (Android) applications, integration of online and offline data, monitoring of electrical equipment facilities, management of ground staff, distribution of tasks. Operations such as tracking and management can respond to urgent emergencies through real-time synchronization protocols, ensure smooth operation of the electrical system, and respond to sudden data surges and needs for emergency troubleshooting and abnormal task monitoring.

16.3.1 Framework of the "Electric-doctor"

The platform adopts the popular method of separating the front and back ends of the technical architecture; that is, the foreground displays information as a view, and initiates an interface call through a RESTful request based on the HTTP protocol.

The front-end part adopts the popular Bootstrap responsive layout framework to display all the IoT information in a multi-dimensional layer on the login management IoT device. By improving development efficiency and saving software development costs, the responsive layout is beneficial to the device. Different automation adaptation effects are achieved with different resolutions. As a demonstration of the management console, we used AdminLTE, an open-source web management front-end template. AdminLTE is a very mature and popular multi-layer presentation template that integrates many excellent front-end components based on jQuery and Bootstrap such as visualization tables, pie charts, maps, and real-time monitoring polylines. There are a lot of presentation delay options in the multidimensional display of IoT devices (Figures 16.5 through 16.8).

The mobile part is based on Google Material Design and development, including task management, device management, personnel management, multimedia, QR code scanning and other functions. The app adopts the Model View Presenter (MVP) decoupling design mode to separate the business logic in View (Activity, Fragment). The data layer uses Glide, Gson, Retrofit, AsyncHttpClient and other framework tools to process data. The interface layer uses AppCompat, Design, CardView, and RecyclerView in the Support Library to implement Material Design and to customize or introduce third-party libraries to implement some interface elements and effects and optimize the startup speed and tablet adaptation.

The back-end part uses SpringMVC as the basic support of the platform. By using various open-source components in Java, such as selecting Hibernate in ORM as a persistent data tool, we can quickly develop the function of background and database interaction. Jackson is used as a quick parsing tool for entity classes.

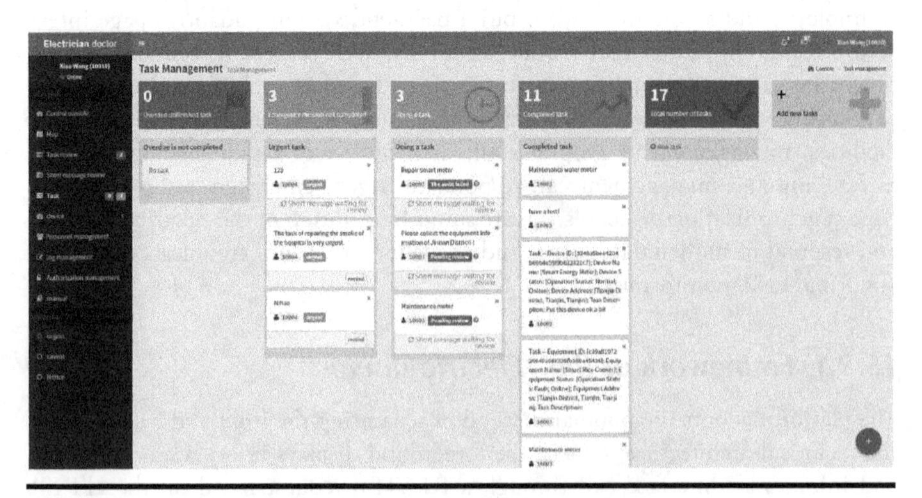

Figure 16.5 Task management interface of "Electric-doctor."

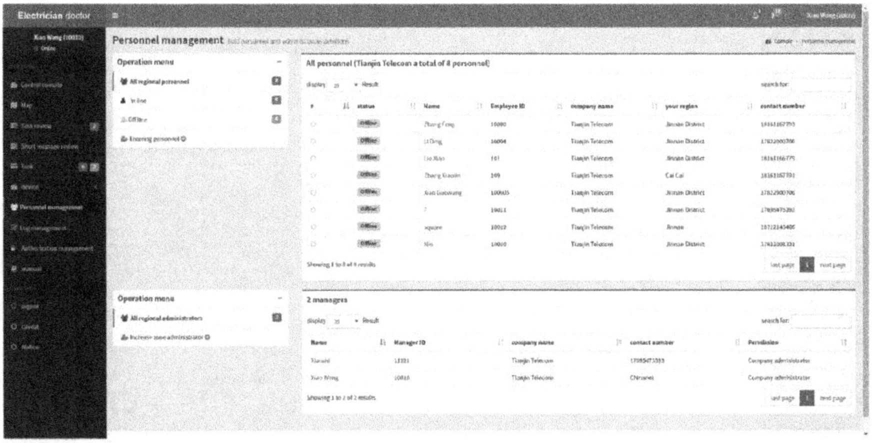

Figure 16.6 Personal management interface of "Electric-doctor."

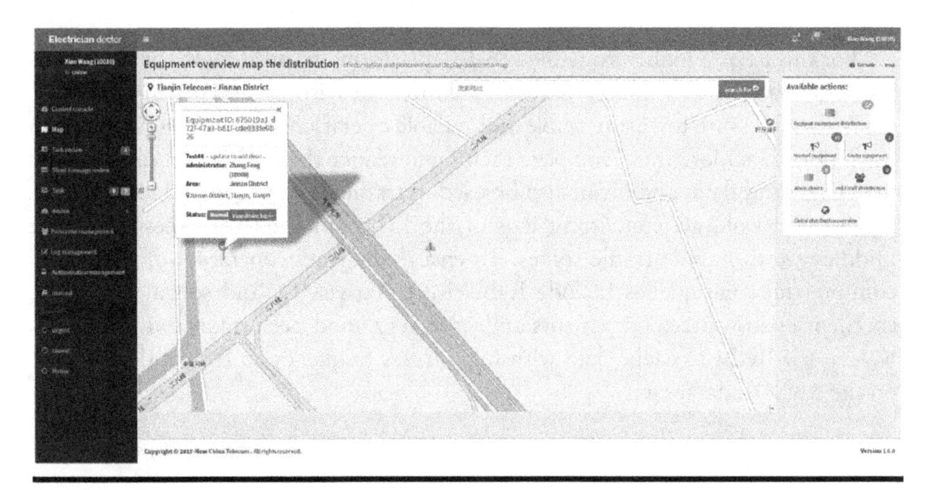

Figure 16.7 Equipment overview interface of "Electric-doctor."

We chose Swagger SpringFox as a code document generation tool. In the background, many requirements may need to be authenticated. To understand the operation of coupling authentication, we use Spring AOP, which is an aspect-oriented programming technique, which separates authentication as a separate module, making the code more scalable. In addition, for some of the users' requirements, such as real-time location display of devices or people, we directly implement this function based on the open API embedding of Baidu map and the real-time caching of the Redis database.

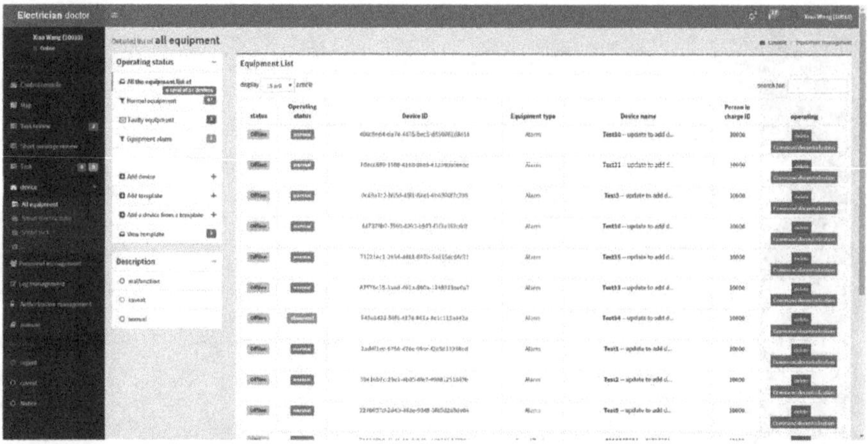

Figure 16.8 All equipment interface of "Electric-doctor."

For this to be a highly available NB-IoT platform, we need more than just conventional coding techniques. We also need to deal with potentially high concurrency data pressure to ensure stable and reliable operation of the platform. We can use Nginx to achieve front-end page caching to reduce the pressure on direct access to background data, and it can also be used in traffic distribution and load balancing for high-volume, concurrent logs of the IoT, combined with message queue middleware to handle traffic spikes. To ensure the stable operation of the server, common message queues include RabbitMQ, ActiveMQ, and so on, which are excellent open-source components and have very good performance. In addition, we can use Redis to cache data with high access frequency to reduce the pressure on the MySQL database.

16.4 Conclusion and Future Work

In this chapter, we introduced the concept of monitoring community safety through the IoT and fog computing. We proposed a hierarchical fog computing architecture based on the fog gateway. We discussed the models for multi-feature risks or threats of community and demonstrated a solution based on Big Data and artificial intelligence for computing. In the future, in order to bring more intelligence to the end user, we will focus on the integration of deep reinforcement learning and federated learning, and optimizing the computing, caching and communications of IoT applications in the fog computing architecture.

References

1. D. Soldani, and A. Manzalini, "Horizon 2020 and beyond: On the 5G operating system for a true digital society," *IEEE Vehicular Technology Magazine*, vol. 10, no. 1, pp. 3242, 2015.
2. E.T. Dresden, and N. Vodafone, "A choice of future M2M access technologies for mobile network operators," *Cellular IoT White Paper*, 2014.
3. J. Pan, and J. McElhannon, "Future edge cloud and edge computing for Internet of Things applications," *IEEE Internet of Things Journal*, vol. 99, pp. 127, 2018.
4. R. Hasan, M. Hossain, and R. Khan, "Aura: An incentive-driven ad-hoc IoT cloud framework for proximal mobile computation offloading," *Future Generation Computer Systems*, vol. 86, pp. 821–835, 2017.
5. H. Gedawy, S. Tariq, A. Mtibaa, and K.A. Harras, "Cumulus: A distributed and flexible computing testbed for edge cloud computational offloading," *Cloudification of the Internet of Things*, pp. 16, 2016.
6. Z. Pooranian, M. Shojafar, P. G. V. Naranjo, L. Chiaraviglio, and M. Conti, "A novel distributed fog-based networked architecture to preserve energy in fog data centers," in *Proceedings of the IEEE 14th International Conference Mobile Ad Hoc and Sensor Systems (MASS)*, Orlando, FL, 2017, pp. 2225.
7. M. Patel, B. Naughton, C. Chan, N. Sprecher, S. Abeta, A. Neal et al., "Mobileedge computing introductory technical white paper," *White Paper, Mobile-Edge Computing, MEC, Industry Initiative*, 2014.
8. L. Gao, T. H. Luan, B. Liu, W. Zhou, and S. Yu, "Fog computing and its applications in 5G," in *Proceedings of the 5G Mobile Communications*, Cham, Switzerland, pp. 571593, 2017.
9. C. Sobin, V. Raychoudhury, G. Marfia, and A. Singla, "A survey of routing and data dissemination in delay tolerant networks," *Journal of Network and Computer Applications*, vol. 67, pp. 128146, 2016.
10. S. Kosta, A. Aucinas, P. Hui, R. Mortier, and X. Zhang, "ThinkAir: Dynamic resource allocation and parallel execution in the cloud for mobile code offloading," in *Proceedings IEEE INFOCOM*, pp. 945953, 2012.
11. E. Ahmed, A. Ahmed, I. Yaqoob, J. Shuja, A. Gani, M. Imran, and M. Shoaib, "Bringing computation closer toward the user network: Is edge computing the solution?," *IEEE Communications Magazine*, vol. 55, pp. 138144, 2017.
12. L. Sun, Y. Li, and R. A. Memon, "An open IoT framework based on microservices architecture," *China Communications*, vol. 14, no. 2, pp. 154162, 2017.
13. A. V. Dastjerdi and R. Buyya, "Fog computing: Helping the Internet of Things realize its potential," *Computer*, vol. 49, no. 8, pp. 112116, 2016.
14. S. Yi, C. Li, and Q. Li, "A survey of fog computing: Concepts, applications and issues," *Proceedings of the 2015 Workshop on Mobile Big Data*, Hangzhou, China, pp. 3742, 2015.
15. B. Mei, W. Cheng, and X. Cheng, "Fog computing based ultraviolet radiation measurement via smartphones," in *Proceedings of the 3rd IEEE Workshop Hot Topics Web Systems and Technologies (HotWeb)*, Washington, DC, pp. 7984, 2015.
16. Y. Liu, J. E. Fieldsend, and G. Min, "A framework of fog computing: Architecture, challenges, and optimization," *IEEE Access*, vol. 5, pp. 25445–25454, 2017.

17. D. Naboulsi, M. Fiore, S. Ribot, and R. Stanica, "Large-scale mobile traffic analysis: A survey," *IEEE Communications Surveys & Tutorials*, vol. 18, no. 1, pp. 124161, 2016.
18. S. Secci, P. Raad, and P. Gallard, "Linking virtual machine mobility to user mobility," *IEEE Transactions on Network and Service Management*, vol. 13, no. 4, pp. 927940, 2016.
19. L. D. Xu, W. He, and S. Li, "Internet of Things in industries: A survey," *IEEE Transactions on Industrial Informatics*, vol. 10, no. 4, pp. 2233–2243, 2014.
20. M. Tang, L. Gao, H. Pang, J. Huang, and L. Sun, "Optimizations and economics of crowdsourced mobile streaming," *IEEE Communications Magazine*, vol. 55, no. 4, pp. 2127, 2017.
21. H. Shi, N. Chen, and R. Deters, "Combining mobile and fog computing: Using CoAP to link mobile device clouds with fog computing," in *Proceedings of the IEEE International Conference on Data Science and Data Intensive Systems* (*DSDIS*), pp. 564571, 2015.

Chapter 17

Tactile Internet over Fiber-Wireless–Enhanced LTE-A HetNets via Artificial Intelligence-Embedded Multi-Access Edge Computing

Amin Ebrahimzadeh and Martin Maier

Contents

17.1 Introduction

While the commercial exploitation of the mobile Internet is enabling users to exchange traditional triple-play (i.e., audio, video, and data) traffic, the emerging *Tactile Internet* envisages to realize *haptic communications*, enabling users to not only see and hear but also touch and manipulate remote physical and/or virtual objects through the Internet [1]. The Tactile Internet, which is driven by recent advances in computerization, automation, and robotization, is expected to significantly augment human-machine interaction, thereby converting today's content delivery networks into skillset/labor delivery networks [2–5]. The Tactile Internet holds promise to create new entrepreneurial opportunities and jobs that are expected to have a profound socioeconomic impact on almost every segment of our everyday life with use cases ranging from augmented/virtual reality (AR/VR) and autonomous driving to health care and smart grid. Many of these industry verticals require very low latency and ultra-high reliability for realizing ultra-responsive interactive applications such as bilateral teleoperation/telepresence. Note, however, that some use cases that do not necessarily require mobility all the time can be realized over fixed broadband networks. This suggests that future fifth generation (5G) cellular networks need to be fully converged networks, allowing for a flexible selection of different fixed and mobile access technologies while sharing core network functionalities [6].

Interactive systems, including in particular AR/VR and teleoperation, demand an ultra-low round-trip latency of 1–10 ms together with high reliability. The high availability and security, ultra-fast and highly-reliable response times, and carrier-grade reliability of the Tactile Internet will add a new dimension to the interaction of humans with machines/robots. To gain a more profound understanding of the Tactile Internet, it may be helpful to compare it to the emerging Internet of Things (IoT) and 5G mobile networks. Although the concept of IoT is far from novel and goes back to 1995, it is only recently that we are experiencing a rapidly increasing growth of interest in IoT from both industry and academia. Figure 17.1 depicts the revolutionary leap of the Tactile Internet in compliance with the International Telecommunication Union–Telecommunication Standardization Sector (ITU-T) Technology Watch Report on the Tactile Internet [7]. While the ultra-fast response time and carrier grade reliability of the Tactile Internet will add a new dimension to human-machine interaction, emerging 5G networks will have to handle an unprecedented growth of mobile data traffic as well as an enormous volume of data from smart sensors and actuators, the empowering elements of the IoT. It is evident that unlike the previous four generations, future 5G mobile networks will result in a seamless integration of wireless fidelity (Wi-Fi) and cellular technologies and standards, paving the way toward realizing so-called heterogeneous networks (HetNets), which mandate the need for addressing the backhaul bottleneck challenge.

The subtle difference between the Tactile Internet and IoT may be best expressed in terms of underlying communications paradigms and enabling end devices. The Tactile Internet involves the inherent human-in-the-loop (HITL) nature of human-to-machine interaction, whereas the IoT is centered around autonomous machine-to-machine (M2M) communications without any human interaction. The Tactile Internet relies on human-to-machine/robot (H2M/R) interaction and thus allows for a human-centric design approach toward creating novel immersive experiences, expanding humans' capabilities through the Internet [3]. A popular misinterpretation about robotics is that intelligent robotic systems will eventually substitute humans in one job after another. This argument may be true for some jobs, but we note that even though advanced robotics can be deployed to automate certain jobs, its greater potential, yet to be further explored, is to complement and augment human capabilities [3]. New jobs and innovative business opportunities that arise

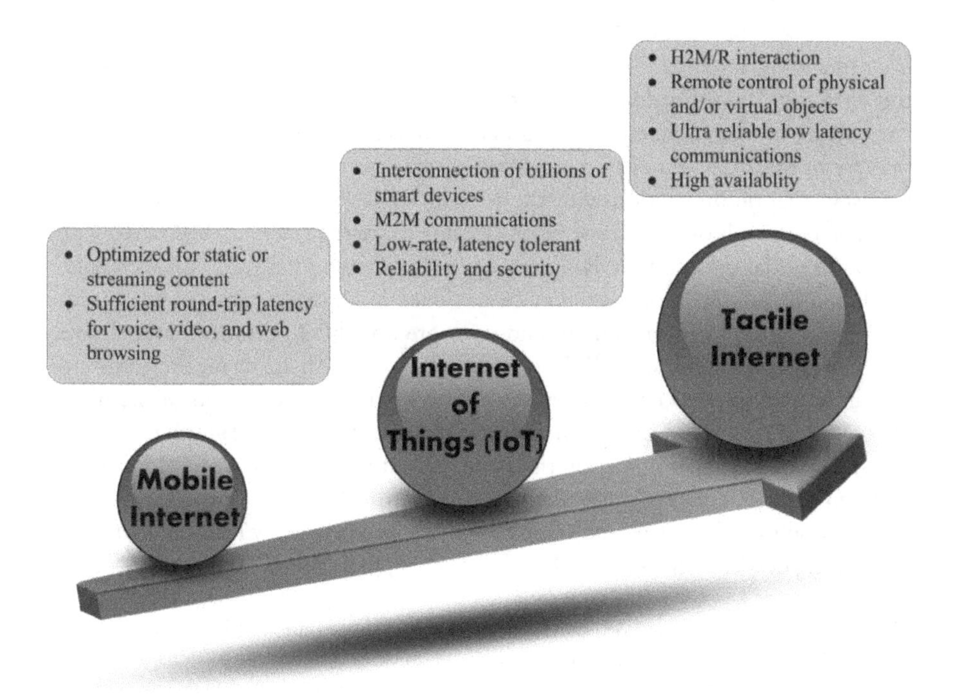

Figure 17.1 Revolutionary leap of the Tactile Internet in compliance with ITU-T Technology Watch Report. (From ITU-T Technology Watch Report, The Tactile Internet, 2014.)

from human-machine symbiosis will emerge in the so-called *missing middle* that refers to the new ways that we will have to bridge the gap between human-only and machine-only activities. This paves the way toward the so-called *third wave of business transformation*, which will be centered around human+machine hybrid activities [8]. An important requirement of developing the missing middle is to fully understand how humans can help machines and how machines can help humans. A recent example of identifying the relative strengths of humans and machines and leveraging them to fill the missing middle can be found at the automobile manufacturer Audi. Having deployed a fleet of Audio Robotic Telepresence (ART) systems, Audi has set forth employee augmentation that not only helps train technicians in diagnostics and repair but also accelerates delivery of service to customers [9].

By augmenting traditional audiovisual and data communications by the haptic modality,[1] touching, sensing, and physically interacting with remote objects becomes a reality. This, in turn, substantially improves human-to-human (H2H) as well as human-to-robot (H2R) interaction. For illustration, Figure 17.2 depicts a typical teleoperation system based on bidirectional haptic communications between a human operator (HO) at one end and a teleoperator robot (TOR) at the other. The ultimate goal of a bilateral teleoperation system is to provide the HO with the impression of feeling immersed and being present in the remote environment. This can be achieved by providing the HO with multi-modal sensory feedback including visual, auditory, and haptic signals [11]. In a teleoperation system, the term degrees-of-freedom (DoF)[2] refers to the number of independent coordinates required to completely specify and control/steer the position, orientation, and velocity of the TOR. Further, a local human system interface (HSI) device is used to display haptic interaction with the remote TOR to the HO. The local control loops on both ends of the teleoperation system ensure the stability and tracking performance of the HSI and TOR. Furthermore, as shown in Figure 17.2, perceptual deadband-based data reduction may be deployed as a lossy compression technique by exploiting the fact that human end users are not able to discriminate relatively small changes in haptic stimuli. In this chapter, we mainly focus on the communication network aspects of teleoperation, paying particular attention to the notion of average end-to-end delay, given its importance in the Tactile Internet. The average end-to-end delay induced by the communication network is the time elapsed between the time a haptic packet arrives at the Media Access Control (MAC) queue of a given source HO/TOR and the time when it is successfully received by the corresponding destination TOR/HO.

---

[1] The term "haptics" refers to both kinesthetic perception (information of forces, torques, position, velocity, and so on, which are sensed by the muscles, joints, and tendons of a human body) and tactile perception (information of surface texture, friction, and so on. sensed by different types of mechanoreceptors in the skin) [10].

[2] Currently available teleoperation systems range from 1-DoF to >20-DoF TORs. For instance, a 6-DoF TOR allows for both translational motion (in three-dimensional space) via force and rotational motion (pitch, yaw, and roll) via torque.

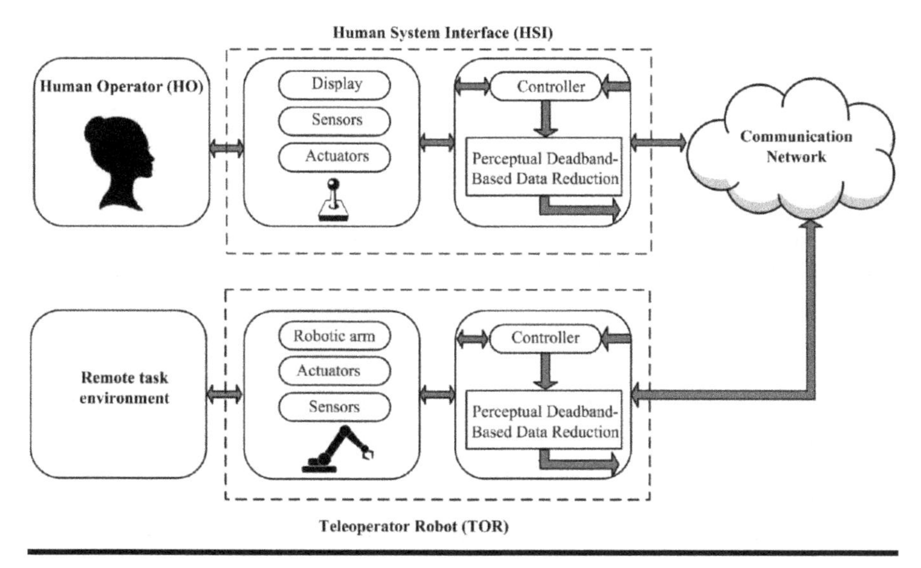

Figure 17.2 Teleoperation system based on bidirectional haptic communications between human operator (HO) and teleoperator robot (TOR) in a remote task environment. (From Maier, M. et al., *IEEE Access*.)

The discussion above indicates that improving quality of experience (QoE) for immersive bilateral teleoperation applications mandates the need for ultra-reliable low-latency communications (URLLCs), which not only provide a low average end-to-end latency but also ensures an upper-bound delay experienced by haptic packets. From a communication network perspective, efforts toward realizing an immersive ultra-responsive teleoperation experience have been pursued in the following domains: (1) communication network and (2) intelligent edge-computing/processing. First, we focus on the communication network for networked teleoperation coexistent with conventional H2H and M2M traffic. Note that delay requirements of HITL-centric bilateral teleoperation systems range from one to hundreds of milliseconds, depending on the application scenarios and dynamicity of the remote environment. If the remote environment is less dynamic, the interaction between the HO and TOR is increased. Conversely, highly dynamic environments may place high demands for ultra-fast response times of as low as 1 ms. Current cellular networks (e.g., 3G or 4G long-term evolution LTE-advanced [LTE-A]) miss this target by at least one order of magnitude. End-to-end latency measurements in an LTE network in a dense urban area for a low-mobility scenario with a proprietary application running on an Android smartphone showed an average end-to-end delay of roughly 47 and 54 ms for low and high cell load scenarios [12]. More recently, the authors of [6] have reported that the achieved average round-trip delay of 3G and 4G networks are 63 and 53 ms, respectively, according to the traces released by UK regulator *Ofcom*. According to carrier Wi-Fi vendor Aptilo Networks, despite the ongoing competition between LTE and Wi-Fi, the two technologies are in fact complementary, as the latency being roughly 10 times less in Wi-Fi leads to a higher user experience, whereas LTE provides long-range outdoor coverage (see http://www.anpdm.com for further details). With the wide deployment of passive optical networks (PONs) providing high capacity and reliability and wireless networks offering ubiquitous and flexible connectivity, interest has been growing in bimodal fiber-wireless (FiWi) networks that leverage the complementary benefits of optical fiber and wireless technologies. FiWi enhanced LTE-A HetNets represent a compelling solution to enable 4G cellular networks to meet the key requirements of low-latency and high-availability [13]. Recently, we evaluated the maximum aggregate throughput, offloading efficiency, and delay performance of FiWi enhanced LTE-A HetNets and have shown that via Wi-Fi offloading and fiber backhaul sharing, an ultra-low latency of 1–10 ms and highly reliable network connectivity can be achieved, especially at low-to-moderate traffic loads.

Next, we turn our attention to intelligent methods to compensate for the communication induced latency. To bridge the gap between the rapid increase of computation-intensive, delay-sensitive applications (e.g., Tactile Internet, AR/VR, interactive gaming) and resource-limited smart mobile devices, mobile cloud computing (MCC) has emerged to reduce the computational burden of mobile devices and widen their capabilities by extending the notion of cloud computing to the mobile environment via computation offloading. MCC allows mobile devices to benefit from powerful computing resources to save their battery power and accelerate task execution although it raises several technical challenges due to the additional communication overhead and poor reliability that remote computation offloading may introduce. To overcome these limitations, mobile edge computing has recently emerged to provide cloud computing capabilities at the edge of access networks, leveraging the physical proximity of edge servers and mobile users to achieve a reduced communication latency and increased reliability. More recently, the European Telecommunications Standards Institute (ETSI) has dropped the word "mobile" and introduced the term *multi-access edge computing* (MEC) in order to broaden its applicability to heterogeneous networks, including Wi-Fi and fixed-access technologies (e.g., fiber) [14]. There is now a growing interest among industry players in extending the cloud to decentralized levels of self-managed entities. This trend toward decentralization has led to the new paradigm of MEC, in which computing and storage resources, variously referred to as cloudlets, micro datacenters, or fog nodes, are placed at the Internet's edge in proximity to wireless end devices in order to achieve low end-to-end latency.

In this chapter, we elaborate on the role of FiWi enhanced 4G networks as the underlying communications infrastructure for enabling the emerging delay-sensitive interactive Tactile Internet applications. In particular, we aim to realize local and/or non-local teleoperation over FiWi enhanced 4G networks, leveraging low-cost data-centric (optical fiber and wireless) Ethernet technologies in both fronthaul and backhaul. As an example of Tactile Internet experience that allows for remote immersion, we study the use case of HITL-centric local and non-local teleoperation in FiWi enhanced LTE-A HetNets with artificial intelligence (AI)-embedded MEC capabilities. Our study focuses on their performance evaluation in terms of low latency and jitter using haptic traces obtained from our teleoperation experiment. We then define the problem of joint prioritized scheduling and assignment of delay-constrained teleoperation tasks onto available skilled human operators. After formulating our multi-objective optimization problem, we propose our so-called context-aware prioritized scheduling and task assignment (CAPSTA) algorithm to achieve satisfactory results by making suitable trade-offs between the contradicting objectives of the problem.

The remainder of this chapter is organized as follows. Section 17.2 reviews related work. In Section 17.3 we describe our AI-embedded MEC-enabled FiWi-enhanced LTE-A HetNets architecture in greater detail. Section 17.4 presents our proposed HITL-centric haptic sample forecasting scheme and verifies trace-driven simulation results. In Section 17.5, we elaborate on our proposed CAPSTA algorithm for solving the problem of scheduling and assignment of teleoperation tasks onto human operators. Section 17.6 concludes the chapter.

17.2 Related Work

In [15], hierarchical edge cloud architectures were shown to be able to cope with larger amounts of mobile computing peak loads. To achieve low per-packet in-order delivery delays for 5G and Tactile Internet applications, the authors of [16] investigated multipath transport protocols across cellular and/or public Wi-Fi networks. A HITL simulation environment for human-machine systems using fuzzy linear regression models for improved cognitive task performance was examined in [17]. An overview of active research challenges in cloud robotics, including low-latency networked robotics, was provided in [18]. Recently, the potential of fog/edge computing as a key enabling technology for IoT, 5G, and embedded AI was described in detail in [19].

There have been numerous research efforts in the area of 5G-enabled Tactile Internet. The authors of [20] presented a game theory–based flexible dynamic network slicing strategy for Tactile Internet applications, where incoming traffic from users can be temporarily offloaded from operator-provided networks to user-provided networks, if needed. In [21] and [22], the radio resource allocation for haptic communications over LTE-A cellular networks was investigated, where an optimization problem was developed and then solved under the assumption of Poisson haptic traffic. The authors of [23] proposed a resource allocation framework in a single-cell sparse code multiple access wireless network with multiple users with the objective of maximizing throughput subject to a given transmit power and delay constraint for haptic users. In [24], a proactive packet dropping mechanism for time division duplexing cellular systems was proposed which, together with optimizing the queue state information– and channel state information–dependent transmission policy, was proven to satisfy given quality-of-service requirements with finite transmit power.

Furthermore, several studies aimed at Tactile Internet-based e-health (e.g., [25,26], and [27]). Toward realizing Tactile Internet capable health-care facilities, the authors of [25] leveraged passive optical local area networks with time–wavelength division multiplexing and predictive dynamic bandwidth allocation. Note, however, that the main focus of this study was only on the optical backhaul without considering any wireless access mechanism. In contrast, [26] studied downlink transmission mechanisms for SmartBANs based on IEEE 802.15.6 technology, considering the 1-ms delay requirement of the Tactile Internet. In [27], the authors developed an analytical framework using an M/G/1 queueing model for estimating the wireless transmission latency from tactile body-worn devices to the wireless access point, whereby the hybrid coordination function controlled channel access (HCCA) MAC protocol is used.

17.3 FiWi Enhanced LTE-A HetNets with AI-Embedded MEC

A generic architecture of FiWi enhanced LTE-A HetNets is illustrated in Figure 17.3. The fiber backhaul consists of a time or wavelength division multiplexing (TDM/WDM) IEEE 802.3ah/av 1/10 Gb/s Ethernet PON (EPON) with a typical fiber range of 20 km connecting the central optical line terminal (OLT) and remote optical network units (ONUs). The backhaul EPON may consist of multiple stages, each separated by a wavelength-broadcasting splitter/combiner or wavelength multiplexer/demultiplexer. We consider three different subsets of ONUs. An ONU may (1) serve fixed (wired) subscribers; (2) connect to a cellular network base station (BS), giving rise to a collocated ONU-BS; or (3) connect to an IEEE 802.11n/ac/s wireless local area network (WLAN) mesh portal point (MPP), resulting in a collocated ONU-MPP. A mobile user (MU), depending on his or her trajectory, may communicate through the cellular network and/or WLAN mesh front-end, which consists of ONU-MPPs, intermediate mesh points (MPs), and mesh access points (MAPs).

The Tactile Internet is expected to enhance conventional audiovisual and data communications by the haptic modality. By doing so, touching, feeling, and physically interacting with remote objects becomes a reality, fundamentally improving H2H as well as H2M/R interaction. We assume that MUs can use their dual-mode 4G/Wi-Fi smartphones to communicate through an ONU-BS or ONU-MAP, whereas the HOs and TORs would communicate only through Wi-Fi.[3]

[3] We make this assumption in light of the fact that most of the state-of-the-art robots operate in Wi-Fi mode only.

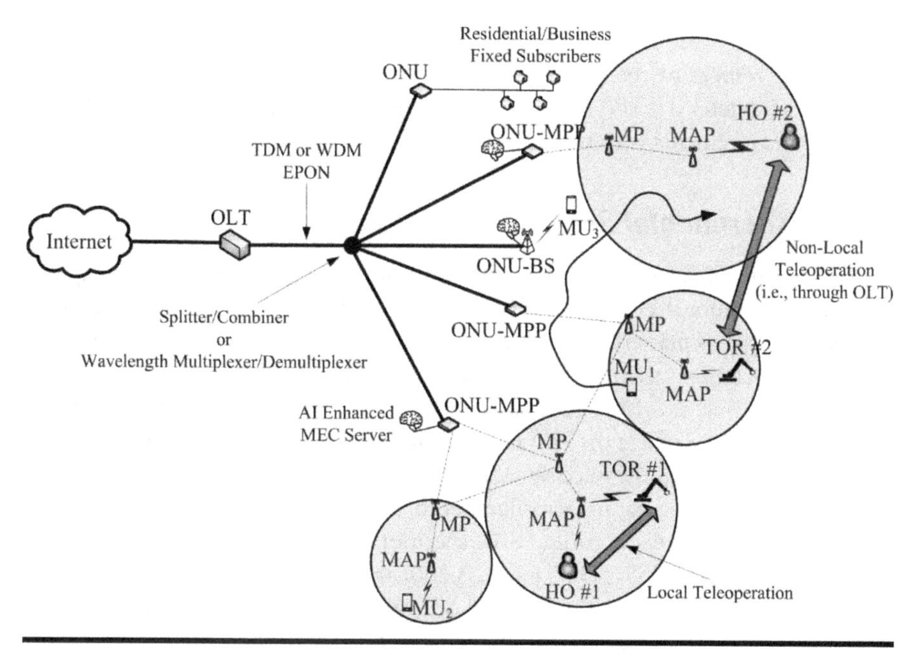

Figure 17.3 Architecture of FiWi-based Tactile Internet with AI-enhanced MEC for local and non-local teleoperation.

Depending on the proximity of the involved HO and TOR, teleoperation can be done either locally or nonlocally, as shown in Figure 17.3. In local teleoperation, the HO and corresponding TOR are associated with the same MAP and exchange their command and feedback samples through this MAP without going through the fiber backhaul. In contrast, non-local teleoperation refers to where the HO and TOR are associated with different MAPs. In such a scenario, the HO and TOR communicate via the backhaul EPON and central OLT. We equip selected ONU-BSs/MPPs with AI-embedded MEC servers collocated at the optical–wireless interface. These entities are responsible for providing the HOs with the required AI-based computing infrastructures to enable them to perform accurate forecasts of delayed haptic samples in the feedback path (to be described Section 17.4.3 in more detail shortly).

17.4 HITL-Centric Edge Sample Forecasting

In this section, we first explain the experimental setup of our haptic traces obtained from teleoperation experiments. Recall from Section 17.1 that we can exploit the human perception of haptics to further reduce the haptic packet rate. More specifically, the well-known Weber's law determines the just noticeable difference (JND), which is the smallest amount of change in the magnitude of a stimulus that humans can detect. Weber's law of the JND gives rise to the so-called *deadband coding*[4] technique, whereby a haptic sample is transmitted only if its change with respect to the previously transmitted haptic sample exceeds a given deadband parameter $d \geq 0$ (given as a percentage) [11]. Our available haptic traces comprise measurements with different values of *d*.

17.4.1 Experimental Setup

Our set of available haptic traces was obtained from the 1-DoF teleoperation experiments at the Technical University of Munich, Germany [28]. Two Phantom Omni[5] devices were used as master (i.e., HO) and slave (i.e., TOR) devices to create a 1-DoF bilateral teleoperation scenario. The communication channel between HO and TOR was emulated using a variable queueing system to generate constant or time-varying delays. The velocity signal at the HO side was sampled before being transmitted to the TOR, which in turn fed the force signal back to the HOR. The experiments were run with different deadband parameter values set to $d \in \{0, 5\%, 10\%, 15\%, 20\%\}$ in both the command and feedback paths. More specifically, the number of samples in the command path was 11,158, 5,327, 4,613, 3,290, and 4,763, whereas in the feedback path, the number of samples was 11,158, 1,532, 508, 349, and 232, respectively. The experiment lasted for 11.158, 11.750, 13.382, 12.606, and 16.254 s, respectively.

17.4.2 Packetization

Typically, haptic samples should be packetized and transmitted immediately once new sensor readings are available to help keep the end-to-end delay as small as possible. This implies a real-time transport protocol (RTP), user datagram protocol (UDP), and Internet protocol (IP) header of 12, 8, and 20 bytes, respectively [11]. For each DoF, the haptic sample of the aforementioned experimental sensor readings comprises 8 bytes. Note that N_{DoF} haptic samples are encapsulated into one RTP/UDP/IP packet, where N_{DoF} denotes the number of DoF in either experiment (i.e., 1 in our case). Thus, the packet size is equal to $40 + 8\ N_{DoF}$ bytes.

17.4.3 AI-Embedded MEC Servers

Recall from Section 17.3 that selected ONU-BSs/MPPs are equipped with AI-embedded MEC servers. The MEC servers rely on the computational capabilities of cloudlets placed at the optical-wireless interface that are enhanced with AI capabilities to create, train, and run artificial neural networks (ANNs). We use

[4] Deadband coding can be interpreted as a lossy compression mechanism exploiting the fact that the human is not able to discriminate arbitrarily small differences in haptic stimuli. The JND threshold for force perception with hand and arm is $7 \pm 1\%$ and for velocity around $8 \pm 4\%$.

[5] Phantom Omni is a widely used human system interface (HSI) device that enables HOs to interact with and manipulate objects by adding three-dimensional navigation to a broad range of applications (e.g., games, entertainment, visualization).

a type of parameterized ANN known as multilayer perceptron (MLP), which is capable of approximating any linear/nonlinear function to an arbitrary degree of accuracy. An MLP with N_h hidden neurons may be viewed as a linear combination of N_h parameterized nonlinear functions called neurons. A neuron is a nonlinear function $g(\cdot)$ of a linear combination of its input variables. Our considered ANN is an MLP with L input variables and one output variable. Let Ξ denote the set of $L{\cdot}N_h + N_h + 1$ weights of the model, that is, $\Xi = \{c_{i,j}: i = 1,...,N_h, j = 1,...,L\} \cup \{c_j': j = 0, 1,...,N_h\}$, which are estimated during the training phase. The output of the MLP is then given by

$$\Psi(\mathcal{A},\Xi)=\sum_{j=1}^{N_h} c'_j g\left(\sum_{i=1}^{L} ci,j\ \mathcal{A}(i)\right)+c'_0 \tag{17.1}$$

where $\mathcal{A} \in \mathbb{R}^L$ denotes the input vector (Figure 17.4). Note that the weights Ξ of the ANN are calculated by the corresponding MEC server and are transmitted to HOs in close proximity.

Our HITL-centric design approach for realizing an immersive and transparent teleoperation experience relies on the combination of deadband coding and MLP-based haptic sample forecasting. Deadband coding is used to reduce the average end-to-end delay, whereas our proposed sample forecasting scheme enables the in-time delivery of expected feedback samples to the HO, thus increasing the reliability of the whole teleoperation system as well as improving the feeling of being present in the remote task environment. Note that selecting larger values of deadband parameter d results in the reduction of the haptic packet rate, which, in turn, results in a decreased average end-to-end delay. Note, however, that this comes at the expense of losing some haptic samples. Besides, deadband coding is in general not sufficient to ensure a reliable teleoperation experience since reducing the average end-to-end delay does not guarantee that the delay experienced by an arbitrary haptic sample be kept under a certain desired threshold, for example, 1–10 ms. To cope with this and to ensure that the HO receives the expected feedback within the desired delay threshold, our proposed multi-sample-ahead-of-time forecasting

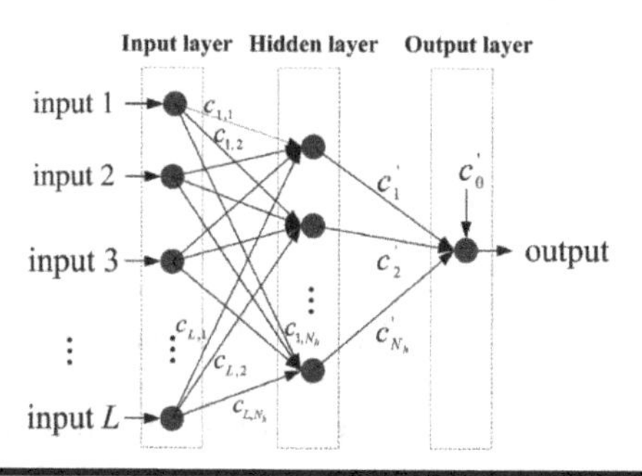

Figure 17.4 Architecture of MLP-ANN deployed at AI-embedded MEC servers.

scheme aims to deliver the forecast sample to the HO if the actual sample does not arrive on time. This in turn increases the reliability of the teleoperation system. Note, however, that the performance of the sample forecasting–based teleoperation system heavily relies on the accuracy of the forecast algorithm.

17.4.4 Haptic Trace Driven Simulations

In our haptic trace-driven simulation, we apply the same default FiWi enhanced LTE-A HetNets parameter settings as in [13]. We assume that MUs as well as HOs and TORs are mostly within Wi-Fi coverage. We consider four ONU-APs, each associated with two MUs, one HO, and one TOR, whereby two MUs communicate with each other via the ONU-AP and the remaining two MUs communicate with uniformly randomly selected MUs associated with a different ONU-AP via the backhaul EPON. Similarly, two of the total of four HO-TOR pairs communicate with each other via the same ONU-AP (i.e., local teleoperation), whereas the other two HOs communicate with two uniformly randomly selected TORs associated with a different ONU-AP via the EPON (i.e., non-local teleoperation). In addition, we consider four ONUs serving fixed (wired) subscribers that communicate with each other. The fixed subscribers and MUs together generate background Poisson traffic at an average packet rate of λ_{BKGD} (given in packets per second). Each point in the presented results is shown with a 95% confidence interval.

Figure 17.5a depicts the average end-to-end delay of HOs in the local teleoperation scenario with and without deadband coding equally applied in command and feedback paths, that is, $d_c = d_f$. As shown in Figure 17.5a, by using deadband coding, the average

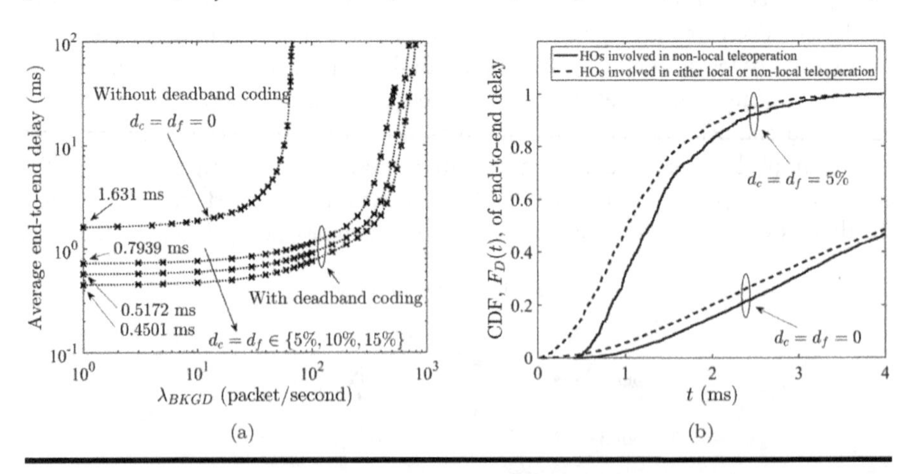

Figure 17.5 (a) Average end-to-end delay of human operators (HOs) for local teleoperation with and without deadband coding; and (b) cumulative distribution function $F_D(t)$ of end-to-end delay of HOs for local and non-local teleoperation with and without deadband coding (λ_{BKGD} = 50 packets/s, fixed). (From Maier, M. et al., *IEEE Access*.)

end-to-end delay can be kept below 1 ms for background traffic loads of up to roughly $\lambda_{BKGD} = 102$ packets/s. Next, we consider the non-local teleoperation scenario for a fixed background traffic set to $\lambda_{BKGD} = 50$ packets/s. Figure 17.5b depicts the cumulative distribution function (CDF), $F_D(t)$, of the end-to-end delay of HOs involved in non-local teleoperation and all HOs together involved in either local or non-local teleoperation for different $d_c = d_f \in \{0\%, 5\%\}$. We observe that with deadband coding ($d_c = d_f = 5\%$), $F_D(t)$ approximately approaches 1 for $t \leq 5$ ms in both local and non-local teleoperation, thus indicating that the end-to-end delay is upper bounded by 5 ms.

Next, we evaluate the performance of our proposed MEC assisted sample forecasting scheme. We use our aforementioned AI-embedded MEC servers for providing the HO with the capability of performing multi-sample-ahead-of-time forecasting of delayed force samples in the feedback path. Feedback samples are considered delayed if they do not arrive at their refresh time instants, which occur every 1 ms due to the typical haptic sampling rate of 1 kHz. By delivering the forecast samples to the HO rather than waiting for the delayed ones, sample forecasters enable HO to perceive the remote environment in real-time at a 1-ms granularity. This results in a tighter immersion and improved QoE for HOs involved in either local or non-local teleoperation. We present the results of the mean absolute percentage error (MAPE), as a Quality-of-control (QoC) metric, between the real force samples sent by the TOR and the forecast force samples delivered to the HO. Figure 17.6 illustrates MAPE versus λ_{BKGD} for scenarios with and without deadband coding.

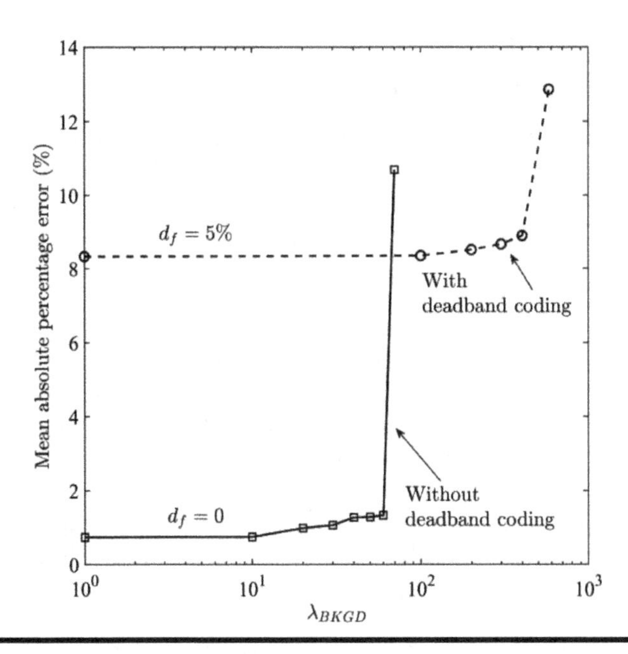

Figure 17.6 Improved quality of experience of HOs using the proposed multi-layer perceptron (MLP) for forecasting delayed feedback samples.

We observe that our sample forecasting scheme is effective without deadband coding by reducing the MAPE to below 2%. This is due to the fact that without deadband coding all force samples are actually fed back by the TOR, which can be used by the sample forecaster to achieve a better forecast accuracy. Furthermore, note that MAPE in both cases without deadband coding (<2%) and with deadband coding (>8%) stays below the typical JND threshold of 10%–20% of humans.

17.5 Scheduling and Assignment of Teleoperation Tasks

In contrast to fully automated robots, semiautonomous robotic systems demand human intervention from time to time via teleoperation and/or telepresence when domain expertise is required to perform a given task. Because these robots will need to request human assistance via teleoperation/presence, mapping these requests to the human operators themselves stands as a difficult optimization problem. Besides, as mentioned earlier, to ensure an immersive teleoperation experience, an extremely low latency of approximately 1–10 ms is required. Therefore, mapping teleoperation tasks to human operators is subject to strict delay constraints. In the following, after formally defining the problem of joint prioritized scheduling and assignment of teleoperation tasks to human operators, we propose our algorithm to minimize the average weighted task completion time, maximum tardiness, and average cost per task.

17.5.1 Problem Definition

We address the problem of joint scheduling and assignment of N delay-constrained teleoperation tasks onto a given fixed number M of HOs. Let $\mathcal{M} = \{O_1, O_2, \ldots, O_M\}$ and $\mathcal{J} = \{J_1, J_2, \ldots, J_N\}$ denote the set of M available HOs and N given tasks, respectively. The execution time of task $J_j \in \mathcal{J}$ is denoted by T_j, which is given by

$$T_j = s_j + w_j, \tag{17.2}$$

where s_j and w_j is the teleoperation session setup time and workload (both in seconds) of task J_j, respectively. Each task $J_j \in \mathcal{J}$ is associated with a due time D_j and weight Ω_j,[6] whereby larger weights correspond to higher priority levels. Even though the tasks should be executed by their specified due times, any incurred tardiness is subject to a cost penalty.

In our study, we consider an *offline* scenario, where all task requests are available at time zero and remain available continuously thereafter. Each task can be executed by only one HO at a time and each HO can execute only one task at a time.

[6] Weight is usually related to the importance, while due time is associated with the urgency of a given task and a prioritized scheduler must prepare a sequence able to first perform high-priority tasks.

We also assume that preemption is not allowed, meaning that tasks cannot be split. This is because if tasks were divided and scheduled in noncontinuous time periods, preemption would incur extra reconfiguration/setup overhead, which is significant when the setup time is non-negligible. Let us denote the start and completion times of task J_j by S_j and C_j, respectively. A feasible assignment/schedule determines when and by which human operator a given task will be executed. Given a feasible schedule, one can calculate the tardiness of task J_j as $\max\{0, C_j - D_j\}$. Our goal is to assign the tasks to the HOs such that the following constraints are satisfied: (1) no more than one task is assigned to an HO at a time, (2) no task is assigned to more than one HO, (3) tasks are not preempted, and (4) the average end-to-end packet delay of a scheduled teleoperation does not exceed a given delay threshold. Also note that the average end-to-end packet delays of any possible HO-TOR pair are characterized by two matrices, one of which represents the command path, whereas the other accounts for the feedback path.

We aim to minimize the following three objective functions: (1) the average weighted task completion time, (2) maximum tardiness, and (3) average operational expenditure (OPEX) per task. We model OPEX as follows:

$$\bar{C} = M \times \in_m + \sum_{j \in \mathcal{J}} \in_h \Omega_j \max\left\{D_j - C_{j,} 0\right\} + \sum_{k \in \mathcal{M}} \sum_{j \in \mathcal{J}} z_{jk} \in_k T_j, \tag{17.3}$$

where $\in_h$ denotes the operational cost per time unit of tardiness, $\in_m$ is the operational cost of activating a teleoperation session, $\in_k$ is the operational cost per time unit of performing a teleoperation task by human operator O_k, and z_{jk} is a binary variable, which equals 0 unless task J_j is assigned to human operator O_k. Note that in Eq. (17.3), the first term represents the cost of setting up M teleoperation sessions, the second term penalizes the tardy tasks proportional to their priority levels (i.e., the tardy tasks with higher priorities are subject to higher penalty), and the third term models the total cost of executing tasks by HOs.

17.5.2 Proposed CAPSTA Algorithm

In this section, we propose our CAPSTA algorithm to find an appropriate trade-off between the conflicting objectives of the problem. Desirable performance of the proposed algorithm, in fact, relies on an accurate estimation of the context parameters (e.g., task parameters, delay matrixes in both command and feedback paths; locations of MUs, HOs, and TORs; incoming H2H/H2M traffic pattern). In the design of our proposed CAPSTA algorithm, we adopt two sorting policies, in both assignment and scheduling stages. The used sorting policies give preference to the high-priority tasks with shorter due times.

Our proposed CAPSTA algorithm starts by partitioning the given task set $\mathcal{J}$ into M subsets. In doing so, the given tasks are sorted in a decreasing order of Ω_j/T_j.

Next, we select the tasks from the sorted set and assign them to the HOs in a round-robin fashion. Note, however, that the assignment of task J_n to human operator O_m is valid only if the estimated average end-to-end delays in both command and feedback paths do not exceed given thresholds. Otherwise, we select the HO that corresponds to the minimum average end-to-end delay in both command and feedback paths. At this point, the assignment subproblem is solved. Next, we turn our attention to the scheduling subproblem. In doing so, among unscheduled tasks, we first select the task with the minimum amount of D_j/Ω_i and then schedule it for the time when the HO first becomes available. This, in turn, results in scheduling in favor of the tasks with larger weights and shorter due times.

17.5.3 Numerical Results

In the following, we evaluate the performance of our proposed CAPSTA algorithm. In our simulations, the task execution time T_j is sampled from a discrete uniform distribution over the range of 10–30 s. The delay threshold D_0 is set to 10 ms. The weight Ω_j is randomly chosen from {1,2,3,4}, thus representing four different priority-levels/classes. We have randomly chosen the due times from $\alpha \cdot \left[1, \left\lceil \frac{1}{M} \sum_{j=1}^{N} T_j \right\rceil\right]$, where $\lceil \cdot \rceil$ denotes the ceiling function. Each point shown in the presented figures is the average of over 50 instances of randomly generated problems and falls within the 95% confidence interval. We compare the performance of our proposed CAPSTA algorithm with a benchmark random assignment and scheduling (RAS) algorithm, which assigns a given teleoperation task to a randomly selected HO from the pool of available HOs [29].

We start by presenting the performance of our proposed CAPSTA algorithm in terms of average weighted completion time (AWCT) and maximum tardiness. Figure 17.7a and b present AWCT and maximum tardiness versus the total number of available HOs, M, respectively. Figure 17.7a indicates that in both RAS and proposed CAPSTA algorithms, as M increases, AWCT decreases exponentially.

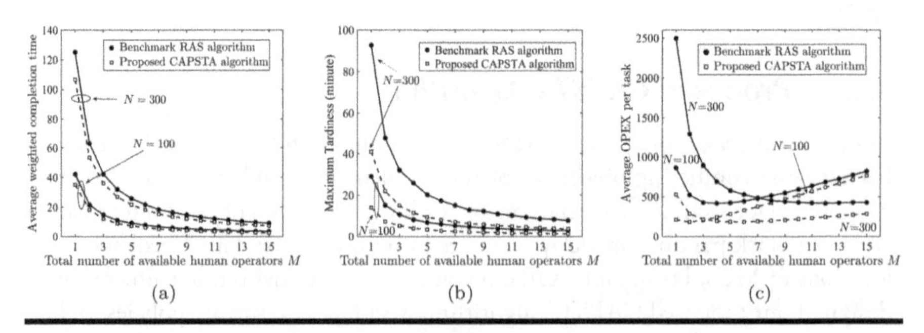

Figure 17.7 (a) Average weighted completion time of tasks versus total number of available human operators, *M*; (b) maximum tardiness of tasks versus total number of available human operators, *M*; and (c) average OPEX per task versus total number of available human operators, *M* ($\epsilon_h = 1, \epsilon_m = 5{,}000$, and $\alpha = 1$ fixed).

More specifically, in our proposed CAPSTA algorithm, increasing M from 1 to 3 results in a 67% reduction, whereas increasing M from 3 to 5 results only in a 41% reduction of AWTC. We also observe that our proposed CAPSTA algorithm achieves a 15%–27% improvement of AWTC compared to that of RAS.

Figure 17.7b demonstrates that the beneficial impact of our proposed CAPSTA algorithm compared to the RAS algorithm is more pronounced in terms of maximum tardiness. As shown in Figure 17.7b, our proposed CAPSTA algorithm achieves a 49%–56% reduction of maximum tardiness compared with that of the RAS algorithm. For $N = 300$, RAS needs 5 HOs to keep the maximum tardiness below 25 minutes, whereas our proposed CAPSTA algorithm needs only 2 HOs to achieve the same performance level. Further, if the decision maker prefers to keep the maximum tardiness below 10 min, the RSA algorithm requires 12 HOs, whereas CAPSTA requires only 5, thus achieving a notable saving in OPEX, to be examined in the following.

The average OPEX per task versus total number of available human operators M for both the proposed CAPSTA and benchmark RAS algorithms is shown in Figure 17.7c. We observe that our proposed CAPSTA algorithm outperforms the RAS algorithm in terms of average OPEX per task, especially when the number of tasks is large, that is, $N = 300$. For $M = 1$, comparing the performance of our proposed CAPSTA algorithm with the benchmark RAS algorithm, we observe a 75.3% and 78.9% reduction of average OPEX per task for $N = 100$ and $N = 300$, respectively. As M increases, the OPEX savings of the CAPSTA algorithm with respect to RAS algorithm decreases until both curves converge. The reason for this is that when the total number of available human operators M is small, the incurred OPEX is mainly due to tardy tasks, which are penalized proportionally to the weighted amount of tardiness. Figure 17.7c also indicates that the appropriate scheduling of our proposed CAPSTA algorithm reduces the number of high-priority tasks that are subject to tardiness, thus achieving a significant reduction of average OPEX per task, compared to that of the benchmark RAS algorithm.

Figure 17.7c gives us further insight into selecting the optimal number of HOs, which should be, on one hand, large enough to reduce the number of high-priority tasks which are subject to tardiness, and, on the other hand, small enough to avoid incurring excessive OPEX due to setting up more teleoperation sessions. For our proposed CAPSTA algorithm, the optimal number of available HOs M^* that minimizes $C(\mathbf{X})$ is two and five for $N = 100$ and $N = 300$, respectively. We note that for our proposed CAPSTA algorithm with $M < 3$, the average OPEX per task for $N = 100$ is less compared to that of $N = 300$. Both curves meet at $M = 4$ and then the OPEX per task for $N = 100$ grows larger than that of $N = 300$. This happens because for $N = 100$, while increasing M doesn't result in a further decrease of tardiness, it does result in an excessive increase of OPEX due to the incurred activation costs of new teleoperation sessions. In contrast, for $N = 300$, a large portion of the tasks are subject to tardiness, thus increasing M reduces the incurred OPEX due to tardiness, which, in turn, partly compensates for the incurred OPEX due to setting up new teleoperation sessions.

17.6 Conclusions

Similar to IoT and 5G networks, the future Tactile Internet demands very low latency and ultra-high reliability, integration of different transmission modes and standards, and coexistence of different types of traffic. Being centered around human-machine interaction, the Tactile Internet will be empowered by the advent of advanced intelligent robotic systems. In this new era of human-machine symbiosis, some human skills may be rendered obsolete, whereas new skills may be needed to build an innovative technology. In this chapter, we took a human-centric design approach and discussed how human and machine can be complementary rather than substituting each other. This chapter elaborated on the commonalities and subtle differences between the Tactile Internet, 5G, and IoT. We then described our envisioned AI-embedded MEC-based FiWi-enhanced LTE-A HetNet architecture for realizing immersive bilateral teleoperation experiences coexistent with conventional H2H/M2M traffic. To satisfy the strict delay and reliability requirements of Tactile Internet applications, we proposed our AI-based multi-sample-ahead-of-time sample forecasting technique, which, together with deadband coding, was shown to successfully reduce the end-to-end delay as well as compensate for delayed samples. Finally, we aimed to solve the problem of assignment and scheduling of teleoperation tasks onto skilled human operators over a FiWi-based networking infrastructure. Our simulation results verified that our proposed CAPSTA algorithm is instrumental in achieving a reduced weighted task completion time, maximum tardiness, and OPEX per executed task.

References

1. M. Simsek, A. Aijaz, M. Dohler, J. Sachs, and G. Fettweis, "5G-Enabled Tactile Internet," *IEEE Journal on Selected Areas in Communications*, vol. 34, no. 3, pp. 460–473, 2016.
2. M. Maier, M. Chowdhury, B. P. Rimal, and D. P. Van, "The Tactile Internet: Vision, Recent Progress, and Open Challenges," *IEEE Communications Magazine*, vol. 54, no. 5, pp. 138–145, 2016.
3. M. Maier, A. Ebrahimzadeh, and M. Chowdhury, "The Tactile Internet: Automation or Augmentation of the Human?" *IEEE Access*, vol. 6, pp. 41607–41618, 2018.
4. M. Dohler, T. Mahmoodi, M. A. Lema, M. Condoluci, F. Sardis, K. Antonakoglou, and H. Aghvami, "Internet of Skills, Where Robotics Meets AI, 5G and the Tactile Internet," in *Proceedings of the European Conference on Networks and Communications (EuCNC)*, 2017, pp. 1–5.
5. M. Dohler, "The Future and Challenges of Communications–Toward a World Where 5G Enables Synchronized Reality and an Internet of Skills," *Internet Technology Letters*, vol. 1, no. 2, pp. 1–3, 2018.

6. M. A. Lema, A. Laya, T. Mahmoodi, M. Cuevas, J. Sachs, J. Markendahl, and M. Dohler, "Business Case and Technology Analysis for 5G Low Latency Applications," *IEEE Access*, vol. 5, pp. 5917–5935, 2017.
7. ITU-T Technology Watch Report, "The Tactile Internet," 2014.
8. P. R. Daugherty and H. J. Wilson, *Human + Machine: Reimagining Work in the Age of AI*. Harvard Business Review Press, Boston, MA, 2018.
9. "Audi and VGo", 2014. Available: http://www.vgocom.com/audi.
10. K. Antonakoglou, X. Xu, E. Steinbach, T. Mahmoodi, and M. Dohler, "Towards Haptic Communications over the 5G Tactile Internet," *IEEE Communications Surveys & Tutorials*, vol. 20, no. 4, pp. 3034–3059, 2018.
11. E. Steinbach, S. Hirche, M. Ernst, F. Brandi, R. Chaudhari, J. Kammerl, and I. Vittorias, "Haptic Communications," *Proceedings of the IEEE*, vol. 100, no. 4, pp. 937–956, 2012.
12. P. Schulz, M. Matthe, H. Klessig, M. Simsek, G. Fettweis, J. Ansari, S. A. Ashraf, B. Almeroth, J. Voigt, I. Riedel, A. Puschmann, A. Mitschele-Thiel, M. Muller, T. Elste, and M. Windisch, "Latency Critical IoT Applications in 5G: Perspective on the Design of Radio Interface and Network Architecture," *IEEE Communications Magazine*, vol. 55, no. 2, pp. 70–78, 2017.
13. H. Beyranvand, M. L'evesque, M. Maier, J. A. Salehi, C. Verikoukis, and D. Tipper, "Toward 5G: FiWi Enhanced LTE-A HetNets With Reliable Low-Latency Fiber Back-haul Sharing and WiFi Offloading," *IEEE/ACM Transactions on Networking*, vol. 25, no. 2, pp. 690–707, 2017.
14. T. Taleb, K. Samdanis, B. Mada, H. Flinck, S. Dutta, and D. Sabella, "On Multi-Access Edge Computing: A Survey of the Emerging 5G Network Edge Cloud Architecture and Orchestration," *IEEE Communications Surveys Tutorials*, vol. 19, no. 3, pp. 1657–1681, 2017.
15. L. Tong, Y. Li, and W. Gao, "A Hierarchical Edge Cloud Architecture for Mobile Computing," in *Proceedings of the IEEE INFOCOM*, 2016, pp. 1–9.
16. A. Garcia-Saavedra, M. Karzand, and D. J. Leith, "Low Delay Random Linear Coding and Scheduling Over Multiple Interfaces," *IEEE Transactions on Mobile Computing*, vol. 16, no. 11, pp. 3100–3114, 2017.
17. J. H. Kim, L. Rothrock, and A. Tharanathan, "Applying Fuzzy Linear Regression to Understand Metacognitive Judgments in a Human-in-the-Loop Simulation Environment," *IEEE Transactions on Human-Machine Systems*, vol. 46, no. 3, pp. 360–369, 2016.
18. B. Kehoe, S. Patil, P. Abbeel, and K. Goldberg, "A Survey of Research on Cloud Robotics and Automation," *IEEE Transactions on Automation Science and Engineering*, vol. 12, no. 2, pp. 398–409, 2015.
19. M. Chiang and T. Zhang, "Fog and IoT: An Overview of Research Opportunities," *IEEE Internet of Things Journal*, vol. 3, no. 6, pp. 854–864, 2016.
20. A. S. Shafigh, S. Glisic, and B. Lorenzo, "Dynamic Network Slicing for Flexible Radio Access in Tactile Internet," in *Proceedings of the IEEE Global Communications Conference (GLOBECOM)*, 2017, pp. 1–7.
21. A. Aijaz, "Towards 5G-enabled Tactile Internet: Radio Resource Allocation for Haptic Communications," in *Proceedings of the IEEE Wireless Communications and Networking Conference*, 2016, pp. 1–6.

22. A. Aijaz, "Toward Human-in-the-Loop Mobile Networks: A Radio Resource Allocation Perspective on Haptic Communications," *IEEE Transactions on Wireless Communications*, vol. 17, no. 7, pp. 4493–4508, 2018.
23. N. Gholipoor, H. Saeedi, and N. Mokari, "Cross-Layer Resource Allocation for Mixed Tactile Internet and Traditional Data in SCMA based Wireless Networks," in *Proceedings of the IEEE Wireless Communications and Networking Conference Workshops*, 2018, pp. 356–361.
24. C. She, C. Yang, and T. Q. S. Quek, "Cross-Layer Transmission Design for Tactile Internet," in *Proceedings of the IEEE Global Communications Conference*, 2016, pp. 1–6.
25. E. Wong, M. P. I. Dias, and L. Ruan, "Predictive Resource Allocation for Tactile Internet Capable Passive Optical LANs," *IEEE/OSA Journal of Lightwave Technology*, vol. 35, no. 13, pp. 2629–2641, 2017.
26. L. Ruan, M. P. I. Dias, and E. Wong, "Towards Tactile Internet Capable E-Health: A Delay Performance Study of Downlink-Dominated SmartBANs," in *Proceedings of the IEEE Global Communications Conference*, 2017, pp. 1–6.
27. Y. Feng, C. Jayasundara, A. Nirmalathas, and E. Wong, "Hybrid Coordination Function Controlled Channel Access for Latency-Sensitive Tactile Applications," in *Proceedings of the IEEE Global Communications Conference*, 2017, pp. 1–6.
28. X. Xu, C. Schuwerk, B. Cizmeci, and E. Steinbach, "Energy Prediction for Teleoperation Systems That Combine the Time Domain Passivity Approach with Perceptual Deadband-Based Haptic Data Reduction," *IEEE Transactions on Haptics*, vol. 9, no. 4, pp. 560–573, 2016.
29. P. Brucker, Scheduling Algorithms. Springer, Berlin, Germany, 2007.

Chapter 18

Smart Power Management Internet of Things System with 5G and LoRa Hybrid Wireless Networks

Yifei Tian, Tengyue Li, Wei Song, Simon Fong, Liangliang Song, and Jinkun Han

Contents

18.1 Introduction

The real-time supervision of electricity consumption and the adaptive adjustment of electrical equipment usage are significant functions of most smart power consumption management applications [1,2]. Multiple innovative technologies from different domains are required to work together in these applications, including embedded Internet of Things (IoT) devices, fast and robust wireless communication, and precise Big Data stream analysis [3,4]. In wireless communication domain, the most significant advantages of protocol Long Range (LoRa) are long propagation distance and low power consumption when compared with other low-power wide-area network (LPWAN) such as narrowband IoT (NB IoT) or LTE Cat M1. To reduce energy consumption without affecting user experience, we developed a smart power management IoT system using fifth generation (5G) and LoRa hybrid wireless networks. Executing automatic operating instructions on electrical appliances leads to a reduction in energy consumption and a modification to electrical generation, which benefits the user side and the demand side, respectively [5,6]. In addition, power management applications promote the open sharing of information on electricity network resources while reforming power production on the supply side [7,8]. The proposed system comprises an integrated sensor module and a data stream mining module. The integrated sensor module delivers power consumption data from distributed IoT sensors to a cloud server, while the data stream mining module analyzes the gathered data and determines the power utilization behavior habits of the distributed users.

Owing to the massive communication requirement between the terminal equipment and cloud management, a 5G wireless network is proposed to increase the data transmission rate [9,10]. The transmission speed characteristic of 5G technology is suitable for delivering latency-sensitive data, particularly in IoT-based applications [11]. By greatly increasing its bandwidth and connection density, the 5G network supports massive communication access for hundreds of millions of connected IoT terminal devices [12,13]. Using a 5G network enables Big Data communication between distributed devices and the IoT server in large areas with no geographical restrictions [14,15]. Although the 4G network has begun to greatly increase its transmission efficiency, a goal of 5G wireless technology is the interconnection of services between high-bandwidth and low-bandwidth networks [16,17]. As another advantage, the 5G wireless network improves the transmission performance of low-latency and high-band width signals to support fast and comprehensive response services [18,19].

In the integrated sensor module, the distributed sensors detect and record environmental information and the energy consumption data of corresponding electrical equipment. To achieve remote monitoring and control of appliances, a LoRa wireless transmission protocol is a feasible method in a downlink network [20]. Compared with other wireless protocols, such as Bluetooth, ZigBee, and wireless fidelity (Wi-Fi), the LoRa protocol has several advantages, such as low cost and energy consumption, which are useful for IoT communication [21,22]. Through 5G wireless networks, a private LoRa network and public wide area networks (WANs) are easily combined into a hybrid framework [23–25]. After delivering sensed data to the cloud platform using LoRa and 5G networks, smart energy-saving decisions based on stream data mining algorithms are obtained [26,27]. In this way, all areas covered by LoRa compose a global IoT power monitoring and management network, sharing power consumption information for smart power management.

The Big Data stream analysis module executes several classification and prediction algorithms on the energy consumption data received from terminal LoRa devices. The massive real-time data generated by our proposed smart power management IoT system can be used to analyze user behavior habits, behavior types, and other such information. Using data stream mining algorithms, this module analyzes and predicts individual energy consumption data that can be used to improve appliance users' experiences, make intelligent control decisions on different electrical equipment, and provide a comprehensive energy analyses for common household appliances. Power consumption predictions and optimal utilization solutions provide the basis for power demand for users as well as the suppliers.

18.2 Smart Power Management for the Internet-of-Things System

Considering the personal habits of using electricity and the power demand predictions of power stations, we developed a smart power management IoT system by integrating smart IoT sensors, 5G WANs and LoRa networks and a cloud platform. Using the LoRa transmission protocol, real-time environment information and energy consumption data of appliances are delivered to private LoRa communication bases. The sensed data in a private network are transmitted to a cloud server via 5G WANs in order to make automatic energy-saving decisions and enable the remote operation of appliances.

As illustrated in Figure 18.1, our proposed system utilizes 5G WANs and LoRa networks to implement communication among IoT devices, LoRa gateways, a cloud platform, an administrator client, and a user client. In order to collect environmental information and power consumption data, a series of different domestic electrical appliances, power resources, and other sensors are installed in individuals' homes. After connecting these appliances to a private LoRa network, the data sensed by different sensors are delivered to LoRa gateways, which are considered LoRa/Internet intermediates between IoT devices and the cloud management platform. In the private LoRa wireless network, IoT devices receive remote operation signals from the cloud platform, administrator clients, and user clients via the LoRa gateways. In addition, some types of IoT terminal devices have computation units to analyze the current environment information and make simple automatic decisions to adjust operation setting. The LoRa gateways deliver received data to the cloud server for data storage and behavior feature analysis, which uses energy Big Data analysis, personalized energy-saving strategy planning, and power demand prediction. The server also receives remote control signals that are transmitted

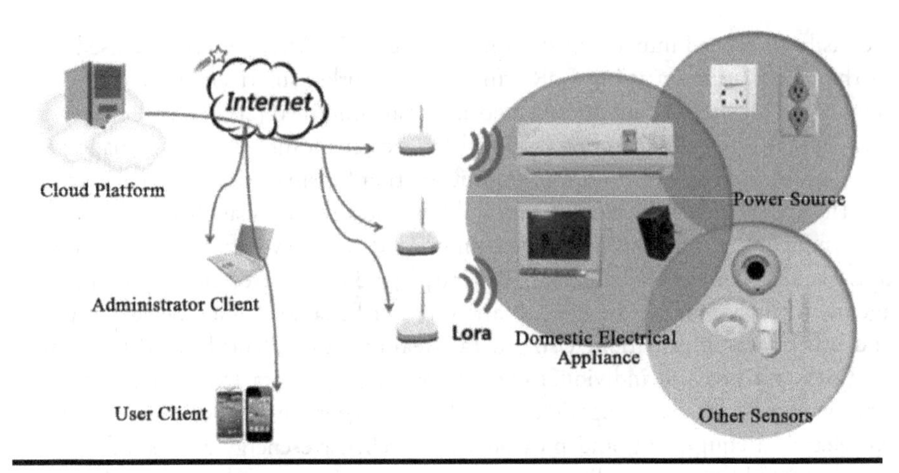

Figure 18.1 Framework of a smart power management IoT system.

from mobile devices via the Internet. Based on the current environmental information displayed on their mobile devices, users are able to remotely control their IoT devices using manual instructions, such as turning on a switch or and changing the operation mode. As intermediate equipment, LoRa gateways receive control signals from the cloud server and forward them to their individual IoT devices as different packets based on IP address. When control requests from mobile devices are received, the cloud server distributes the control instructions to the relevant gateways connected to the destination LoRa terminal devices.

By integrating the private LoRa and public networks, a smart power management IoT system overcomes low-bandwidth transmission problems in difficult scenarios under weak signals, such as in tall skyscrapers and underground garages. If IoT equipment attempts to connect to a private network by delivering a joining request, the LoRa gateways confirm the connection request and generate the configured parameters. After generating these parameters, the IoT equipment begins to record electric consumption data and transmit them to the server twice every minute. Activation by personalization (ABP) and over-the-air activation (OTAA) methods are applied to establish a connection among IoT devices, LoRa gateways, and the cloud platform. The ABP is an insecure connection pathway and is only suitable for use in private networks. The OTAA is more secure than the ABP; thus, it better guarantees the confidentiality and safety of user data delivered over the 5G and LoRa hybrid wireless networks.

As shown in Figure 18.2, several end-node parameters are initialized as random values (LoRa RSSI) to generate a password for encrypting data, which is a significant step in the working principle of OTAA method for implementing data security during the transmission process. By using formatted, packed, encapsulated, and encrypted user data and control instructions, the security and privacy of our proposed IoT system can be ensured.

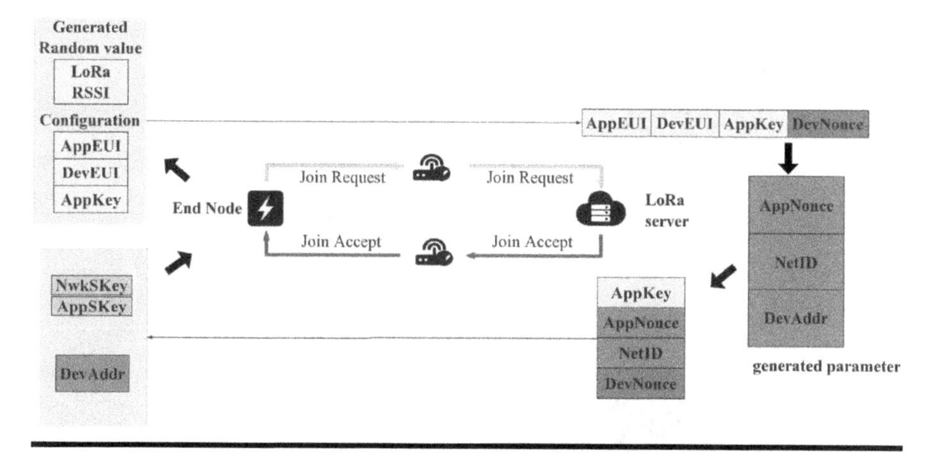

Figure 18.2 Flowchart of LoRa network establishing a connection.

To implement automatic energy-saving decisions without impacting users' experiences, our power management cloud platform analyzes users' habits and real-time environmental information. Through the total energy consumption data collected from different IoT devices installed in the same area, our cloud platform supports the precise prediction of electrical consumption for the following period; this allows the power station to adjust its electricity generation strategy accordingly. In addition, the cloud platform is treated as a data transmission medium for transferring energy consumption data and control instructions between IoT devices and users' mobile terminals.

To support a comfortable user experience during automatic energy-saving decisions, a user's habit analysis model is embedded into our cloud platform to classify the usage habit characteristics of different types of IoT-based electrical devices. Using an IoT-embedded refrigerator as an example, an energy-saving scheme can be automatically generated according to the real-time changing trends of ambient temperature and air humidity around the device and the user's historical habits recorded in our cloud platform. Using the power consumption features on time series, a large amount of historical power data provides sufficient training data for medium- and long-term forecasting of total power consumption in different areas, which facilitates peak avoidance and electrical generation planning.

18.3 Experiments

In this section, we analyze the performance of the proposed IoT system for smart power management using several data mining algorithms. The IoT system was implemented using an Arduino board, ATT7053BU board, LoRa transmitter, and LoRa gateway, as shown in Figure 18.3a–d, respectively. An ATT7053BU module was attached to the Arduino board, which was connected to a computer. This system tested the electricity consumption data originating from the UART through a USB local area network (LAN).

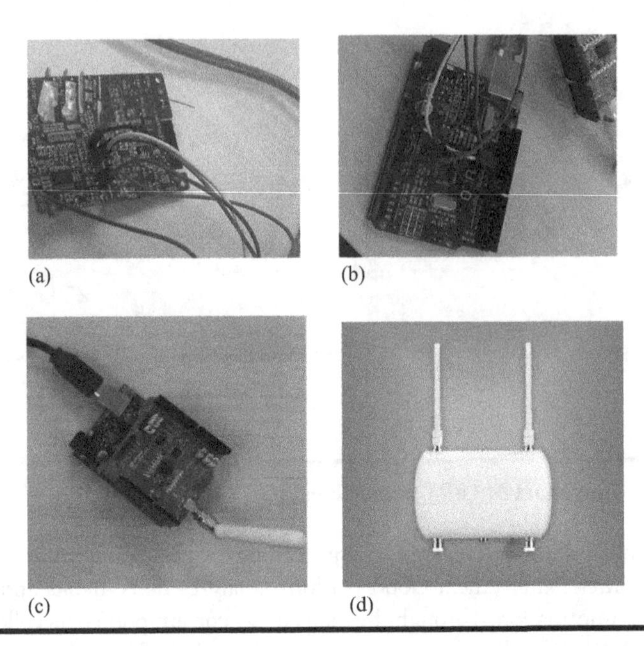

Figure 18.3 Main devices utilized in proposed system: (a) ATT7053BU board, (b) Arduino board, (c) LoRa transmitter, and (d) LoRa gateway.

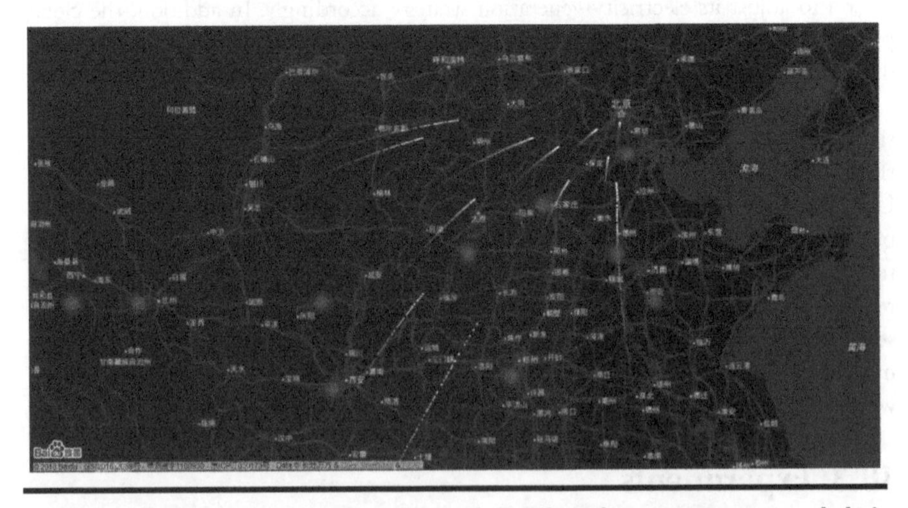

Figure 18.4 Data visualization images of distributed LoRa gateways and their communication.

Figure 18.4 presents the data visualization of the distributed LoRa gateways and their communication on a Baidu map. The map displays the status of real-time interactions between the server and gateways. For example, Figure 18.5 shows the electricity consumption datasets of a refrigerator tested in 1 week, including voltage, current, electric capacity, and total energy. The figure shows that the voltage

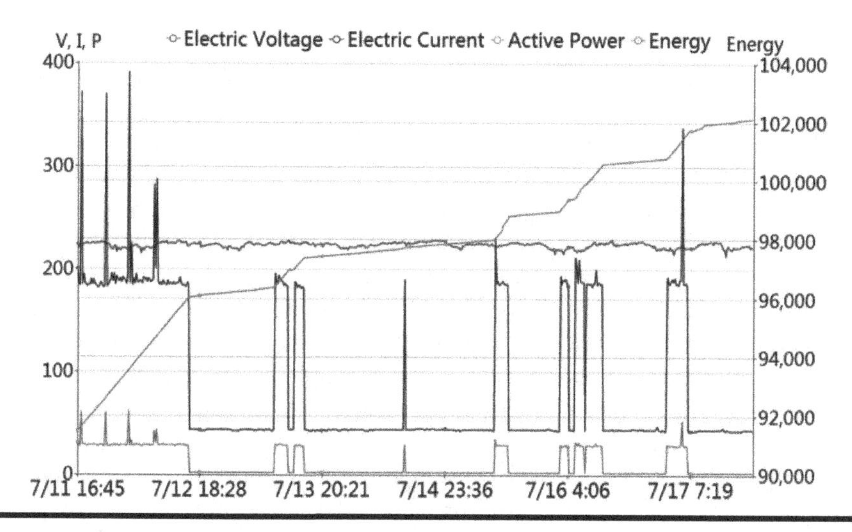

Figure 18.5 The electricity consumption datasets of a refrigerator tested over 1 week.

fluctuated slightly, but it remained largely at 220 V. The electric capacity and current increased when the refrigerator was used. Therefore, the system obtained user features of power utilization so as to determine periods of peak consumption during a day.

Using the gathered power consumption datasets, the objective of data mining algorithms was to investigate the suitability of two groups of algorithms: batch mode data mining, where the model needed to be fully rebuilt by loading all the data, and incremental data stream mining, where the data arrived segment by segment and the model updated progressively. A simple feature selection scheme called correlation feature selection (cfs) was applied on these two groups of algorithms. The impact of cfs on the algorithms in terms of forecasting performance was quantified by mean absolute errors (MAEs), root mean square errors (RMSEs) and time taken to complete the forecasting over the whole time series.

The design of the experiment was to run specific data mining or data stream mining algorithms, one at a time, for forecasting the future 200 steps ahead at the end of the time series. The data or data stream mining algorithms learned the underlying mapping between the attributes and the target class (which is the main total energy), as well as learning recurring patterns from the past of the time series. In other words, we attempted to induce a fitting curve by using the data or data stream mining algorithm that followed the original curve. At each data point, based on what has been learned, future values were derived from the fitting curve as predicted values. The batch mode algorithms included classical algorithms such as Linear Regression, Multilayer Perception, Regression optimized by support vector machine (SMOReg), Holt-Winter and Random Tree. The incremental algorithms included Stochastic Gradient Decent (SGD), on-line regression trees with options (ORTO), Fading Target Mean, Adaptive Model Rules, was rule-based learning algorithm for regression problems on streams (AMRulesRegressor) and Random Rules.

All the data stream mining algorithms achieved lower error rates than those of data mining algorithms when it came to forecasting future values over the data streams. In general, data mining algorithms incurred error rates at a magnitude of three to four digits. In particular, the algorithm AMRulesRegressor outperformed all others with its MAE = 9.03, which was far lower than the worst performer, Multilayer Perception, at MAE = 3,269.96. For speed, data stream mining algorithms generally were several times faster than the batch mode algorithms. In particular, Fading Target Mean was the fastest among all, requiring only 142 s to execute the forecast compared with the worst case, the SVM optimized regression, which took 16,416 s.

The predictions of the energy consumptions in the next 40 steps are shown in Figure 18.6. The predicted values or the trend as successive values, vary quite largely. Holt-Winter was too optimistic with a downward trend. Neural network had a curve with the largest degree of fluctuation, probably due to its sensitivity in learning non-linear relations between attributes and the target. The rest of the algorithms gave predicted values that are quite consistent with each other.

By implementing a simple correlation test using Pearson coefficient computation, interestingly, there were several attributes that were very well correlated to the prediction target—the total main energy. Several attributes were found to be quite correlated to the predicted target. They were Energy, HP_in_Tot (heat pump inside unit), HP_out_Tot (heat pump outside unit), and

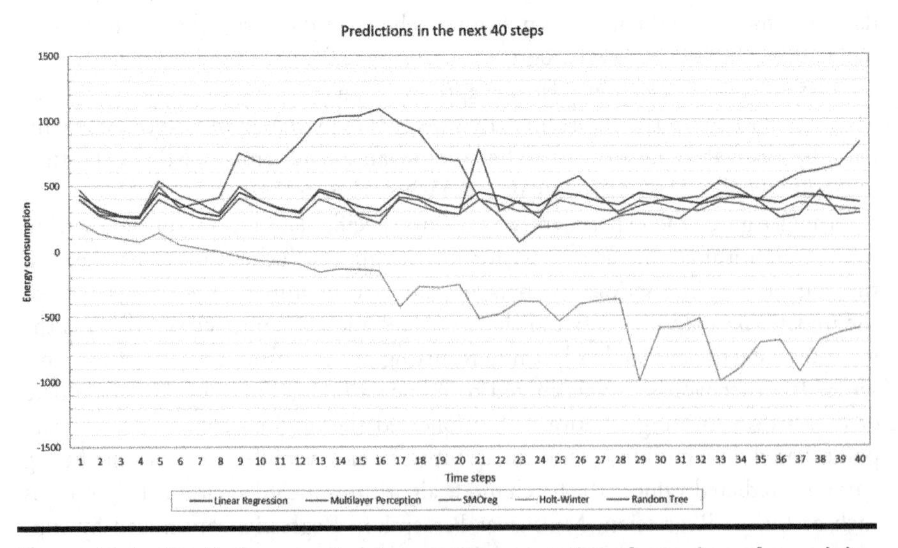

Figure 18.6 Prediction results of energy consumptions by various data mining algorithms.

wash_Tot (energy, washing machine). Common sense told us that the heater and washing machines in a house were among the appliances that used most of the electricity. By knowing the correlations and whose attributes were most related to the predicted target, it offered a clue in enhancing the forecasting performance by the algorithms. Feature selection where a subset of attributes that were most correlated to the predicted target should be included in the forecasting. The cfs was applied over the dataset. It eliminated the redundant attributes, leaving only the attributes that were well correlated to the predicted target. The forecasting experiment was repeated after cfs was applied to the dataset. The forecasting performance results were tabulated in Table 18.1.

Several interesting phenomena can be identified in Table 18.1, after the application of cfs. The error rates were not affected by the reduced number of attributes for all the algorithms except for cfs-ORTO, cfs-AMRulesRegressor, and cfs-RandomRules. For cfs-ORTO, there was a significant improvement as the error rates dropped almost in half and the time taken shortened from 1,979 to 8 s. For the cfs-AMRulesRegressor and cfs-RandomRules algorithms, the error rates rose after cfs was applied. Cfs was not helping those two algorithms in improving forecasting accuracy. However, cfs did indeed help the data stream mining algorithms reduce processing time from hundreds or thousands of seconds to merely

Table 18.1 Performance Results of Data Mining and Data Stream Mining Algorithms in Forecasting Smart House Total Energy Consumption with Feature Selection Applied

Batch-Mode Data Mining Algorithms with Feature Selection	*MAE*	*RMSE*	*Time (s)*
cfs-Linear Regression	156.6418	219.2652	171
cfs-Multilayer Perception	3,269.963	5,878.646	3353
cfs-SMOReg	166.3897	250.7913	14686
cfs-Holt-Winter	2,154.039	129,594.3	3521
cfs-Random Tree	195.4466	286.6616	1983
Incremental Data Stream Mining Algorithms with Feature Selection			
cfs-SGD	435.73	502.89	0.38
cfs-ORTO	270.58	395.8	8
cfs-FadingTargetMean	161.37	217.12	0.2
cfs-AMRulesRegressor	28.51	44.28	10.3
cfs-RandomRules	63.66	91.97	116

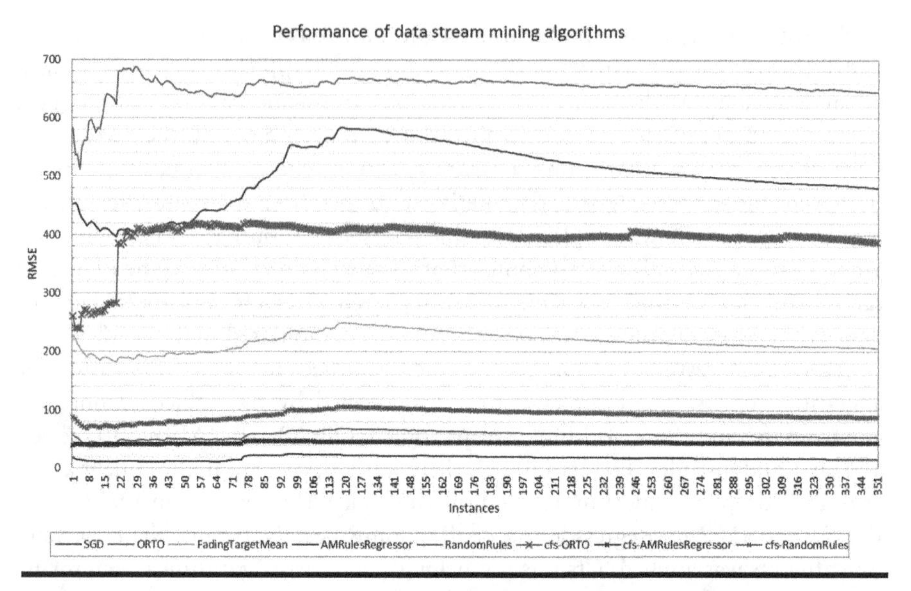

Figure 18.7 Prediction results of energy consumptions by various data stream mining algorithms with and without the correlation feature selection.

single-digit seconds. This made it a strong approach in shaping the data stream mining process fast and accurate online forecasting. Considering both factors of accuracy and speed, cfs-AMRulesRegressor was by far the optimal combination, with a MAE rate of 28.51 and time of 10.3 s. The longitudinal performance of each combination of algorithm and cfs is charted in Figure 18.7. AMRulesRegressor and Random Rules data stream mining outperformed the others, especially those algorithms of batch mode. They took a long time, and the forecasting errors were very high compared with those of data stream mining algorithms. As a concluding remark, this experiment shows that data stream mining algorithms were far more favorable than data mining algorithms, when it came to forecasting task over a long data stream. Cfs feature selection by correlation had enhancing effects on certain data stream mining algorithms, but it was useless compared with the data mining algorithms in forecasting.

18.4 Conclusion

Automatic monitoring and adaptive management of electric energy in real time are important solutions for reducing the energy consumption of individuals and industries. In transmitting real-time power consumption data collected by different types of terminal devices in an IoT-based system, LoRa and 5G wireless network technologies were employed to increase the data transmission rate. All of the energy data were transmitted to a cloud platform, where the data features were analyzed and individual operating instructions were generated with the goal of saving energy based on users' habits. Owing to the power consumption features, several data stream mining algorithms were used in our cloud platform to extract data characteristics. To illustrate different usage models based on individual features detected in the cloud platform, we used data visualization to represent energy consumption in real time. We developed several batch-mode and incremental-mode data mining algorithms to forecast total energy consumption in smart house system, all of which supports effective criteria for autonomous adjustment of electricity generation from power station.

Acknowledgment

This research was supported by National Natural Science Foundation of China (61503005), the Great Wall Scholar Program (CIT&TCD20190304), Beijing Natural Science Foundation (4184086), Beijing Young Topnotch Talents Cultivation Program (No. CIT&TCD201904009), North China University of Technology "The Belt and Road" Countries Talent Training Base Project, and "Yuyou" Project of North China University of Technology.

References

1. Y. Strengers and L. Nicholls, Convenience and Energy Consumption in the Smart Home of the Future: Industry Visions from Australia and Beyond, *Energy Research & Social Science*, Vol. 32, pp. 86–93, 2017.
2. M. Collotta and G. Pau, An Innovative Approach for Forecasting of Energy Requirements to Improve a Smart Home Management System Based on BLE, *IEEE Transactions on Green Communications and Networking*, Vol. 1, no. 1, pp. 112–120, 2017.
3. R. Yang and L. Wang, Multi-objective Optimization for Decision-Making of Energy and Comfort Management in Building Automation and Control, *Sustainable Cities and Society*, Vol. 2, no. 1, pp. 1–7, 2012.
4. B. L. R. Stojkoska and K. V. Trivodaliev, A Review of Internet of Things for Smart Home: Challenges and Solutions, *Journal of Cleaner Production*, Vol. 140, pp. 1454–1464, 2017.
5. K. Y. Liu, J. Peng, H. Li et al., Multi-device Task Offloading with Time-Constraints for Energy Efficiency in Mobile Cloud Computing, *Future Generation Computer Systems*, Vol. 64, pp. 1–14, 2016.
6. B. Mack and K. Tampe-Mai, An Action Theory-Based Electricity Saving Web Portal for Households with an Interface to Smart Meters, *Utilities Policy*, Vol. 42, pp. 51–63, 2016.
7. D. S. Markovic, D. Zivkovic, I. Branovic et al., Smart Power Grid and Cloud Computing, *Renewable and Sustainable Energy Reviews*, Vol. 24, pp. 566–577, 2013.
8. S. Abras, S. Pesty, S. Ploix et al., An Anticipation Mechanism for Power Management in a Smart Home using Multi-Agent Systems, In *2008 3rd International Conference on Information and Communication Technologies: From Theory to Applications*, Damascus, Syria, April 7–11, 2008.

9. P. Schulz, M. Matthé, H. Klessig et al., Latency Critical IoT Applications in 5G: Perspective on the Design of Radio Interface and Network Architecture, *IEEE Communications Magazine*, Vol. 55, pp. 70–78, 2017.
10. M. Agiwal, N. Saxena, and A. Roy, Towards Connected Living: 5G Enabled Internet of Things (IoT), *IETE Technical Review*, 2018. doi:10.1080/02564602.2018.1444516.
11. N. Saxena, A. Roy, B. J. R. Sahu et al., Efficient IoT Gateway over 5G Wireless: A New Design with Prototype and Implementation Results, *IEEE Communications Magazine*, Vol. 55, pp. 97–105, 2017.
12. M. Zeinali, J. Thompson, C. Khirallah et al., Evolution of Home Energy Management and Smart Metering Communications towards 5G, In *2017 8th International Conference on the Network of the Future* (*NOF*), London, UK, November 22–24, 2017.
13. L. Xu, R. Collier, and G. M. P. O'Hare, A Survey of Clustering Techniques in WSNs and Consideration of the Challenges of Applying Such to 5G IoT Scenarios, *IEEE Internet of Things Journal*, Vol. 4, No. 5, pp. 1229–1248, 2017.
14. S. Husain, A. Kunz, A. Prasad et al., Mobile Edge Computing with Network Resource Slicing for IoT, In *IEEE World Forum—IoT* (*WF-IoT*), Singapore, 2018.
15. K. Lei, S. Zhong, F. Zhu et al., An NDN IoT Content Distribution Model With Network Coding Enhanced Forwarding Strategy for 5G, *IEEE Transactions on Industrial Informatics*, Vol. 14, pp. 2725–2735, 2018.
16. R. Hussain, A. T. Alreshaid, S. K. Podilchak et al., Compact 4G MIMO Antenna Integrated with a 5G Array for Current and Future Mobile Handsets, *IET Microwaves, Antennas & Propagation*, Vol. 11, pp. 271–279, 2017.
17. R. Ma, K. H. Teo, S. Shinjo et al., A GaN PA for 4G LTE-Advanced and 5G: Meeting the Telecommunication Needs of Various Vertical Sectors Including Automobiles, Robotics, Health Care, Factory Automation, Agriculture, Education, and More, *IEEE Microwave Magazine*, Vol. 18, pp. 77–85, 2017.
18. J. A. del Peral-Rosado, J. A. López-Salcedo, and G. Seco-Granados, Impact of Frequency-Hopping NB-IoT Positioning in 4G and Future 5G Networks, In *2017 IEEE International Conference on Communications Workshops* (*ICC Workshops*), Paris, France, May 21–25, 2017.
19. K. E. Skouby, and P. Lynggaard, Smart Home and Smart City Solutions Enabled by 5G, IoT, AAI and CoT Services, In *2014 International Conference on Contemporary Computing and Informatics* (*IC3I*), Mysore, India, November 27–29, 2014.
20. W. Song, N. Feng, Y. Tian et al., A Deep Belief Network for Electricity Utilisation Feature Analysis of Air Conditioners Using a Smart IoT Platform, *Journal of Information Processing System*, Vol. 14, No. 1, pp. 162–175, 2018.
21. R. S. Sinha, Y. Wei, and S. H. Hwang, A Survey on LPWA Technology: LoRa and NB-IoT, *ICT Express*, Vol. 3, pp. 14–21, 2017.
22. A. Carlsson, I. Kuzminykh, R. Franksson et al., Measuring a LoRa Network: Performance, Possibilities and Limitations, In *Internet of Things, Smart Spaces, and Next Generation Networks and Systems*, pp. 116–128, 2018.
23. Y. H. Song, J. Lin, M. Tang, and S. F. Dong, An Internet of Energy Things Based on Wireless LPWAN, *Engineering*, Vol. 3, No. 4, pp. 460–466, 2017.
24. Y. Z. Li, X. Q. Yan, L. Y. Zeng, and H. L. Wu, Research on Water Meter Reading System Based on LoRa Communication, In *2017 IEEE International Conference on Smart Grid and Smart Cities*, pp. 23–26, 2017.

25. N. Hayati and M. Suryanegara, The IoT LoRa System Design for Tracking and Monitoring Patient with Mental Disorder, In *IEEE International Conference on Communication*, pp. 135–139, 2017.
26. J. Wang, C. Ju, Y. Gao, A. K. Sangaiah, and G. Kim, A PSO Based Energy Efficient Coverage Control Algorithm for Wireless Sensor Networks, *Computers Materials & Continua*, Vol. 56, No. 3, pp. 433–446, 2018.
27. J. Wang, J. Cao, S. Ji, and J. H. Park, Energy Efficient Cluster-based Dynamic Routes Adjustment Approach for Wireless Sensor Networks with Mobile Sinks, *Journal of Supercomputing*, Vol. 73, No. 7, pp. 3277–3290, 2017.

Index

Note: Page numbers in italic and bold refer to figures and tables, respectively.